The Chemical Biology
of Thrombin

T0331371

The Chemical Biology of Thrombin

Roger L. Lundblad

CRC Press
Taylor & Francis Group
Boca Raton London

CRC Press is an imprint of the
Taylor & Francis Group, an **informa** business

First edition published 2022
by CRC Press
6000 Broken Sound Parkway NW, Suite 300, Boca Raton, FL 33487–2742

and by CRC Press
2 Park Square, Milton Park, Abingdon, Oxon, OX14 4RN

© 2022 Taylor & Francis Group, LLC

CRC Press is an imprint of Taylor & Francis Group, LLC

ISBN: 978-1-138-05590-2 (hbk)
ISBN: 978-1-032-20330-0 (pbk)
ISBN: 978-1-315-16564-6 (ebk)

DOI: 10.1201/b22204

Typeset in Times
By Apex CoVantage, LLC

This book is dedicated to Jennifer Lee Bergstrom, who has provided moral support for the past 30 years, for which I am very thankful.

Contents

Preface

I will be honest and say that while I enjoyed the topic, having been in a relationship with thrombin for more than 50 years, working through the pandemic has been a challenge. Library access has been exceedingly difficult, and I was not able to access a large amount of material. As a result, this work is not as complete as I would have liked. I do apologize to many investigators whose work is not included in this book. Regardless, I do feel that this is a useful piece of work. Despite a long history with thrombin, I learned much about this protein.

Acknowledgements

It is a pleasure to acknowledge the continued help of Professor Bryce Plapp of the University of Iowa with various aspects of the book; his moral support was essential. I also want to acknowledge the contributions of Sherry Thomas, Dr. Chuck Crumly, Kyle Meyer, and their colleagues at Taylor & Francis as well as those at Apex CoVantage in the development and production of the final product.

About the Author

Roger L. Lundblad is a native of San Francisco, California. He received his undergraduate education at Pacific Lutheran University and his PhD degree in biochemistry at the University of Washington. After postdoctoral work in the laboratories of Nobel Laureates Stanford Moore and William Stein at the Rockefeller University, he joined the faculty of the University of North Carolina at Chapel Hill in 1968, attaining the rank of full professor in 1976. He joined the Hyland Division of Baxter Healthcare in 1990. Currently, Dr. Lundblad is an independent consultant and writer in biotechnology in Chapel Hill, North Carolina, as well as an adjunct professor of pathology at the University of North Carolina at Chapel Hill. Dr. Lundblad is the former editor-in-chief of *Biotechnology and Applied Biochemistry* and *Internet Journal of Genomics and Proteomics*. He is also the co-editor of three books: *Chemistry and Biology of Thrombin*, *Chemistry and Biology of Heparin*, and *The Handbook of Biochemistry and Molecular Biology*, fourth and fifth editions. He is also the author of more than 100 papers in refereed journals and 12 books, as well as holding three patents.

1 The History of Thrombin

The fact that blood, when removed from the circulatory system, forms a "clot" has been known for centuries.[1] There were early studies exploring the mechanism of blood coagulation leading to two hypotheses. One school of thought considered the coagulation of blood as a process of solidification caused by cooling or exposure to air or other substances. The other school of thought was based on the concept of the "intrinsic" formation of a factor in blood, which resulted in the formation of fibrin as insoluble material that would stop bleeding. A consideration of these hypotheses is beyond the scope of this work, and the reader is directed to several discussions of early work on blood coagulation.[2–7]

Although there is some legitimate controversy on the matter,[8] it would seem that Andrew Buchanan deserves the credit for the discovery of a "substance" intrinsic in blood that could form an insoluble substance when added to blood or other "fibrogeneous" fluid such as hydrocele or ascites.[9,10] It was noted that hydrocele or ascites did not clot without the addition of their fibrin ferment. Buchanan isolated a "washed blood-clot" from blood which had been added to 6–10 volumes of water, stirred for 5 minutes, and allowed to stand for 24 hours. The washed blood-clot was isolated by filtration through a coarse linen cloth. The isolated washed blood-clot retained coagulant activity when added to blood or hydrocele. While blood would clot in the absence of exogenous material, hydrocele and other "fibriniferous" fluids would clot on the addition of the washed blood-clot. The isolated washed blood-clot could be dried and stored. It was noted that the addition of "spirit of wine" (distilled wine, mostly likely what would be considered brandy today) preserved activity. Gamgee[11] repeated and extended Buchanan's experiments, presenting evidence that the washed blood-clot was fibrin (*Faserstoff*) and did retain coagulant activity, which he described as fibrin ferment. Subsequent studies with somewhat better characterized reagents confirmed that thrombin did bind to fibrin[12,13] and that such binding was associated with the conversion of fibrinogen to fibrin. Later studies showed that thrombin generated in plasma could be adsorbed to fibrin during the process of fibrinogen clotting.[14] More sophisticated studies performed 60 years later showed that the majority of thrombin generated *in situ* under dynamic conditions is incorporated into the fibrin thrombus.[15] The inclusion of the tripeptide, GlyProArgPro which inhibits fibrin polymerization, increased the amount of thrombin-antithrombin detected. Early work had designated fibrin as antithrombin I.[16–18] Thrombin bound to fibrin does retain enzymatic activity.[19–23] This work with purified proteins supports the earlier the observations by Buchanan and Gamgee[9–11] that crude fibrin isolated could contain bound thrombin, which would still be active. Wilner and coworkers[19] showed that streptokinase digested the complex of thrombin with fibrin, with the release of active thrombin, which was likely still bound to soluble fibrin degradation products.[20] Thrombin bound to blood platelets also retained activity.[24,25] Platelets were also shown to have antithrombin activity by Tullis and coworkers.[26,27] Gamgee[11]

DOI: 10.1201/b22204-1

was able to elute the fibrin ferment from the isolated washed blood-clot in what might have been the first purification of thrombin; in later work, Schmidt was able to form thrombin from prothrombin isolated from plasma.[28-31] And Howell,[32] also in later work, isolated "thrombin" from porcine fibrin (obtained by the whipping of porcine blood). "Thrombin" was eluted with NaCl (8 gm/L; 0.14 M) and the turbid solution subjected to repeated extraction with chloroform until a clear solution was obtained. However, the properties of the product raise some question, as it was said to be stable to boiling. There are reasons to suspect the quality of the early fibrin ferment preparations and the relationship of such presentations to thrombin. However, this author acknowledges that it is very difficult to totally inactivate thrombin. The product obtained by Howell could be dialyzed (collodion tubes [nitrocellulose membranes][33-35]) in water but was said to be unstable in the absence of salt. The material could be precipitated with alcohol or ammonium sulfate (half-saturation). The material also provided a positive biuret reaction and a positive Millon reaction (tyrosine); it was also positive for tryptophan. Howell's product did behave as a globulin (insoluble in water, soluble in dilute salt solution). Gamgee[11] had previously shown that his fibrin ferment also behaved as globulin (insoluble in water) and was inactivated by heat, suggesting that the isolated material was a protide[36] (protein). There was some suggestion by Gangee[11] that the isolated material was an enzyme; hence, the term ferment[37] was used to describe the activity. In later work it will be seen that the suffix -ase is attached in a combination word, such as in thrombokinase, to indicate designation of the material as an enzyme. As noted by Gamgee,[11] Buchanan's work had been largely ignored, possibly obscured by the work of Alexander Schmidt who, for good reason, is frequently given the credit for the discovery of thrombin (fibrin ferment) for work first appearing in 1861,[28-30] culminating with a "magnum opus" in 1892.[31] It is accepted that Schmidt was the first to use the term thrombin to describe fibrin ferment activity based on his observation on the formation of a intravascular thrombus following administration of thrombin. In his early work, Schmidt questioned as to whether fibrin ferment was an enzyme favoring a concept where fibrin ferment combined with fibrinogen in the presence of a paraglobulin to form fibrin. In later work,[33] Schmidt did promote the enzymatic nature of fibrin ferment, and together with others including Hammarsten,[38] developed the concept of soluble fibrin as an intermediate in the formation of a fibrin clot. As will be discussed later, there were several theories for the formation of fibrin. There was a lively discussion of the nature, formation, and mechanism of action of fibrin ferment during this period of time (1860–1930).[5,6,39,40] As will be shown later, discussion over the identity of thrombin as an enzyme continued into middle of the 20th century.[41-50] Chargaff,[41] about whom I will write more later, noted: "Many of these discussions degenerated into mere exercises in terminological dialetics." I observe that blood coagulation has been cursed by terminology issues, leading to the formation of an international committee on nomenclature, which met in exotic places to assign Roman numerals to the various factors. I do grant that factor XI is a more useful term than "plasma thromboplastin antecedent," so there is some merit in the Roman numeral system. On the other hand, there was some grousing by an eminent hemophilia specialist that a patient with Christmas disease would not remember that he is deficient in factor IX or factor VIII, just that he had hemophilia. I would further observe that nomenclature is a problem that continues in science and may be getting worse, eliciting continuing comparison to the biblical tale of the Tower of Babel.[51–53] For someone like myself who has diverse interests, a Babel fish[54] may be needed.

In 1905, Paul Morawitz[2] attempted to rationalize the information of the past several centuries into a model for the coagulation of blood. He postulated that four factors were important for the formation of a blood clot: prothrombin (preferment; thrombogen), calcium salts, fibrinogen, and thrombokinase. The Morawitz hypothesis has been referred to as the four-factor model.[55] Thrombokinase was a factor considered to be an enzyme; hence, the term kinase. derived from the disintegration of platelets and leukocytes. Schmidt[31] was a leader in promoting the role of cells in the process of blood coagulation. Schmidt and others proposed the existence of a "zymoplastic" substance that was responsible for the conversion of preferment to ferment. It is of interest to be able to take this early data and fit the observations into current concepts. Today we know that platelets can contribute factor V activity[56,57] and neutrophils can contribute tissue factor.[58–60] Thus, with some "enhancement" of the concept developed by Morawitz,[2] there is ample support for his suggested mechanism, which is consistent with the formation of the complex known as prothrombinase.[61–63] Thrombokinase was described as a concept during the 19th century as a substance important in the formation of fibrin ferment from prothrombin (thrombogen).[2,64,65] Thrombokinase is also known as zymoplastic substance or cytozyme.[66] The nature of this material is not clear, but it is not unreasonable that the material is tissue thromboplastin, known today as tissue factor. Mellanby[67] demonstrated that thrombokinase formed fibrin ferment (thrombin) from prothrombin in a manner similar to the conversion of trypsinogen to trypsin by enterokinase. It is of interest that the time course for the evolution of fibrin ferment is sigmoidal, similar to that observed with subsequent work on prothrombin activation.[68] It is of interest that Wooldridge[69] suggested the fibrin ferment was a product of plasma and not derived from cells. Wooldridge obtained "salted plasma," which was free of observable cellular elements (canine blood collected into an equal volume of 10% common salt, presumed to be sodium chloride and subjected to centrifugation [conditions not defined]). Such plasma clots on dilution with five volumes of water. The serum obtained from clotting of the diluted plasma and the undiluted plasma were each treated with a large excess of absolute alcohol and allowed to stand for a considerable period of time (weeks). The precipitate obtained from each was taken up in water. The material obtained from the diluted plasma (serum) clotted magnesium sulfate plasma, while the material obtained from undiluted plasma did not. Magnesium sulfate was one of the several inorganic salts used as an anticoagulant in early studies.[70,71] Wooldridge concluded that the substances responsible for the formation of fibrin ferment were intrinsic to plasma. However, he also suggested that fibrin ferment was free of protides. I would note a consideration of their data suggests that the results could have been obtained with a very low concentration of thrombin; data was not available as to protein concentration. It is possible that the protein concentration was too low to be detected by the various precipitation techniques.

While questions regarding thrombin as an enzyme continued into the middle of the 20th century, this did not stop continuing work on the mechanism of thrombin formation. In support of earlier work, Ferguson and coworkers[50] identified prothrombin as a precursor of thrombin, which was activated by a thromboplastic enzyme in the presence of calcium ions and phospholipid. Other researchers obtained a prothrombin preparation that was activated by a thromboplastin derived from tissue in the presence of calcium ions;[72] activation of prothrombin also occurred in the presence of strontium ions but was slower. A number of prothrombin preparations were obtained that demonstrated activation in the presence of calcium ions. Ken Robbins, who was later a leader in fibrinolytic therapeutics,[73-75] obtained a prothrombin (prothrombase) preparation by isoelectric precipitation (pH 5.3) from diluted plasma,[76,77] following the work of Mellanby.[78] This product formed thrombin (thrombase) in the presence of calcium ions, suggesting the presence of factors VII and X (impurities, as suggested by Chargaff[5]). Milstone, in studies[79-83] at Yale University, brought considerable clarity to the activation of prothrombin by work on thrombokinase and thromboplastin. Thrombokinase was later shown to be identical to factor Xa.[84] Thromboplastin was an evolving concept into the 1940s,[85] and its role was not understood until the identification of blood coagulation factor VII (proconvertin, pro-serum prothrombin conversion accelerator, pro-SPCA).[86] Some years later, the late Yale Nemerson, also at Yale University, brought further clarity to the nature of tissue factor.[87] Milstone can be given credit in establishing the enzymatic activation of prothrombin to form thrombin, as well as the need for the enzymatic conversion of a precursor (prokinase) to the active component (kinase), which would then convert prothrombin to thrombin. A careful consideration of the data suggests that the preparations of thrombokinase (factor Xa) and thrombin were likely contaminated. Milstone did report the activation of chymotrypsinogen by his thrombokinase preparations,[82] while others reported the activation of chymotrypsinogen by thrombin.[88] Milstone did report the separation of thrombokinase from thrombin in commercial thrombin.[83] Subsequent work showed that thrombokinase was identical to factor Xa and autoprothrombin C.[89,90] Milstone was somewhat hesitant in accepting the identity of thrombokinase with factor Xa.[91] But Milstone should be given credit for establishing firmly the enzymatic nature of prothrombin activation to thrombin by thrombokinase, as well as the recognition that thrombokinase also existed in a precursor form. Milstone's work and the work of a number of others led to the development of the cascade or waterfall hypothesis for the process of blood coagulation.[92,93]

In 1935, Eagle[94] demonstrated that prothrombin could be isolated as a euglobulin from citrated or oxalated plasma after dilution and dispersion of CO_2. The infusion of CO_2 lowers the pH, permitting isoelectric precipitation of prothrombin together with other proteins, including components of what is known as a prothrombin complex concentrate (factors II, VII, IX, and X). This material could be activated to form thrombin and the rate of activation could be accelerated either by platelets or cephalin (the ether extract of an acetone powder of porcine brain). Eagle observed that the platelet-free plasma failed to clot and suggested that the slow rate of thrombin formation permitted inhibition by antithrombin before there could be substantial fibrin formation. Eagle also summarized the state of understanding of prothrombin activation (classical theories as to the mechanism of coagulation), concluding that his work supported the concept that thrombin was a cause of fibrinogen clotting and not a product of fibrinogen clotting. I find it of interest that Eagle[94] also commented on the nomenclature issues in the study of blood coagulation stating, "If we disregard the nosological differences, and see the basic observations." This is a most useful concept which is as applicable today as it was then. The data obtained in these early studies can be fitted quite nicely into current concepts. As an example, consider Chargaff's statement that the "autoactivation of prothrombin" in the presence of calcium ions can be ascribed to the presence of impurities.[5] While information is not available, the various prothrombin preparations isolated from plasma at this time contained substantial amounts of factor VII (proconvertin; pro-serum prothrombin conversion accelerator), factor IX (Christmas factor), and factor X (Stuart Prower factor), as in some of the current prothrombin complex concentrates. Factors VII, IX, and X had yet to be described. Metathrombin is an old term used to describe the thrombin-antithrombin complex, where metathrombin in serum dissociates to form active thrombin and modified antithrombin. Early researchers considered metathrombin dissociation a potential mechanism for thrombin formation during coagulation.[95–97] Before leaving Harry Eagle for the moment, it is important to point out that Eagle was responsible for a major development in cell culture technology: Eagle's minimal essential media.[98–100]

Another observation that can be clarified by present knowledge is that of Freund in 1886,[101] who observed that a glass rod coated with oil did not promote coagulation of blood, while an uncoated rod promoted blood coagulation. Today we recognize the role of contact activation (factor XII, Hageman factor) in the initiation of blood coagulation.[102] Current thinking suggests that the tissue factor pathway (extrinsic blood coagulation) is the dominant process in the hemostatic response, with the intrinsic pathway important for the propagation or stabilization phase. I remember the late Yale Nemerson telling me that siliconized glassware delayed the discovery of the importance of the tissue factor pathway by a least a decade.

A major problem in the early work in blood coagulation was a lack of understanding of the role of divalent cations in coagulation.[103] Early work used a variety of inorganic salts, such as sodium chloride or magnesium sulfate, at high concentrations such that dilution or addition of fibrin ferment promoted coagulation.[48,69,71] Peptone (a mixture of peptides obtained from the partial digestion of a protein or proteins by proteases) was also used as an anticoagulant.[104] Barrett[104] brought clarification to the mechanism of the peptone effect. It had been argued that the effect was directed at the conversion of fibrinogen to fibrin. Barrett studied the effect of peptone on the formation of fibrin in plasma using *Echis carrinatus* (Ecarin) instead of thrombin to form fibrin. In course of this work, Barrett did observe the effect of fibrinogen concentration on the quality of the fibrin clot. Barrett also observed that peptone affected the polymerization of fibrin and hence the ability to observe a clot. A more recent contribution by Marcum[105] has established that the *in vivo* effect of peptone is the mobilization of heparin. It has been suggested that J.R. Green was responsible for the observation that calcium ions accelerated the process of blood coagulation,[70] with the observation that the addition of calcium sulfate accelerated the coagulation of blood. A major step forward was the discovery of oxalate as an anticoagulant[106–109] and the importance of calcium ions, as well as other metal ions such as strontium, in the process of blood coagulation.[109] Arthus[109] also compared the coagulation of blood to the coagulation of milk, which is an enzymatic reaction catalyzed by rennin.[110] Citrate replaced oxalate as an anticoagulant for use in transfusion medicine.[111,112] Mellanby[67] used avian blood for his early studies, as it did not clot for some time after removal from the source (I suspect chicken; the term fowl was used in the text). This phenomena can be understood today by the absence of factor XII (Hageman factor) and factor XI (plasma thromboplastin antecedent) in avian blood.[102,113–116] Bird (cockerel) plasma continued to be used because of this quality of not requiring anticoagulant (heparin).[117] This cited study described early work on fibrin sealant during WWII, which involved Peter Medawar (Sir Peter Medawar; Nobel Laureate in Physiology or Medicine, 1960. for work in immunology). While this work was very promising, the development of an effective fibrin sealant (fibrin glue) required a higher fibrinogen concentration than that available in plasma. Nevertheless, this early work demonstrated the value of a fibrin product in tissue repair. Medawar was one of a number of famous scientists involved in work during WWII on thrombin and fibrinogen. Notable among this group was Erwin Chargaff, who went on to make fundamental discoveries in the structure of DNA that were critical to the development of the double helix structure.[118] In humans, the absence of factor XII does not result in hemorrhagic diathesis,[119] and factor IX deficiency results in a mild deficiency.[120]

The identification of thrombin as a serine protease took a considerable period of time following the early observations of Buchanan.[8–11] First it was necessary to establish that thrombin was an enzyme. As stated by Eagle in 1935,[44] while the preponderance of evidence suggested that thrombin is an enzyme, there was still some question. Copley also suggested that thrombin was an enzyme.[45] In his work,[45] Copley did make the argument that the action of thrombin on fibrinogen was a specific interaction. Copley also studied the inactivation of thrombin by a substance (antithrombin) present in serum, forming a product which others described as a metathrombin.[2] Alfred L. Copley provided some early contribution to blood coagulation, but is best known for his subsequent development of biorheology[46] and hemorheology.[47] Copley is also well-known as an artist for his abstract expressionist paintings under the name of L. Alcopley.[48]

The concept that the formation of a complex between thrombin and fibrinogen was responsible for the fibrin clot continued into the early part of the 20th century.[12,13] Our understanding of the relationship between protein in enzymes was still developing at this point. It was only with the work of James Sumner in 1926 on the crystallization of urease[121–123] that there was solid data supporting the chemical nature of enzymes as proteins. Even with the publication of Sumner's work, it took considerable time for the concept of enzymes as proteins to gain acceptance.[123] Simoni and coworkers[123] presented an excellent discussion of Sumner and his work as part of the JBC Centennial in 2002. There was work that complicated the recognition of thrombin as a proteolytic enzyme. Notable was the observation that certain organic compounds could "clot" fibrinogen.[124,125] Ninhydrin, chloramine-T (N-chlorotosylamide; sodium; chloro-[4-methylphenyl]sulfonylazanide), and several naphthoquinone sulfonates were reported to react with fibrinogen to yield an insoluble product resembling a fibrin clot.[125] Reducing agents such as sodium bisulfite and glutathione inhibited the action of ninhydrin or the naphthoquinone sulfonates on fibrinogen; sodium bisulfite, but not glutathione, inhibited the clotting of horse fibrinogen by bovine thrombin. These investigators suggested that thrombin could be an oxidase. There were others who suggested that thrombin formed a clot from fibrinogen by a process of disulfide interchange or oxidation.[126–129] While subsequent work showed that sulfhydryl groups/disulfide bonds were not important in the formation of a fibrin clot, allosteric disulfides have been identified as being important in the process of hemostasis.[130,131] Some years after his early work on the clotting of fibrinogen which suggested that thrombin was an oxidase, Chargaff[5] defined thrombin as a *proteotropic* enzyme that causes an irreversible change in the structure of fibrinogen by an undefined mechanism. Mellanby[132] defined thrombin (thrombase) as a *proteoelastic* enzyme[133] that cleaved fibrinogen into fibrin and a serum globulin, which remained in solution. Mellanby observed that the clotting time of plasma with thrombase was inversely proportional to the amount of thrombase. He also observed that a portion of the thrombase was removed from solution on the clotting of plasma, which could be recovered by treatment of the clot with NaCl solution (approximately 1.4 M NaCl). Other investigators suggested that thrombin was a denaturase.[134–136] Challenging earlier work which suggested that thrombin was a hydrolytic enzyme, Strughold and Wöhlisch[134] were not able to demonstrate the hydrolysis of fibrinogen by thrombin. These experiments compared the action of thrombin to that of trypsin-kinase (trypsin). They suggested that thrombin was instead a denaturing enzyme. In subsequent experiments,[135] Wöhlisch and Jühling advanced denaturase in experiments comparing the action of thrombin with papain in the clotting of fibrinogen. They suggested that the denaturing process in fibrinogen clotting could involve protein hydrolysis. These investigators showed that heat treatment (incomplete thermal denaturation) enhanced the action of either thrombin or papain in the clotting of fibrinogen. This observation was one of several diverse observations leading to the demonstration that there was an intermediate (profibrin) in the formation of the fibrin clot. The observation that papain clotted fibrinogen has been of continuing interest. There was some thought that the papain clotting of fibrinogen involved the action of factor XIIIa (fibrin-stabilizing factor). Russ Doolittle showed that papain directly clotted fibrinogen, but there were differences in the fibrin structure.[137] Protein denaturation was a concept evoked by several investigators in explaining the conversion of fibrinogen to fibrin. The formation of insoluble protein was a criteria for protein denaturation associated with protein unfolding at this period of time.[138] Haurowitz[138] did advance the concept of a denaturase to explain the process of proteolysis of native proteins, where cleavage of the first peptide bond increases susceptibility of other peptide bonds to cleavage. Haurowitz argued that the first step in the proteolysis of a native protein could be considered to be a "denaturase." Discussion on the role of denaturation in the proteolysis of native proteins continues. Linderstrøm-Lang and colleagues[139] had previously argued that the proteolysis of native proteins relied upon the proposed equilibrium between the native form and denatured form of a protein, where the denatured form would be susceptible to proteolysis.[140]

Boyles and coworkers[50] presented other evidence to support thrombin as an enzyme. They showed that the final yield of fibrin (by weight) was independent of thrombin concentration and that the action of thrombin continued after visible clot formation. The effect of acacia on decreasing the fibrinogen clotting time was described as "fibrinoplastic." They also separated the effect of calcium chloride as a "specific salt effect" from that of sodium chloride as a "non-specific salt effect."

However, these investigators did use sodium chloride to study the "profibrin" concept. A number of investigators had advanced the profibrin concept as an intermediate step in the formation of fibrin (*Fasserstoff*). Boyles and coworkers[50] showed that heating of fibrinogen in sodium chloride (12 minutes/48°C/0.86 M NaCl) enhanced the rate of fibrin clot formation with no effect on fibrin yield after cooling and decreasing NaCl concentration by dilution with water. They interpreted this as a partial denaturation of fibrinogen, which provided a profibrin intermediate in the formation of fibrin. Boyles and coworkers[44] also showed the inactivation of thrombin by trypsin in a process they described as thrombolysis. While thrombin was being accepted as an enzyme, it was still not accepted as a protease into the 1940s.[5,95,97]

Lyons[141] demonstrated the presence of an intermediate, "fibrinogen b," in the clotting of fibrinogen by thrombin. He also suggested a separate thrombin species, thrombin A and thrombin B, were responsible for the separate steps in the conversion of fibrinogen to fibrin via fibrinogen B. In 1935 Eagle showed that thrombin did not form a stoichiometric complex with fibrinogen/fibrin but was rapidly incorporated into the fibrin product during clotting.[94] Thrombin could be expressed from the fibrin clot that was formed with a crude euglobulin precipitate. Calcium is not thought to have an effect on the catalytic activity of thrombin, but may affect fibrinogen polymerization.[142–145] Bailey and Bettelheim reviewed the status of thrombin as an enzyme in 1955, showing that there was still a lack of consensus on the issue.[146] However, this was rapidly changing, as there were studies appearing showing the release of peptide material from fibrinogen during clotting by thrombin.[147–150]

Understanding of the enzymatic nature of thrombin was frustrated by the lack of sufficiently purified proteins (thrombin and fibrinogen), the lack of technologies, and the lack of basic science information on the isolation and characterization of proteins. There were early attempts to purify prothrombin by salt precipitation and isoelectric precipitation. It is recognized that Mellanby made a major contribution to prothrombin purification in 1930 by further fractionating the isoelectric precipitate by solubilization in calcium bicarbonate.[78] The next major advance in the purification of prothrombin occurred in the laboratory of H.P. Smith in the Department of Pathology at the University of Iowa.[151–153] This group extended the work of Mellanby[78] by taking the solubilized isoelectric precipitate and adsorbing the prothrombin with magnesium hydroxide and eluting it with bicarbonate. After dialysis and removal of insoluble material, the bovine prothrombin was obtained with a second isoelectric (pH 5.3) precipitation. Thrombin was obtained from this material by activation with tissue thromboplastin (bovine lung) in the presence of calcium ions.[154] Thrombin was purified from the activation mixture by acetone precipitation. This material had 300–540 Iowa units per mg of dried material; 1 Iowa unit is equal to 0.83 NIH units. This would give a value of specific activity of 250–400 NIH units/mg, which would suggest a purity of 20% compared to the current purified material.[155] Seghatchian reviewed thrombin units in 1984.[156] As noted, 1 NIH unit is equal to 1.2 Iowa units, 1 WHO unit is equal to 0.667 Iowa units or 0.564 units, and 1 WHO unit represents 2.48×10^{-2} nmol.

This work involved Walter Seegers who, after some time at Parke-Davis, joined the faculty of the School of Medicine at Wayne State University and was a leader in research on prothrombin and thrombin for generations. He was also famous for his unique contributions to nomenclature, as well as the "Seeger's Symposia." Kenneth Brinkhous was also part of this research group and later joined the Department of Pathology at the University of North Carolina at Chapel Hill, where he was a leader in hemophilia research for generations. The method developed for thrombin purification by the Iowa group[157,158] was the basis of the topical thrombin developed by Parke-Davis.[159,160] I recall a conversation with Brinkhous in Chapel Hill shortly after my arrival there. He told me that he (meaning the Iowa group) thought that every barber shop would stock their thrombin to take care of cuts from shaving customers. I would think that most of the readers today will not recall a day when men routinely were shaved in barber shops. Brinkhous stated the rationale for topical thrombin in 1947,[161] noting the amount of topical thrombin was substantially more than could be contributed by blood at the site of application and that the time required for thrombin formation was much greater than that required for the formation of a fibrin clot. There is no question that the Iowa group made major contributions to thrombin, which Seegers continued at Wayne State. I would note that the work on prothrombin purification at Iowa ended in 1940 with the start of WWII but was carried on by Walter Seegers and coworkers in later work.[162,163] The purified bovine prothrombin was not stable, but was stabilized in the presence of oxalated bovine plasma.[163]

One of the more interesting observations on the formation of thrombin from prothrombin was the discovery of citrate activation. Seegers and colleagues observed that prothrombin formed thrombin in the presence of 25% sodium citrate.[164–166] This was a useful observation as it did provide a process for the production of thrombin. As would be expected from data obtained in earlier studies of prothrombin activation, later work from the Seegers laboratory showed the activation of factor X occurred in these preparations of prothrombin, forming Xa,[167] which could then activate prothrombin under these reaction conditions. Completing the story, David Aronson, one of the more underrated investigators in blood coagulation, showed that factor VII was required for the citrate activation of prothrombin via the activation of factor X.[168] Scheraga and Laskowski[169] classified the various thrombin preparations as citrate thrombin as that derived from citrate activation, biothrombin as that derived from thromboplastin activation, and trypsin thrombin as that obtained by the activation of prothrombin by trypsin. The activation of prothrombin by trypsin likely occurs via action of factor X, which could be present as a contaminant in prothrombin.[170,171]

WWII resulted in the development of a unique organization: the Center for Blood Research at Harvard Medical School, which developed a scheme for the fractionation of human plasma. The Center for Blood Research was organized by Edwin J. Cohn out of the Department of Physical Chemistry at Harvard Medical School. The primary focus of the program was the production of albumin for the treatment of shock.[172] The albumin was intended to replace plasma (frozen and dried).[173] Plasma had been advocated for the treatment of shock in 1918,[174,175] but was placed on a solid basis during WWII.[173] One note on the work on shock from 1918: Peyton Rous, who was active in transfusion medicine,[174] was later responsible for the discovery of tumor-inducing viruses, for which he was awarded the Nobel Prize in 1916.[176,177] The choice of albumin was based on its importance in maintaining the osmotic pressure in the circulation, as well the ability to use small packaging. There were significant logistical challenges in this project; there was no blood banking system, and technology both for plasma acquisition and plasma fractionation had to be developed. There was a program to use bovine albumin, but that was discontinued in 1943 after serious adverse reactions, including one fatality. The program to develop human albumin as a plasma substitute started in 1940 and was closed in October 1944. It is unfortunate that the infrastructure developed for blood procurement was not continued, and there was still no organized blood banking program when the Korean War started in 1950. Fatty acids bind to albumin, and it is notable that J. Murray Luck at Stanford University was involved in the study of the role of fatty acids in stabilizing human albumin. Professor Luck had a distinguished research career but may be best known as the founding editor of *Annual Reviews of Biochemistry*.

There was use of by-products of plasma fractionation, with an emphasis on immune serum globulin (currently known as intravenous immune globulin [IVIG]), fibrin foam, and fibrin film.[178] Under the leadership of Edwin Cohn, a method (the Cohn method)[179,180] was developed for the fractionation of human plasma that still provides the technical basis for the plasma fractionation industry.[181] Cohn had assembled a group of scientists of considerable talent, including John Edsall, who was later the editor of the *Journal of Biological Chemistry* for several decades. Edsall was an excellent physical biochemist who, with Cohn, developed the use of organic solvents (ethyl alcohol) to separate/fractionate plasma proteins into various fractions, which are still known as Cohn fractions.[182] After WWII, Edsall became editor of the *Journal of Biological Chemistry* in 1958 and continued to make major contributions to protein chemistry. While the major focus of the plasma fractionation effort was the production of albumin for transfusion medicine, the program also produced prothrombin, thrombin, and fibrinogen.[183] These materials were produced as part of the WWII war effort. This particular paper by Edsall and colleagues contains little experimental information. As it was part of the WWII war effort, the research was classified. Brief reports of the research can be accessed at the Library of Congress. However, this does not diminish the importance of this work or, for that matter, any of the various papers published by the Cohn group and associates in this issue (volume 23, issue 4) of the *Journal of Clinical Investigation*.

A crude thrombin preparation was obtained by Parfentjev from an ammonium sulfate fraction derived from rabbit plasma.[184] This material is dialyzed for several days and concentrated by isoelectric precipitation. The Parfentjev process was further developed by another group at Harvard Medical School[185] for use as a therapeutic.[186] The Harvard group appears to have modified the original Parfentjev process by stirring the first ammonium sulfate precipitation at 37.5°C for 16–18 hours. Ammonium sulfate can promote the "autoactivation" of prothrombin.[187] A similar material was obtained from bovine, porcine, and human plasma.[188]

The end of WWII brought an increase in the study of thrombin. While human thrombin saw use as a therapeutic during WWII and for some time after WWII, use stopped after the discovery of hepatitis virus in human thrombin products[189,190] and the dismantling of the blood coagulation and processing system developed during WWII. The bovine topical product has dominated the therapeutic market for the past 60 years.

Kenneth Bailey's laboratory published a number of papers some 60+ years ago[146–149] when I was in junior high school in Glendale, California, which firmly established that thrombin was a proteolytic enzyme in experiments showing the release of peptide material from fibrinogen during clotting by thrombin. Subsequently, Jules Gladner and Kolman Laki[191] showed that thrombin was inactivated by reaction with diisopropylphorphorophosphate (DFP; 2-[fluoro(propan-2-yloxy)phorphoryl]oxy-propane)). This was the first suggestion that thrombin was a serine protease[192] similar to chymotrypsin[193,194] and trypsin.[195] DFP had been developed as a nerve gas during WWII[196] and shown to inhibit acetylcholine esterase (AChE; cholinesterase).[197] These observations were followed by studies which showed that chymotrypsin, trypsin, and thrombin are similar enzymes.[198–203] The primary points of similarity were the hydrolysis of tosyl-arginine methyl esters (TAME)[204,205] and the common active site sequence of amino acids, GlyAspSerGly, for trypsin, chymotrypsin, and thrombin.[200] There was also increasing evidence for the presence of an active site for histidine residue in thrombin, chymotrypsin, and trypsin.[198] The nature of the fibrinogen-fibrin transition was also being clarified during this time period, with the concept of limited proteolysis being used to explain the action of thrombin on fibrinogen.[206] This concept was based on the clotting of milk by rennin[207] and the formation of plakalbumin from ovalbumin by subtilisin.[208–210] The term plakalbumin is derived from the Greek for plate, reflecting the plates formed by plakalbumin crystals. The concept of limited proteolysis appears to have been advanced by Desnuelle in 1955,[211] as cited by Mihalyi in his comprehensive work on proteolysis.[212]

There was an increase in the technologies available to study proteins between 1950 and 1960. Ion-exchange resins developed during WWII were being used to separate peptides and proteins.[213] A technique for the determination of amino acids by column chromatography was also developed during this time frame,[214,215] as well as a method, the Edman degradation, for the determination of the sequence of amino acids.[216,217] Rasmussen took advantage of the availability of ion-exchange resins to purify thrombin with Amberlite IRC-50.[218] Rasmussen observed that despite an isoelectric point between 4.7 (low) and 6 (high), thrombin behaved as a cation at pH 7.0 in sodium phosphate buffer. His choice of IRC-50 (cation exchange resin) was based on previous observations that thrombin bound to glass (which bears a diffuse negative charge) at pH 7.0.[219,220] Rasmussen did suggest that regardless of isoelectric point, thrombin may not be an acid protein. This was an earlier indication of the presence of anion-binding exosites. Miller[221] subsequently showed that impurities in either biothrombin or citrate thrombin could be removed by precipitation with Rivanol (ethacridine lactate; 7-ethoxyacridine-3,9-diamine; 2-hydroxypropanoic acid). A more efficient procedure used adsorption and subsequent elution of crude thrombin from CG-50 (cation exchange) resin previously equilibrated with ethacridine lactate. Ethacridine lactate has been used in the purification of antibody fragments produced in *Escherichia coli*, where it precipitates host cell protein.[222] In earlier work, separation by precipitation of free ligand (thyroxin) from bound ligand in a radioimmunoassay was accomplished with ethacridine lactate.[223] Miller[224] extended his work on thrombin purification with the purification of biothrombin with chromatography on CG-50 resin in either cacodylate buffer or tris buffer at pH 7.0. John Fenton, in two later publications, improved the Miller process for the purification of human thrombin on CG-50, starting with human Cohn III paste.[225,226] Prothrombin was purified from Cohn III paste by barium citrate precipitation and elution and activated with thromboplastin. The crude thrombin was purified by chromatography on CG-50 at pH 8.0. It is fair to say that the provision of thrombin by John Fenton enabled a great amount of work on thrombin for the next several decades. At the same time I developed a procedure for the purification of bovine thrombin using sulfoethyl-sephadex.[227] At that time, we needed the thrombin for work on factor VIII,[228] but thrombin became of major interest to my laboratory for the next decade. My choice of Sulfoethyl-Sephadex was based on a conversation with Stanford Moore, who mentioned they had substituted Sulfoethyl-Sephadex for IRC-50 in work on pancreatic ribonuclease. I would note that at this time neither John Fenton nor I picked up on the Rasmussen comment about thrombin as an acid protein. However, Fenton and others shortly later developed the exosite (see Chapter 3) concept for thrombin.[229,230]

It is fair to say the accomplishments in the purification of thrombin in the 1960–1975 time frame ended the early phase of thrombin research, with the state of the work summarized in the proceedings of a conference held at the Mayo Clinic on March 31–April 1, 1977.[231] The conference contained diverse contributions showing progress in structural analysis and interactions with other proteins and with cells. All of this was made possible largely through the availability of a purified thrombin. Work since that conference has focused on the interaction with cells with the discovery of PAR (protease activated receptors) receptors. While redundant, the term PAR receptor is commonly used. The PAR receptors were discovered by Shaun Coughlin and colleagues at the University of California in San Francisco.[232,233] Staffan Magnusson[234] and others solved the primary structure of thrombin,[235,236] confirming homology with other serine proteases.[203] The crystal structure of thrombin and the production of prothrombin/thrombin by recombinant DNA technology was in the future, as was the elucidation of thrombin actions outside of the vascular bed.

REFERENCES

1. Tsoucalas, G., Karamanou, M., Papaioannou, T.G., and Sgantzos, M., Theories about blood coagulation in the writings of ancient Greek medico-philosophers, *Curr. Pharm. Des.* 23, 1275–1278, 2017.
2. Morawitz, P., Die Chemie der Blutgerinnung, *Ergebn. Physiol.* 4, 307–405, 1905 (Morawitz, P., *The Chemistry of Blood Coagulation*, tr. R.C. Hartmann and P.F. Guenther, C.C. Thomas, Springfield, Illinois, 1958).
3. Hewson, W., *An Experimental Inquiry into the Properties of Blood*, T. Cadell, London, UK, 1771.
4. Lister, J., On the coagulation of blood, *Proc. Roy. Soc. Lond.* 12, 580–611, 1863–1863.
5. Chargaff, E., The coagulation of blood, *Adv. Enzymol.* 5. 31–65, 1945.
6. Brücke, E., An essay on the coagulation of blood, *Brit. Foreign Med. Chir. Rev.* 19, 183–212, 1857.
7. Owen, C.A., *A History of Blood Coagulation*, ed. W.L. Nichols and E.J.W. Bowie, Mayo Foundation for Medical Research, Rochester, Minnesoate, USA, 2001.
8. Marcum, J.A., Defending the priority of 'remarkable research': the discovery of fibrin ferment, *Hist. Philos. Life Sci.* 20, 51–76, 1998.

9. Buchanan, A., Contributions to the physiology and pathology of the animal fluids, *London Medical Gazette* 18, 50–54, 1835.

10. Buchanan, A., On the coagulation of blood and other fibriniferous liquids, *London Medical Gazette* volume 1 (new series), 617, 1845 (as cited in reference 4).

11. Gamgee, A., Some old and new experiments on the fibrin-ferment, *J. Physiol.* 2, 145–157, 1878.

12. Rettiger, L.A., The coagulation of blood, *Am. J. Physiol.* 24, 406–435, 1909.

13. Howell, W.H., The preparation and properties of thrombin, together with observations on antithrombin and prothrombin, *Am. J. Physiol.* 26, 453–473, 1910.

14. Wilson, S.J., Quantitative studies on antithrombin, *Arch. Int. Med.* 69, 647–661, 1942.

15. Zhu, S., Lu, Y., Sinne, T., and Diamond, S.L., Dynamics of thrombin generation and flux from clots during whole human blood flow over collagen/tissue factor surfaces, *J. Biol. Chem.* 291, 23027–23035, 2016.

16. Fell, C., Ivanovic, N., Johnson, S.A., and Seegers, W.H., Differentiation of plasma antithrombin activities, *Proc. Soc. Exptl. Biol. Med.* 85, 194–202, 1954.

17. Mosesson, M.M., Updates of antithrombin I (fibrin), *Thromb. Haemost.* 98, 105–108, 2007.

18. Abildgaard, U., Antithrombin: early prophecies and present challenges, *Thromb. Haemost.* 98, 97–104, 2007.

19. Wilner, G.D., Danitz, M.P., Mudd, M.S., Hsieh, K.H., and Fenton, J.W., 2nd, Selective immobilization of alpha-thrombin by surface-bound fibrin, *J. Lab. Clin. Med.* 97, 403–411, 1981.

20. Francis, C.W., Markham, R.E., Jr., Barlow, G.H., *et al.*, Thrombin activity of fibrin thrombi and soluble plasmic derivatives, *J. Lab. Clin. Med.* 102, 220–230, 1983.

21. Kumar, R., Béguin, S., and Hemker, H.C., The influence of fibrinogen and fibrin on thrombin generation—evidence for feedback activation of the clotting system by clot bound thrombin, *Thromb. Haemost.* 72, 713–721, 1994.

22. Weitz, J.I., Hudoba, M., Massal, D., Maraganore, J., and Hirsh, J., Clot-bound thrombin is protected from inhibition by heparin-antithrombin III but is susceptible to inactivation by antithrombin III -independent inhibitors, *J. Clin. Invest.* 86, 385–391, 1990.

23. Pospisil, C.H., Stafford, A.R., Fredenburgh, J.C., and Weitz, J.I., Evidence that both exosites on thrombin participate in its high affinity interaction with fibrin, *J. Biol. Chem.* 278, 21584–21591, 2003.

24. Wu, H.F., White, G.C., II., Workman, E.F., Jr., Jenzano, J.W., and Lundblad, R.L., Affinity chromatography of platelets on immobilized thrombin: retention of catalytic activity by platelet-bound thrombin, *Thromb. Res.* 67, 419–427, 1992.

25. Lundblad, R.L. and White, G.C., 2nd, The interaction of thrombin with blood platelets, *Platelets* 16, 373–385, 2005.

26. Chao, F.C., Tullis, J.L., Kenney, D.M., Conneely, G.S., and Doyle, J.R., Concentration effects of platelets, fibrinogen and thrombin on platelet aggregation and fibrin clotting, *Thromb. Diath. Haemorrh.* 32, 216–231, 1974.

27. Watanabe, K., Chao, F.C., and Tullis, J.L., Platelet antithrombins: role of thrombin binding and the release of platelet fibrinogen, *Br. J. Haematol.* 35, 123–133, 1977.

28. Schmidt, A., Ueber den Faserstoff und die Ursache seiner Gerinnung. *Arch. Anat. Physiol.* (Reichert-Dubois-Reymonds), 545, 675, 1861.

29. Schmidt, A., Neue Untersuchungen über die Fasterstoffgerinnung. *Pflügers Arch. ges. Physiol.* 6, 413–538, 1872.

30. Schmidt, A., *Die Lehre von den fermentativen Gerinnungserscheinungen in den eiweissartgen thierischen Körperflüssigkeiten*, C. Matthieson, Dorpat, Germany, 1876.

31. Schmidt, A., *Zur Blutlehre*, Vogel, Leipzig, Germany, 1892.

32. Howell, W.H., The preparation and properties of thrombin, together with observations on antithrombin and prothrombin, *Am. J. Physiol.* 26, 453–473, 1910.
33. Bigelow, S.L., Collodion membranes, *J. Am. Chem. Soc.* 29, 1576–1589, 1907.
34. Kunitz, M. and Simms, H.S., Dialysis with stirring, *J. Gen. Physiol.* 11, 641–644, 1928.
35. Bugher, J.C., Characteristics of collodion membranes for ultrafiltration. *J. Gen. Physiol.* 36, 431–448, 1953.
36. Protide is term used to describe a protein; the term is rarely used to in the current literature.
37. Harden, A., The alcoholic ferment of yeast juice, *Proc. Roy. Soc. B Biol. Sci.* 77(519), 405–420, 1905.
38. Hammarsten, O., Zur Lehre von der Faserstoffgerinnung. *Pflüg Arch. ges. Physiol.* 14, 211–283, 1877.
39. Brücke, E., Ueber die Ursache der Gerinnung dews Bluts, *Arch. Pathol. Anat. Physiol. Klin. Med.* 12, 81–100, 1857.
40. Beck, E.A., The discovery of fibrinogen and its conversion to fibrin, in *Fibrinogen*, ed. K. Laki, Chapter 2, pp. 25–37, Marcel Dekker, New York, USA, 1968.
41. Beck, E.A., Historical development of the prothrombin concept, in *Prothrombin and Related Coagulation Factors*, ed. H.C. Hemker and J.J. Veltkamp, Leiden University Press, Leiden, Netherlands, 1975.
42. Edsall, J.T., Ferry, R.M., and Armstrong, S.H., Jr., Chemical, clinical and immunological studies on the products of human plasma fractionation. XV. The proteins concerned in the blood coagulation mechanism, *J. Clin. Invest.* 23, 557–565, 1944.
43. Mellanby, J., Thrombase: its preparation an properties, *Proc. Roy. Soc. London B* 113, 93–106, 1933.
44. Eagle, H., Studies on blood coagulation. II. The formation of fibrin from thrombin and fibrinogen, *J. Gen. Physiol.* 18, 547–554, 1935.
45. Copley, A.L., On the specificity of thrombin action, *Am. J. Physiol.* 126, 310–315, 1935.
46. Copley, A.L., Fluid mechanics and biorheology, *Biorheology* 27, 3–19, 1990.
47. Copley, A.L., On the practice of basic and clinical hemorheology at blood transfusion centers, *Biorheology* 17, 5–7, 1980.
48. www.moma.org/artists/104
49. Ferguson, J.H. and Glazko, A.J., Heparin, *J. Lab. Clin. Med.* 26, 1559–1564, 1941.
50. Boyles, P.W., Ferguson, J.H., and Muehlke, P.H., Mechanisms involved in fibrin formation, *J. Gen. Physiol.* 34, 493–513, 1950.
51. Bean, W.B., Tower and Babel, *Arch. Intern. Med.* 110, 375–381, 1962.
52. Nelson, D.R., Gene nomenclature by default, or BLASTing to Babel, *Hun. Genomics* 2, 196–201, 2005.
53. Murray, P.J., Allen, J.E., Biswas, S.K., *et al.*, Macrophage activation and polarization: nomenclature and experimental guidelines, *Immnity* 41, 14–20, 2014.
54. Adams, D., *Hitchhiker's Guide to the Galaxy*, Harmony Books, New York, USA, 1979.
55. Gray, C.R., On the coagulation of the blood: an elaboration of Lord Lister's hypothesis and the four-factor model of Morawitz, *Med. Hypotheses* 33, 63–68, 1990.
56. Viskup, R.W., Tracy, P.B., and Mann, K.G., The isolation of human platelet factor V, *Blood* 69, 1188–1195, 1987.
57. Camire, R.M., A new look at blood coagulation factor V, *Curr. Opin. Hematol.* 18, 338–342, 2011.
58. Maugeri, N. and Manfredi, A.A., Tissue factor expressed by neutrophils: another piece in the vascular inflammation puzzle, *Semin. Thromb. Hemost.* 41, 728–736, 2015.

59. Lisman, T., Platelet-neutrophil interaction as drivers of inflammatory and thrombotic disease, *Cell Tissue Res.* 371, 567–576, 2018.
60. von Brühl, M.L., Stark, K., Steinhart, A., *et al.*, Monocytes, neutrophils, and platelets cooperate to initiate and propagate venous thrombosis in mice *in vivo, J. Exp. Med.* 209, 819–835, 2012.
61. Barton, P.G., Jackson, C.M., and Hanahan, D.J., Relationship between factor V and activated factor X in the generation of prothrombinase, *Nature* 214, 923–924, 1967.
62. Nesheim, M.E., Taswell, J.B., and Mann, K.G., The contribution of bovine Factor V and Factor Va to the activity of prothrombinase, *J. Biol. Chem.* 254, 10952–10962, 1979.
63. Rosing, J., Tans, G., Govers-Riemslag, J.W., Zwaal, R.F., and Hemker, H.C., The role of phospholipids and factor Va in the prothrombinase complex, *J. Biol. Chem.* 255, 274–283, 1980.
64. Howell, W.H., The nature and action of the thromboplastic (zymoplastic) substance of the tissue, *Am. J. Phys.* 31, 1–21, 1912.
65. Pickering, J.W. and Hewitt, J.A., LVIII., Studies of the coagulation of blood. Part II. Thrombin and antithrombin, *Biochem. J.* 16, 587–598, 1922.
66. Bordet, J. and Delange, L., Sur de nature du cytozyme, Recherhe sur le coagulation du sang, *Ann. Inst. Pasteur* 27, 341–356, 1913.
67. Mellanby, J., The rate of formation of fibrin ferment from prothrombin by the action of thrombokinase and calcium chloride, *J. Physiol.* 51, 396–403, 1917.
68. Foster, W.B., Nesheim, M.E., and Mann, K.G., The factor Xa-catalyzed activation of factor V, *J. Biol. Chem.* 258, 13970–13977, 1983.
69. Wooldridge, L.C., On the origin of fibrin ferment, *Proc. Royal. Soc. Lond* 36(1883–1884), 417–420, 1884.
70. Green, J.P., On certain points connected with the coagulation of blood, *J. Physiol.* 8, 354–371, 1887.
71. Wright, J.H. and Minot, G.R., The viscous metamorphosis of the blood platelets, *J. Expt. Med.* 26, 395–409, 1917.
72. Seegers, W.H., Lommis, E.C., and Vandenbelt, J.M., Preparation of prothrombin products. Isolation of prothrombin and its properties, *Arch. Biochem.* 6, 85–91, 1945.
73. Robbins, K.C., The human plasma fibrinolytic system: regulation and control, *Mol. Cell. Biochem.* 20, 149–157, 1978.
74. Robbins, K.C., Barlow, G.H., Nguyen, G., and Samama, M.M., Comparison of plasminogen activators, *Semin. Thromb. Hemost.* 13, 131–138, 1987.
75. Robbins, K.C., Fibrinolytic therapy: biochemical mechanisms, *Sem. Thromb. Hemost.* 17, 1–6, 1991.
76. Robbins, K.C., Prothombase and thrombase. I. Preparation and properties, *Arch. Biochem.* 6, 69–74, 1945.
77. Robbins, K.C., Prothrombase and thrombase. II Chemical composition, relationships, and chemistry of the activation of prothrombase, *Arch. Biochem.* 6, 75–84, 1945.
78. Mellanby, J., Prothombase: its preparation and properties, *Proc. Royal Soc. B* 107, 271–285, 1930.
79. Milstone, J.H., Fractionation of plasma globulin for prothrombin, thrombokinase, and accessory thromboplastin, *J. Gen. Physiol.* 35, 67–87, 1951.
80. Milstone, J.H., Separation of thrombin from thrombokinase by continuous flow paper electrophoresis, *Proc. Soc. Exptl. Biol. Med.* 101, 660–662, 1959.
81. Milstone, J.H., Oulianoff, N., and Milstone, V.K., Activities associated with thrombokinase derived from bovine plasma, *Proc. Soc. Expt. Biol. Med.* 119, 804–807, 1965.

82. Milstone, J.H. and Milstone, V.K., Activation of chymotrypsinogen by preparations of thrombokinase derived from bovine plasma, *Thromb. Diath. Haemorrh.* 17, 388–400, 1967.

83. Kerwin, D.M. and Milstone, J.H., Removal of thrombokinase from commercial thrombin, *Thromb. Haemost.* 17, 247–255, 1967.

84. Henrikson, R.A. and Jackson, C.M., The chemistry and enzymology of bovine factor X, *Sem. Thromb. Hemost.* 1, 284–309, 1973.

85. Smith, H.P. and Flynn, J.K., The coagulation of blood, *Annu. Rev. Physiol.* 10, 417–444, 1948.

86. Alexander, B., Clotting factor VII (proconvertin): synonymy, properties, clinical and clinicolaboratory aspects, *N. Engl. J. Med.* 260, 1218–1222, 1959)

87. Nemerson, Y., My life with tissue factor, *J. Thromb. Haemost.* 5, 221–223, 2007.

88. Engel, A. and Alexander, B., Activation of chymotrypsinogen A by thrombin preparations, *Biochemistry* 5, 3591–3598, 1966.

89. Seegers, W.H., Physiology of blood coagulation: advances, problems and present status, *Pflügers Arch. ges. Physiol. Menschen Tiere* 299, 226–246, 1968.

90. Kisiel, W. and Hanahan, D.J., The action of factor Xa, thrombin, and trypsin on human factor II, *Biochim. Biophys. Acta* 329, 221–237, 1973.

91. Milstone, J.H., Blood coagulation—a basic science review, *Conn. Med.* 28, 653–656, 1964.

92. Davie, E.W. and Ratnoff, O.D., Waterfall sequence for intrinsic blood clotting, *Science* 145, 1310–1312, 1964.

93. MacFarlane, R.G., An enzyme cascade in the blood clotting mechanism, and its function as a biochemical amplifier, *Nature* 202, 498–499, 1964.

94. Eagle, H., Studies on blood coagulation: I. The role of prothrombin and of plaletets in the formation of thrombin, *J. Gen. Physiol.* 18, 531–545, 1935.

95. Ferguson, J.H., Blood coagulation, thrombosis, and hemorrhagic disorders, *Annu. Rev. Physiol.* 8, 231–262, 1946.

96. Morawitz, P., Die Chemie der Blutgerinnung, *Ergeb. Physiol.* 4, 307–422, 1905.

97. Lein, J., *J. Cell. Comp. Physiol* 30, 43–77, 1947.

98. Eagle, H., Propagation in a fluid medium of a human epidermoid carcinoma, strain KB, *Proc. Soc. Exp. Biol. Med.* 89, 362–364, 1955.

99. Dunn, C.D., Jarvis, J.H., and Greenman, J.M., A quantitative bioassay for erythropoietin using mouse fetal liver cells, *Exp. Hematol.* 3, 65–78, 1975.

100. Cinatl, J., Jr., Cinatl, J., Gerein, V., Kornhuber, B., and Doerr, H.W., The establishment of mouse L-929 cells in protein-free Eagle's minimal essential medium. *J. Biol. Stand.* 16, 249–257, 1988.

101. Freund, E., Ein Beitrag zur Kenntnis der Blutgerinnung, *Med. Jahrb. Wein* 1, 46, 1886 as cited in Morawitz in reference no. 15.

102. Soulier, J.-P., Wartelle, O., and Ménaché, D., Hageman trait and PTA deficiency: the role of contact of blood with glass, *Br. J. Haematol.* 5, 121–138, 1959.

103. Mellanby, J., Calcium and blood coagulation, in *Perspectives in Biochemistry*, ed. J. Needham and D.E. Green, pp. 286–295, Cambridge University Press, London, UK, 1938.

104. Barratt, J.O.W., XXXII, The anticoagulant activity of peptone *in vitro*, *Biochem. J.* 22, 230–235, 1928.

105. Marcum, J.A., The discovery of heparin revisited: the peptone connection, *Perspect. Biol. Med.* 39, 610–625, 1996.

106. Arthus, N.M. and Pagés, C., Novelle theorie chimique de la coagulation du sang, *Arch. Physiol. Norm. Pathol.* 2. 737–746, 1890.

107. Arthus, M., *Coagulation des liquides organique sang lymph, transudents*, Rueff, Paris, France, 1894.
108. Bordet, J. and Delange, L., Coagulation du sang et la genese de la thrombin, *Ann. Inst. Pasteur* 26, 737–766, 1912.
109. Arthus, M., Paralléle de la coagulation due sang et de la caséfication du lait, *Comp. Rend. Soc. Biol.* 45, 435–437, 1893.
110. Berridge, N.J., Rennin and the clotting of milk, *Adv. Enzymol.* 15, 423–448, 1954.
111. Weil, R., Sodium citrate in the transfusion of blood, *JAMA* 64, 425–426, 1915.
112. Greenwalt, T.J., A short history of transfusion medicine, *Transfusion* 37, 550–563, 1993.
113. Shimada, T., Kato, H., Iwanaga, S., Iwamori, M., and Nagai, Y., Activation of factor XII and prekallikrein with cholesterol sulfate, *Thromb. Res.* 38, 21–31, 1985.
114. Frost, C.L., Naudé, R.J., and Oelofsen, W., and Jacobson, B., Comparative blood coagulation in ostrich, *Immunopharmacology* 45, 75–81, 1999.
115. Ambrus, J.L., Ambrus, C.M., and Robin, J.C., Hemophilia D factor (factor XII), red-cell deformability and platetet aggregates in raptors, *J. Med.* 35, 19–25, 2004.
116. Ponczek, M.B., Gailani, D., and Doolittle, R.F., Evolution of the contact phase of verte-brate blood coagulation, *J. Thromb. Haemost.* 6, 1876–1883, 2008.
117. Young, J.Z. and Medawar, P.B., Fibrin suture of peripheral nerve. Measurement of the rate of regeneration, *Lancet* 237, 126–128, 1940.
118. Watson, J.D., *The Annotated and Ilustrated Double Helix*, Simon & Shuster, New York, USA, 2012.
119. Ratnoff, O.D. and Margolius, A., Jr., Hageman trait: an asymptomatic disorder of blood coagulation, *Trans. Associ. Am. Physicians* 68, 149–154, 1955.
120. Wheeler, A.P. and Gailani, D., Why factor XI deficiency is a clinical concern, *Expert Rev. Hematol.* 9, 629–637, 2016.
121. Sumner, J.B., The isolation and crystallization of the enzyme urease: preliminary paper, *J. Biol. Chem.* 69, 435–441, 1926.
122. Sumner, J.B., Kirk, J.S., and Howell, S.F., The digestion and inactivation of crystalline urease by pepsin and by papain, *J. Biol. Chem.* 98, 543–552, 1932.
123. Simoni, R.D., Hill, R.D., and Vaughn, M., Urease, the crystalline enzyme and the proof that enzymes are proteins. The work of James B. Sumner, *J. Biol. Chem.* 277, 23e, 2002.
124. Chargaff, E. and Ziff, M., Coagulation of fibrinogen by simple organic substances as a model of thrombin action, *J. Biol. Chem.* 138, 787–788, 1941.
125. Chargaff, E. and Bendich, A., On the coagulation of fibrinogen, *J. Biol. Chem.* 149, 93–110, 1943.
126. Baumberger, J.P., Some evidence in support of a sulfhydryl mechanism of blood clot-ting, *Am. J. Physiol.* 133, 206–207, 1941.
127. Jeener, R., Sur le role des groupes thiol dans la coagulation due plasma sanguin, *Experentia* 3, 243–244, 1947.
128. Carter, J.R. and Warner, E.D., Importance of the disulfide (-S-S-) linkage in the blood-clotting mechanism, *Am. J. Physiol.* 173, 109–114, 1953.
129. Jensen, E.V., Sulfhrydryl-disulfide interchange, *Science* 130, 1319–1323, 1959.
130. Zucker, M., Seligsohn, U., Yeheskel, A., and Mor-Cohen, R., An allosteric disulfide bon is involved in enhanced activation of factor XI by protein disulfide isomerase, *J. Thromb. Haemost.* 14, 2202–2211, 2016.
131. Zhou, B., Hogg, P.J., and Gräter, F., One-way allosteric communication between the two disulfide bonds in tissue factor, *Biophys. J.* 112, 78–86, 2017.
132. Mellanby, J. Thrombase-Its preparation and properties, *Proc. Roy. Soc. London Series B*, 133, 93–106, 1933.

133. Fearon, W.C., Protein metabolism in the animal body, *Irish J. Med. Sci.* 3, 303–315, 1924.

134. Strughold, H., and Wöhlisch, E., Ist das thrombin ein proteolytische Ferment?, *H. -S. Zeit. physiol. Chem.* 223, 267–280, 1934.

135. Wöhlisch, E. and Jühling, E., Das Thrombin als fibrinogen-denaturase und seine Beziehungen zum Papain, *Biochem. Z.* 297, 353–368, 1938.

136. Apitz, K., Über Profibrin. II Die Bildung von Profibrin bei der Denaturierung des Fibrinogens, *Zeitschrift für gesamte Exptl. Med.* 102, 202–211, 1938.

137. Doolittle, R.F., Clotting of mammalian fibrinogen by papain: a re-examination, *Biochemistry* 53, 6687–6694, 2014.

138. Haurowitz, F., *Chemistry and Biology of Proteins*, Academic Press, New York, USA, 1950.

139. Linderstrøm-Lang, K., Hotchkiss, R.D., and Johansen, G., Peptide bonds in globular proteins, *Nature* 142, 996, 1938.

140. Anson, M.L. and Mirsky, A.E., The equilibria between native trypsin and inactive denatured trypsin, *J. Gen. Physiol.* 17, 393–398, 1934.

141. Lyons, R.N., Thrombin components in blood coagulation, *Aust. J. Exptl. Biol. Med. Sci.* 29, 383–392, 1951.

142. Boyer, M.H., Shainoff, J.R., and Ratnoff, O.D., Acceleration of fibrin polymerization by calcium ions, *Blood* 39, 382–387, 1972.

143. Endres, G.F. and Scheraga, H.A., Equilibria in the fibrinogen-fibrin conversion. IX. Effects of calcium ions on the reversible polymerization of fibrin monomer, *Arch. Biochem. Biophys.* 153, 266–278, 1972.

144. Mihalyi, E., Review of some unusual effects of calcium binding to fibrinogen, *Biophys. Chem*, 112, 131–140, 2004.

145. Weisel, J.H. and Litvinov, R.I., Mechanism of fibrin polymerization and clinical implications, *Blood* 121, 1712–1719, 2013.

146. Bailey, K. and Bettelheim, F.R., The nature of the fibrinogen-thrombin reaction, *Brit. Med. Bull.* 11, 50–53, 1955.

147. Bailey, K., Bettelheim, F.R., Lorand, L., and Middlebrook, W.R., Action of thrombin in the clotting of fibrinogen, *Nature* 167, 233–234, 1951.

148. Bettelheim, F.R. and Bailey, K., The products of the action of thrombin on fibrinogen, *Biochim. Biophys. Acta* 9, 578–579, 1952.

149. Bailey, K. and Bettleheim, F.R., The clotting of fibrinogen. I. The liberation of peptide material, *Biochim. Biophys. Acta* 18, 495–503, 1955.

150. Bettelheim, F.R., The clotting of fibrinogen. II. Fractionation of peptide material liberated, *Biochim. Biophys. Acta* 19, 121–130, 1956.

151. Smith, H.P., Warren, E.D., and Brinkhous, K.M., Prothrombin deficiency and the bleeding tendency in liver injury (chloroform intoxication), *J. Exptl. Med.* 66, 801–811, 1937.

152. Seegers, W.H., Smith, H.P., Warner, E.D., and Brinkhous, K.M., The purification of prothrombin, *J. Biol. Chem.* 123, 751–754, 1938.

153. Seegers, W.H. and Smith, H.P., The purification of prothrombin, *J. Biol. Chem.* 126, 677–678, 1941.

154. Seegers, W.H., Brinkhous, K.M., Smith, H.P., and Warner, H.D., The purification of thrombin, *J. Biol. Chem.* 126, 91–95, 1938.

155. Workman, E.F., Jr. and Lundblad, R.L., On the preparation of bovine thrombin, *Thromb. Haemost.* 39, 193–200, 1978.

156. Seghatchian, M.J., Enzymology of thrombin: purification procedures, characterization, surface binding, unitage and standardization, in *The Thrombin*, ed. R. Machovich, Chapter 6, pp. 103–130, CRC Press, Boca Raton, FL, USA, 1984.

157. Seegers, W.H., Warner, E.D., Brinkhous, K.M., and Smith, H.P., The use of purified thrombin as hemostatic agent, *Science* 89, 86, 1939.

158. Brinkhous, K.M., Thrombin and its clinical applications, *South. Surg.* 13, 397–402, 1947.

159. Smith, H.P., Blood coagulation product and method for obtaining same, US Patent 2,398,077, April 9, 1946.

160. Smith, H.P., Prothrombin product and process of preparation of same, US Patent 2,408,536, October 1, 1946.

161. Brinkhous, K.M., Thrombin and its clinical applications, *South. Surg.* 13, 397–402, 1947.

162. Seegers, W.H., Loomis, E.C., and Vandenbelt, J.M., Preparation of prothrombin product. Isolation of prothrombin and its preparation, *Arch. Biochem.* 6, 85–91, 1945.

163. Ware, A.G., Guest, M.M., and Seegers, W.H., Stability of prothrombin, *Amer. J. Physiol.* 150, 58–64, 1947.

164. Seegers, W.H., McClaughry, R.I., and Fahey, J.L., Some properties of purified prothrombin and its activation with sodium citrate, *Blood* 5, 421–433, 1950.

165. Seegers, W.H., History of prothrombin activation in sodium citrate solution, *Thromb. Diath. Haemost.* 19, 610–611, 1968.

166. Seegers, W.H., Marciniak, E., and Cole, E.R., Autocatalysis in prothrombin activation, *Amer. J. Physiol.* 203, 397–400, 1962.

167. Teng, C.M. and Seegers, W.H., Activation of factor X and thrombin zymogens in 25% sodium citrate solution, *Thromb. Res.* 22, 203–212, 1981.

168. Aronson, D.L. and Mustafa, A.J., The activation of human factor X in sodium citrate: the role of factor VII, *Thromb. Haemost.* 36, 104–114, 1976.

169. Scheraga, H.A. and Laskowski, M., Jr., The fibrinogen-fibrin conversion, *Adv. Prot. Chem.* 12, 2–131, 1957.

170. Bajaj, S.P. and Mann, K.G., Simultaneous purification of bovine prothrombin and factor X. Activation of prothrombin by trypsin-activated factor X, *J. Biol. Chem.* 248, 7729–7741, 1973.

171. Suttie, J.W. and Jackson, C.M., Prothrombin structure, activation, and biosynthesis, *Physiol. Rev.* 57, 1–70, 1977.

172. Kendrick, D.B., The bovine and human albumin program, in *Blood Program in World War II*, Chapter 12(XII), pp. 325–357, Office of the Surgeon General, Department of the Army, Washington, DC, USA, 1964.

173. Kendrick, D.B., The plasma program, in *Blood Program in World War II*, Chapter 9(IX), pp. 265–323, Office of the Surgeon General, Department of the Army, Washington, DC, USA, 1964.

174. Rous, P. and Wilson, G.W., Fluid substitutes for transfusion after hemorrhage. First communication, *JAMA* 70, 219–229, 1918.

175. Mann, F.C., Further experimental study of surgical shock, *JAMA* 71, 1184–1188, 1918.

176. Koelbing, H.M., Scientific autographs. IV. Peyton Rous (1879–1970) and his Novel Prize, *Agents Actions* 1, 211–214, 1970.

177. Rubin, H. The early history of tumor virology: Rous, RIF, and RAV, *Proc. Natl. Acad. Sci. USA* 108, 14389–14396, 2011.

178. Kendrick, D.B., Byproducts of plasma fractionation, in *Blood Program in World War II*, Chapter XIII (13), pp. 359–369, Office of the Surgeon General, Department of the Army, Washington, DC, USA, 1964.

179. Cohn, E.J., Oncley, J.L., Strong, L.E., Hughes, W.L., and Armstrong, S.H., Chemical, clinical, and immunological studies on the products of human plasma fractionation. I. The characterization of the protein fractions of human plasma, *J. Clin. Invest.* 23, 417–432, 1944.

180. Edsell, J.T., The plasma proteins and their fractionation, *Adv. Prot. Chem.* 3, 384–479, 1947.
181. Burnouf, T. and Seghatchian, J., "Go no Go" in plasma fractionation in the world's emerging economies: still a question asked 79 years after the COHN process was developed!, *Transfus. Apher. Sci.* 51, 113–119, 2014.
182. Edsall, J.T., The plasma proteins and their fractions, *Adv. Protein Chem.* 3, 169–225, 1947.
183. Edsall, J.T., Ferry, R.M., and Armstrong, S.H., Chemical, clinical, and immunological studies on the products of human plasma fractionation, XV, The proteins concerned in the blood coagulation mechanism, *J. Clin. Invest.* 23, 557–565, 1944.
184. Parfentjev, I.A., A globulin fraction in rabbit's plasma possessing a strong clotting property, *Am. J. Med. Sci.* 202, 578–584, 1941.
185. Taylor, F.H.I., Lozner, E.I., and Adams, M.A., The thrombic activity of a globulin fraction derived from rabbit plasma, *Am. J. Med. Sci.* 202, 585–593, 1941.
186. Lozner, E.L., MacDonald, H., and Taylor, F.H.L., The use of rabbit thrombin as a local hemostatic, *Am. J. Med. Sci.* 202, 593–598, 1941.
187. Seegers, W.H., Heene, D.L., and Marciniak, E., Activation of purified prothrombin in ammonium sulfate solutions: purification of autoprothrombin C, *Thromb. Diath. Haemorrh.* 15, 1–11, 1966.
188. Adams, M.A. and Taylor, F.H.L., The thrombic activity of a globulin fraction of the plasma proteins of beef, swine and human blood, *Am. J. Med. Sci.* 205, 538–504, 1943.
189. Lessess, M.F. and Hamolsky, M.W., Epidemic of homologous serum hepatitis apparently caused by human thrombin, *J. Am. Med. Assoc.* 147, 727–730, 1951.
190. Porter, J.E., Shapiro, M., Maltby, G.L., *et al.*, Human thrombin as vehicle of infection in homologous serum hepatitis, *J. Am. Med. Assoc.* 153, 17–19, 1953.
191. Gladner, J. and Laki, K., The inhibition of thrombin by diisopropyl phosphorofluoridate, *Arch. Biochem. Biophys.* 62, 501–503, 1956.
192. Hedstrom, L., Serine protease mechanism and specificity, *Chem. Rev.* 102, 4501–4524, 2002.
193. Jansen, E.F., Nutting, F., and Balls, A.K., Mode of inhibition of the proteinase and esterase activities by diisopropyl fluorophosphate; crystallization of inhibited chymotrypsin, *J. Biol. Chem.* 179, 189–199, 1949.
194. Schaffer, N.K., May, S.C., Jr., and Balls, A.K., Mode of inhibition of chymotrypsin by diisopropyl fluorophosphoryl chymotrypsin, *J. Biol. Chem.* 179, 201–204, 1947.
195. Dixon, G.H. and Neurath, H., The reaction of DFP with trypsin, *Biochim. Biophys. Acta* 20, 572–574, 1956.
196. Gros, D. and Harvey, D.M., The effect and treatment of nerve gas poisoning, *Am. J. Med.* 14, 52–63, 1953.
197. Koelle, G.B., Gilman, A., and Binzer, B.D., The relationship between cholinesterase inhibition and the pharmacological action of di-isopropyl fluorophosophate (DFP), *J. Pharmacol. Extpl. Therapeut.* 87, 421–434, 1946.
198. Mounter, L.A., Alexander, H.C., 3rd, Tuck, K.D., and Dixon, L.T.H., The pH dependence and dissociation constants of esterases and proteases treated with diisopropyl fluorophosphate, *J. Biol. Chem.* 226, 867–872, 1957.
199. Laki, K., Gladner, J.A., Folk, J.E., and Kominz, D.P., The mode of action of thrombin, *Thromb. Haem. Haemorrh.* 2, 205–217, 1958.
200. Gladner, J.A. and Laki, K., The active site of thrombin, *J. Am. Chem. Soc.* 80, 1263–1264, 1958.

201. Hartley, B.S., Proteolytic enzymes, *Annu. Rev. Biochem.* 29, 45–72, 1960.
202. Mounter, L.A., Shipley, B.A., and Mounter, M.-E., The inhibition of hydrolytic enzymes by organophosphorous compounds, *J. Biol. Chem.* 238, 1979–1983, 1963.
203. Magnusson, S., Homologies between thrombin and other serine proteinases, *Biochem. J.* 110, 25P–26P, 1968.
204. Ronwin, E., Thrombin properties, *Can. J. Biochem.* 35, 743–758, 1957.
205. Hummel, B.C.W., A modified spectrophotometric determination of chymotrypsin, trypsin, and thrombin, *Can. J. Biochem.* 37, 1393–1399, 1959.
206. Laki, K., Gladner, J.A., and Folk, J.E., Some aspects of the fibrinogen-fibrin transition, *Nature* 187, 758–761, 1960.
207. Berridge, N.J., Rennin and the clotting of milk, *Adv. Enzymol.* 15, 423–448, 1954.
208. Linderstrøm-Lang, K. and Ottesen, M., A new protein from ovalabumin, *Nature* 157, 807–808, 1947.
209. Eeg-Larsen, N., Linderstrøm-Lang, K., and Ottesen, M., Transformation of ovalbumin into plakablumin, *Arch. Biochem.* 19, 340–344, 1947.
210. Ottesen, M., The transformation of ovalbumin into plakalbumin; a case of limited proteolysis, *C. R. Trav. Lab. Carlsberg. Chim.* 38, 211–270, 1958.
211. Desnuelle, P and Rovery, M., Sur quelques protéolyses limileés d'intérêt biologique, in *Proc. Third Intern. Cong. Biochem.*, ed. C. Liébecq, p. 78, Academic Press, New York, USA, 1956.
212. Mihalyi, E., *Application of Proteolytic Enzymes to Protein Structure Studies*, CRC Press, West Palm Beach, FL, USA, 1978.
213. Moore, S. and Stein, W.H., Column chromatography of peptides and proteins, *Adv. Prot. Chem.* 11, 1011–1236, 1956.
214. Spackman, D.H., Stein, W.H., and Moore, S., Automatic recording apparatus for use in the chromatography of amino acids, *Anal. Chem.* 30, 1190–1206, 1958.
215. Hirs, C.H.W., Moore, S., and Stein, W.H., The sequence of amino acid residues in performic acid-oxidized ribonuclease, *J. Biol. Chem.* 235, 633–647, 1960.
216. Edman, P., A method for the determination of amino acid sequence in peptides, *Arch. Biochem.* 22, 475–476, 1949.
217. Edman, P., Phenylthiohydantoins in protein analysis, *Ann. N. Y. Acad. Sci.* 88, 602–610, 1960.
218. Rasmussen, P.S., Purification of thrombin by chromatography, *Biochim. Biophys. Acta* 16, 157–158, 1955.
219. Waugh, D.F. and Livingstone, B.J., Kinetics of the interaction of bovine fibrinogen and thrombin, *J. Phys. Chem.* 55, 1206–1218, 1951.
220. Waugh, D.F., Anthony, L.J., and Ng, N., The interaction of thrombin with borosilicate glass surfaces, *J. Biomed. Mater. Res.* 9, 511–536, 1975.
221. Miller, K.D., Rivanol, resin and the isolation of thrombin, *Nature* 184, 450, 1959.
222. Persson, J. and Lester, P., Purification of antibody and antibody-fragment from *E. coli* homogenate using 6,9-diamino-2-ethoxyacridine lactate as precipitation agent, *Biotechnol. Bioeng.* 87, 424–434, 2004.
223. Bhupal, V. and Mani, R.S., Separation of bound and free ligand by ethacridine (Rivenol) in thyroxin radioimmunoassay, *Clin. Chem.* 29, 1937–1940, 1983.
224. Miller, K.D. and Copeland, W.H., Human thrombin: isolation and stability, *Exptl. Mol. Pathol.* 4, 431–437, 1965.
225. Fenton, J.W., II., Campbell, W.P., Harrington, J.C., and Miller, K.D., Large scale preparation and preliminary characterization of human thrombin, *Biochim. Biophys. Acta* 229, 26–32, 1971.

226. Fenton, J.W., 2nd, Fasco, M.J., Stackrow, A.B., *et al.*, Human thrombins. Production, evaluation, and properties of α-thrombin, *J. Biol. Chem.* 252, 3587–3598, 1977.

227. Lundblad, R.L., A rapid method for the purification of bovine thrombin and the reaction of the purified enzyme with phenylmethylsulfonyl fluoride, *Biochemistry* 10, 2501–2506, 1971.

228. Vogel, C.N., Parfitt, H.E., Jr., Kingdon, H.S., and Lundblad, R.L., Preparation of modified bovine factor VIII with enhanced biological activity using insoluble-trypsin columns, *Thromb. Diath. Haemorrh.* 30, 229–234, 1973.

229. Fenton, J.W., 2nd, Olson, T.A., Zabinksi, M.P., and Wilner, G.D., Anion-binding exosite of human α-thrombin and fibrin(ogen) recognition, *Biochemistry* 27, 7106–7112, 1988.

230. Fenton, J.W., 2nd, Witting, J.I., Pouliott, C., and Fareed, J., Thrombin anion-binding exosite interactions with heparin and various polyanions, *Ann. N. Y. Acad. Sci.* 556, 158–165, 1989.

231. *Chemistry and Biology of Heparin*, ed. R.L. Lundblad and J.W. Fenton, 2nd ed., and Mann, K.G., Ann Arbor Science, Ann Arbor, MI, USA, 1977.

232. Vu, T.K., Hung, D.T., Wheaton, V.I., and Coughlin, S.R., Molecular cloning of a functional thrombin receptor reveals a novel proteolytic mechanism of a receptor activation, *Cell* 64, 1057–1068, 1991.

233. Hung, D.T., Vu, T.H., Nelken, N.A., and Coughlin, S.R., Thrombin-induced events in non-platelet cells are mediated by the unique proteolytic mechanism established for the cloned platelet thrombin receptor, *J. Cell. Biol.* 116, 827–832, 1992.

234. Magnusson, S., Thrombin and prothrombin, in *The Enzymes*, ed. P.D. Boyer, Vol. 3, pp. 277–321, Academic Press/Elsevier, New York, USA, 1971.

235. Butkowski, R.J., Elion, J., Downing, M.R., and Mann, K.G., Primary structure of human prethombin 2 and α-thrombin, *J. Biol. Chem.* 262, 4947–4957, 1977.

236. Thompson, A.R., Enfield, D.L., Ericsson, L.H., Legaz, M.E., and Fenton, J.W., 2nd, Human thrombin: partial primary structure, *Arch. Biochem. Biophys.* 178, 356–367, 1977.

2 The Formation of Thrombin

Thrombin can be found and formed in both the intravascular and extravascular space. Thrombin has a critical role in the hemostatic response within the intravascular space. As such, there is a requirement for the rapid generation of thrombin and a mechanism for rapid inactivation in the intravascular space. However, while the function of thrombin in the extravascular space is likely regulatory in nature and rapid generation is not required, it is not clear if a need for rapid inactivation is required in hemostasis.

The primary action of thrombin in blood is the activation of platelets; the "activation" of factors VIII, V, and XIII; and the clotting of fibrinogen. Dave Aronson has compared the relative sensitivities of these substrates to thrombin (see Table 2.1).[1]

The primary mechanism for the formation of thrombin for hemostasis in the intravascular space uses the prothrombinase complex; there may be some other mechanisms, such as the MASP protease mechanism, which might participate in disease states. There are a number of potential mechanisms for thrombin formation in the extravascular space. A compilation of these methods, including various snake venoms, is presented in Table 2.2. A more detailed consideration of snake venoms that activate prothrombin is presented in Table 2.3. The potential of snake venoms to provide hemostasis in congenital bleeding disorders was recognized by R.G. Macfarlane in 1934.[2,3] As products obtained from snake venoms have become more sophisticated and safe from neurotoxins, there has been continued interest in providing local hemostasis.[4-6] Snake venoms include diverse biological activities such as disintegrins.[7]

Prothrombin, the precursor of thrombin, is one of the several vitamin K–dependent proteins present in blood plasma. The vitamin K–dependent proteins are characterized by requirement for vitamin K for synthesis of functional protein containing

TABLE 2.1
Relative Sensitivity of Thrombin Substrates in Plasma

Reaction	Sensitivity
Factor VIII activation	100
Fibrinopeptide A release	10
Platelet aggregation	10
Fibrinopeptide B release	1
Factor V activation	1

Source: Aronson, D.L., *Thromb. Haemost.* 36, 9–13, 1976

DOI: 10.1201/b22204-2

TABLE 2.2
Formation of Thrombin (Fibrinogen clotting activity)

Mechanism	Reference
Prothrombinase: The complex formed between factor Xa, factor V(Va), calcium ions, and phospholipid (platelets).[1,2] The product of this reaction has been referred to as biothrombin[3] to distinguish it from citrate thrombin.[4] The term prothrombinase is attributed to P.A. Owen[5,6] but it is not clear that he was referring to the current complex instead of to factor VI, which was related to factor V (labile factor).[7] The canonical activation of prothrombin can occur via two different pathways, both of which lead to the formation of α-thrombin. One pathway occurs with the cleavage at Arg271-Thr272, leading to the formation of a polypeptide consisting of fragment 1 and fragment 2 and prethrombin 2; prethrombin 2 is then cleaved at Arg320-Ile321 to yield α-thrombin. Another pathway can occur via cleavage at Arg320-Il321, yielding meizothrombin, which undergoes autocatalytic cleavage at Arg2271-Thr272 to yield α-thrombin. Meizothrombin can also autocatalytically cleave at Arg155-Ser156 to release fragment 1 with the formation of desF1-meizothrombin.	1–7
Polyanion activation of prothrombin. The activation of "prothrombin" occurs in high concentrations of sodium citrate (25% trisodium citrate, approximately 0.85 M, pH ~9). Activation has also been observed in the presence of sulfate. This process requires the presence of factor VII and factor X.[8] The product of this reaction has been referred to as citrate thrombin.[9]	8,9
Activation of prothrombin in the presence of polylysine.[10] This reaction is inhibited by calcium ions.[11] Prothrombin activation by histone B4 has also been reported.[12,13] The effect of histone B4 is likely different from the effect of polylysine, as the reaction proceeds in the presence of calcium ions.	12,13
Poly(ethylene)glycol (PEG) has been reported to support the activation of prothrombin by factor Xa. A concentration of 20% PEG-4000 was optimal.[14] The rate of activation was slow (days). PEG does affect the clotting of fibrinogen by thrombin,[15] but did not affect the hydrolysis of the peptide nitroanilide substrate, S-2238 (H-D-PhePipArgpNA).[16–18] The effect of PEG was dependent on the molecular weight, with PEG-1000 having no effect and a weak effect observed with PEG-2000; PEG-4000 was used in these studies.	14–18
Staphylocoagulase is a protein secreted by *Staphylococcus* which forms a complex with prothrombin, which will clot fibrinogen, and hydrolyzes synthetic substrates.[19] Staphylocoagulase is thought be important in the pathogenicity of *Staphylococcus aureus*.[20] Staphylocoagulase forms a complex with prothrombin by inserting a terminal amino group into the latent active site of thrombin present in prothrombin.[21] The free amino terminal amine is necessary for the formation of a salt bridge with an aspartic acid, permitting formation of an enzyme-active site.[22,23] The staphylocoagulase-prothrombin complex active site histidine can be labeled with a chloromethyl derivative.[24,25] Dabigatran (oral novel anticoagulant [NOAC] directed at thrombin; Pradaxa) reduces *S. aureus* virulence in a murine model.[26]	19–26
Von Willebrand factor–binding protein (VWbp) from *S. aureus* was shown to have homology to staphylocoagulase.[27] The recombinant protein was demonstrated to clot human and porcine plasma. Subsequent work showed that VWbp bound to prothrombin, and via a process described as "conformational activation," the amino terminal sequence is inserted into the nascent active site of thrombin in the prothrombin molecule.[28] The proteolytic cleavage of prothrombin is not required for the development of fibrinogen clotting activity.[29]	27–29

Mechanism	Reference
The autoactivation of prothrombin and prothrombin activation intermediates has been reported.[30] Susceptibility to autoactivation is suggested to depend on conformation (open versus closed).[31] Autoactivation is promoted by histone H4.[12,13]	12,13,30,31
Some snake venoms can activate prothrombin.[32] There are a variety of mechanisms. Snake venom prothrombin activators can be divided into four categories.[33–35] There are metalloproteinases,[35] which form meizothombin in the presence (B) or absence of calcium ions (A) and serine proteinases,[36] both of which require the presence of calcium ions and phospholipid. Group D requires the addition of factor V(Va) for optimal activity, while group C does not require exogenous factor V(Va); group C activator contains a factor Xa–like subunit and a factor Va–like subunit.[36] A further discussion of snake venoms is presented in Table 2.3.	32–36
Prothrombin can be cleaved by subtilisin (Bacillus subtilis) to yield a derivative with catalytic activity.[37] Cleavage occurs between alanine 490 and asparagine 491 to yield a two-chain derivative, which the authors designate as σPre2. This derivative has activity in the hydrolysis of S-2238 (k_{cat}/K_m is 165-fold lower than that seen for α-thrombin and was inactivated by D-Phe-Pro-Arg-chloromethyl ketone (PPACK) at a rate similar to α-thrombin (1.6–2.0 $\mu M^{-1}sec^{-1}$). The rate of release of fibrinopeptide A was 300-fold lower than that observed with α-thrombin; the rate of release of fibrinopeptide B was 100-fold lower. A subtilisin-like serine protease from an Acremonium sp. was reported to cleave prothrombin to yield a meizothrombin-like product.[38] Cleavage occurs at Tyr316-Ile317, close to the factor Xa cleavage site at Arg320-Ile321.	37,38
Fibrinogen-like protein 2 (Fgl2; fibroleukin). Membrane-bound Fgl2 is also referred to as cell membrane prothrombinase[39] and activates prothrombin in the presence of calcium ions and factor Va.[40] Recombinant Fgl2, when reconstituted into a phospholipid vesicle, is able to activate prothrombin in the presence of calcium ion and factor V.[41] Soluble Fgl2 is secreted by T-lymphocytes[42] and has immunosuppressive activity.[43]	39–43
Mannose-binding lectin (MBL)–associated serine proteases (MASPs) have been reported to activate prothrombin in addition to their role in complement. MBL is a C-type lectin that functions in innate immunity.[44] It is a pattern-recognition receptor that recognizes carbohydrate arrays on the surface of pathogenic organisms. MASPs are activated on the binding of MBL to targets, resulting in the activation of complement and the coagulation system.[45] Activation also occurs with ficolins.[1] Both MASP-1[46] and MASP-2[47] have been reported to activate prothrombin. The action of MASP-1 has been studied in some detail.[48] It is of interest that the reaction occurs in the absence of divalent cations. Both meizothrombin-like and prothrombin-like intermediates are observed in the MASP-1 activation of prothrombin.	44–48
Tryptase, a trypsin-like serine protease from mast cells, has been reported to activate prothrombin.[49] These experiments were performed with rat trypsin; later studies showed that human tryptase did not activate prothrombin,[50] although another study showed that human tryptase did activate prothrombin.[51] However, a later study[52] concluded that the human tryptase activated prothrombin.	49–52

(*Continued*)

TABLE 2.2
(Continued)

Mechanism	Reference
Early studies suggested that trypsin could activate prothrombin.[53] Later studies supported the activation of prothrombin by insoluble trypsin.[54] The use of insoluble trypsin reduced the destruction of the product thrombin. However, it was most likely that the earlier studies on the activation of prothrombin by trypsin were mediated by the activation of factor X present in the prothrombin preparations.[55] It is of interest that the presence of platelets or lipid (lecithin) enhanced the yield of thrombin from bovine thrombin.[56]	53–56
Der p 1 is a cysteine protease allergen produced by the house dust mite,[57–59] which has been shown to activate prothrombin.[60,61]	57–61

References
1. Rosing, J., Tans, G., Govers-Riemslag, J.W., Zwall, R.F., and Hemker, H.C., The role of phospholipid and factor V in the prothrombin complex, *J. Biol. Chem.* 255, 274–283, 1980.
2. Krishnaswamy, S., Church, W.R., Neisheim, M.E., and Mann, K.G., Activation of human prothrombin by human prothrombinase. Influence of factor Va on the reaction mechanism, *J. Biol. Chem.* 262, 3291–3299, 1987.
3. Seegers, W.H. and Landaburu, R.H., Esterase and clotting activity derived from purified prothrombin, *Am. J. Physiol.* 191, 167–173, 1957.
4. Seegers, W.H., McClaughry, R.I., and Fahey, J.L., Some properties of purified prothrombin and its activation with sodium citrate, *Blood* 5, 421–433, 1950.
5. Quick, A.J., The coagulation of blood and hemostasis, *Annu. Rev. Physiol.* 12, 257–264, 1950.
6. Owen, C.A., Jr., Magath, J.P., and Bollman, J.L., Prothrombin conversion factors in blood coagulation, *Am. J. Physiol.* 166, 1–11, 1951.
7. Owren, P.A., The fifth coagulation factor ('Factor V'). Preparation and properties, *Biochem. J.* 43, 136–139, 1948.
8. Aronson, D.L. and Mustafa, A.J., The activation of human factor X in sodium citrate: the role of factor VII, *Thromb. Haemost.* 36, 104–114, 1976.
9. Aizawa, P., Winge, S., and Karlsson, G., Large-scale preparation of thrombin from human plasma, *Thromb. Res.* 122, 560–567, 2008.
10. Miller, K.D., The nonenzymic activation of prothrombin by polylysine, *J. Biol. Chem.* 235, PC63–PC-64, 1960.
11. Vogel, C.N., Butkowski, R.J., Mann, K.G., and Lundblad, R.L., Effect of polylysine on the activation of prothrombin, Polylysine substitutes for calcium ions and factor V in the activation of prothrombin by factor Xa, *Biochemistry* 15, 3265–3269, 1976.
12. Barranca-Medina, S., Pozzi, N., Vogt, B.D., and Di Cera, E., Histone H4 promotes prothrombin autoactivation, *J. Biol. Chem.* 288, 35749–35757, 2013.
13. Pozzi, N. and Di Cera, E., Dual effect of histone H4 on prothrombin activation, *J. Thromb. Haemost.* 14, 1814–1818, 2016.
14. Kaetsu, H., Mizuguchi, I., Hamamoto, T., *et al.*, Large-scale preparation of human thrombin: polyethylene glycol potentiates the factor Xa-mediated activation of prothrombin, *Thromb. Res.* 90, 101–109, 1998.
15. Fenton, J.W., 2nd and Fasco, M.J., Polyethylene glycol 6,000 enhancement of the clotting of fibrinogen solutions in visual and mechanical assays, *Thromb. Res.* 4, 809–817, 1974.
16. Pozgay, M., Szabó, G., Bajusz, S., *et al.*, Study of the specificity of thrombin with tripeptidyl-*p*-nitroanilide substrates, *Eur. J. Biochem.* 115, 491–495, 1981.
17. Lottenberg, R., Hall, J.A., Fenton, J.W., 2nd, and Jackson, C.M., The action of thrombin on peptide *p*-nitroanilide substrates: hydrolysis of Tos-gly-pro-arg-pNA and D-Phe-Pip-Arg-pNA by human α- and γ-thrombin and bovine α- and β-thrombins, *Thromb. Res.* 28, 313–332, 1982.

18. Lottenberg, R., Hall, J.A., Blinder, M., Blinder, E.P., and Jackson, C.M., The action of thrombin of peptide *p*—nitroanilide substrates. Substrate selectivity and examination of hydrolysis under different reaction conditions, *Biochim. Biophys. Acta* 742, 539–557, 1983.
19. Hemker, H.C., Bas, B.M., and Muller, A.D., Activation of a pro-enzyme by a stoichiometric straphylocoaguase, *Biochim. Biophys. Acta.* 379, 180–188, 1975.
20. Peetermans, M., Verhame, P., and Vanassche, T., Coagulase activity by *Staphylococcus aureus*: A potential target for therapy?, *Semin. Thromb. Haemost.* 41, 433–444, 2015.
21. Friedrich, R., Panizzi, P., Kawabata, S., Bode, W., and Bock, P.E., Structural basis of reduced staphylocoagulase-mediated bovine prothrombin activation, *J. Biol. Chem.* 281, 1188–1195, 2006.
22. Magnusson, S. and Hofmann, T., Inactivation of thrombin by nitrous acid, *Can. J. Biochem.* 48, 432–437, 1970.
23. de Haën, C., Neurath, H., and Teller, D.C., The phylogeny of trypsin-related serine proteases and their zymogens. New methods for the investigation of distant evolutionary relationships, *J. Mol. Biol.* 92, 225–259, 1975.
24. Panizzi, P., Friedrich, R., Fuentes-Prior, P., *et al.*, Novel fluorescent prothrombin analogs as probes of staphylcoagulase-prothrombin interactions, *J. Biol. Chem.* 281, 1169–1178, 2006.
25. Panizzi, P., Friedrich, R. Fuentes-Prior, P., *et al.*, Fibrinogen substrate recognition by Staphylocoagulase prothrombin complexes, *J. Biol. Chem.* 281, 1179–1187, 2006.
26. Vanassche, T., Verhaegen, J., Peetermans, W.E., *et al.*, Inhibition of staphylothrombin by dabigatran reduces *Staphylococcus aureus* virulence *J. Thromb. Haemost.* 9, 2436–2446, 2011.
27. Bjerktorp, J., Jacobsson, K., and Fryberg, L., The von Willebrand factor-binding protein (vWbp) of *Staphylococcus aureus* is a coagulase, *FEMS Microbiol. Lett.* 234, 309–314, 2004.
28. Kroh, H.K., Panizzi, P., and Bock, P.E., Von Willebrand factor-binding protein is a hysteretic conformational activator of prothrombin, *Proc. Natl. Acad. Sci. USA* 106, 7786–7791, 2009.
29. Thomer, L., Schneewind, O., and Missiakas, D., Multiple ligands of von Willebrand factor-binding protein (vWbp) promote *Staphylococcus aureus* clot formation in human plasma, *J. Biol. Chem.* 288, 28283–28293, 2013.
30. Pozzi, N., Chen, Z., Zapata, F., *et al.*, Autoactivation of thrombin precursors, *J. Biol. Chem.* 288, 11601–11610, 2013.
31. Pozzi, N., Chen, Z., Gohara, D.W., *et al.*, Crystal structure of prothrombin reveals conformational flexibility and mechanism of activation, *J. Biol. Chem.* 288, 22734–22744, 2013.
32. Markland, F.S., Jr., Snake venoms, *Drugs* 54 (Suppl 3), 1–10, 1997.
33. Serrano, S.M., The long road of research on snake venom serine proteinases, *Toxicon* 62, 19–26, 2013.
34. Kini, R.M. The intriguing world of prothrombin activators from snake venom, *Toxicon* 45, 1133–1145, 2005.
35. Bos, M.H. and Carmire, R.M., Procoagulant adaptation of blood coagulation prothrombinase-like enzyme complex in Australian elapid venom, *Toxins* (Basel) 2, 1554–1567, 2010.
36. Kini, R.M. and Koh, C.Y., Metalloproteinases affecting blood coagulation, fibrinolysis and platelet aggregation from snake venoms: definition and nomenclature of interaction sites, *Toxins* (Basel) 8, E284, 2016.
37. Portorollo, G., Acquasaliente, L., Peterle, D., *et al.*, Non-canonical proteolytic activation of human prothrombin by subtilisin from *Bacillus subtillis* may shift the procoagulant-anticoagulant equilibrium toward thrombosis, *J. Biol. Chem.* 292, 15161–15179, 2017.
38. Liu, C., Matsushita, Y., Shimizu, Y., Makimura, K., and Hasumi, K., Activation of prothrombin by two subtilisin-like serine proteases from *Acremonium* sp., *Biochem. Biophys. Res. Commun.* 358, 356–362, 2007.

Fibrinogen-like protein 2
39. Rabizadeh, E., Cherny, I., Lederfein, D., *et al.*, The cell-membrane prothrombinase, fibrinogen-like protein 2, promotes angiogenesis and tumor development, *Thromb. Res.* 136, 118–124, 2015.
40. Ding, J.W., Ning, Q., Lin, M.F., Fulminant hepatic failure in murine hepatitis 3 infections: tissue-specific expression of a novel *fgl2* prothormbinase, *J. Virol.* 71, 9223–9230, 1997.

41. Chan, C.W., Chan, M.W., Liu, M., *et al.*, Kinetic analysis of a unique direct prothrombinase, fgl2, and identification of a serine residue critical for the prothrombinase activity, *J. Immunol.* 168, 5170–5177, 2002.

42. Marazzi, S., Blum S., Hartmann, R., *et al.*, Characterization of human fibroleukin, a fibrinogen-like protein secreted by T lymphocytes, *J. Immunol.* 161, 138–147, 1998.

43. Melnyk, M.C., Shelev, I., Zhang, J., *et al.*, The prothrombinase activity of FGL2 contributes to the pathogenesis of experimental arthritis, *Scand. J. Rheumatol.* 40, 269–278, 2011.

44. Messias-Reason, I. and Boldt, A.B.W., Introduction, in *Mannose-Binding Lectin in the Innate Immune Systems*, Chapter 1, pp. 1–23 (references 45–76), Nova Science Publishers, Hauppauge, NY, USA, 2009.

45. Garred, P., Genster, N., Pilely, K., *et al.*, A journal through the lectin pathway of complement-MBL and beyond, *Immunol. Rev.* 274, 74–97, 2016.

46. Krarup, A., Wallis, R., Presanis, J.S., *et al.*, Simultaneous activation of complement and coagulation by MBL-associated serine protease 2, *PLoS One* 2(7), e623, 2007.

47. Jenny, L., Dobó, J., Gál, P., and Schroeder, V., MASP-1 of the complement system promotes clotting via prothrombin activation, *Mol. Immunol.* 65, 398–405, 2015.

48. Jenny, L., Dobó, J., Gál, P., and Schroeder, V., MASP-1 induced clotting—the first model of pro-thrombin activation by MASP-1, *PLoS One* 10(12), e0144633, 2015.

49. Kido, H., Fukusen, N., Katunuma, N., Morita, T., and Iwanaga, S., Tryptase from rat mast cells converts bovine prothrombin to thrombin, *Biochem. Biophys. Res. Commun.* 132, 613–619, 1985.

50. Harvima, I.T., Harvima, R.J., Pentillä, M., *et al.*, Effect of human mast cell tryptase on human plasma proenzymes, *Int. Arch. Allergy Immunol.* 90. 104–109, 1989.

51. Dietz, S.C., Auerswald, F.A., and Fritz, H., A new highly sensitive enzyme assay for human tryptase and its use for identification of tryptase inhibitors, *Biol. Chem. Hoppe-Seylers* 371(Suppl), 65–73, 1990.

52. Katunuma, N. and Kido, H., Biological functions of serine proteases in mast cells in allergic inflammation, *J. Cell. Biochem.* 38, 291–301, 1988.

53. Eagle, H. and Harris, H.N., Studies in blood coagulation. V. The coagulation of blood by proteolytic enzymes, *J. Gen. Physiol.* 20, 543–560, 1935.

54. Rimos, A., Alexander, B., and Katchalski, E., Action of water-insoluble trypsin derivatives on pro-thrombin and related clotting factors, *Biochemistry* 5, 792–798, 1966.

55. Radcliffe, R.D. and Barton, P.G., Comparison of the molecular forms of activated bovine factor X. Products of activation with Russell's Viper Venom, insoluble trypsin, sodium citrate, tissue factor, and the intrinsic system, *J. Biol. Chem.* 248, 6788–6798, 1973.

56. Landaburu, R.H., Barnhart, M.T., and Seegers, W.H., Prothrombin activation with trypsin as an enzyme, *Am. J. Physiol.* 298–302, 1961.

57. Schulz, O., Sewell, H.F., and Shakib, E., Proteolytic cleavage of CD25, the α subunit of the human T cell interleukin 2 receptor by Der p 1, a major mite allegen with cysteine protease activity, *J. Expl. Med.* 187, 271–275, 1998.

58. López-Rodríguez, J.C., Manosalvo, J. Cobrero-García, J.D., Human glutathione-S-transferase pi promotes the cysteine protease activity of the Der p 1 allergen from the house dust mite through a cysteine redox mechanism, *Redox Biol.* 26, 101256, 2019.

59. Reithofer, M., and John-Schmid, B. Allergens with proteolytic activity from house dust mites, *Int. J. Mol. Sci.* 18, 1386, 2017.

60. Zhang, J., Chen, J., Allen-Philbey, K., *et al.* Innate generation of thrombin and intracellular oxidants in airway epithelium by allergen Der p 1, *J. Allergy Clin. Immunol.* 138, 1224–1227, 2016

61. Zhang, J., Chen, J., Mangat, S.C. *et al.*, Pathways of airway oxidant formation by house dust mite allergen and viral RNA converge through myosin, panexons and Toll-like receptor 4, *Immun. Inflamm. Dis.* 6, 276–296, 2018.

TABLE 2.3
Activation of Prothrombin by Snake Venoms[a]

Venom	Comment	Reference
Echis carinatus (saw-scaled viper; ecarin)	*E. carinatus* (saw-scaled viper) is the best known of the venoms that can accelerate blood coagulation.[1] There may be several active components in *E. carinatus* venom,[2] but the best known is an enzyme (ecarin) that activates prothrombin.[3,4] Ecarin activates prothrombin to meizothrombin,[5] which can then be autocatalytically converted to thrombin.[5-7] *E. carinatus* venom, unlike factor Xa, does not require carboxylated prothrombin for activation to thrombin and was used to detect des-carboxy-prothrombin (vitamin K deficiency).[8,9] Ecarin is also used to characterize abnormal prothrombins[11,12] and to measure dabigatran concentration in blood.[13,14] The *E. carinatus* venom prothrombin activator would be considered a group A (group IA), in that no cofactors are required for the action on prothrombin. The inactivation of a snake venom metalloproteinase from *E. carinatus* has been reported.[15]	1–15
Oxyuranus scutellatus (Taipan snake venom)	The prothrombin activator was purified from *O. scutellatus* venom and shown to generate the same products from prothrombin as those seen with factor Xa, fragment 1, fragment 2, and thrombin.[16] The reaction was enhanced by the presence of phospholipid and calcium ions, but not by factor V. Subsequent work[17] from these investigators showed that the presence of 2.0 M NaCl eliminated the ability of the venom activator to activate prothrombin, but the ability to activate meizothrombin was retained. The hydrolysis of S-2222, a peptide nitroanilide substrate for factor Xa, was enhanced in the presence of 2.0 M NaCl. Evidence was obtained to suggest that the venom activator has a molecular weight in excess of 300 kDa. Later work from another laboratory[18] showed that the *O. scutellus* prothrombin activator has a molecular weight of 300,000 and was composed of several subunits of 110 kDa, 80 kDa, and 60 kDa. The 60-kDa unit consisted of two-disulfide–linked 30-kDa subunits. Treatment of the purified activator with sodium thiocyanate, a chaotropic salt, resulted in loss of the ability to activate prothrombin, which could be restored with factor V. It was suggested that the 60-kDa mass was the catalytic subunit, while the 110-kDa to 80-kDa mass had factor V function. The *O. scutellatus* venom activator would be considered a group C (formerly group III) activator in that calcium ions and phospholipid are required cofactors. There is an excellent review of the factor V function of this group of snake venoms (Australian elapid venoms).[19]	16–19
Bothrops atrox	A prothrombin activator with molecular weight 70 KDa was purified from *B. atrox* venom by a combination of gel filtration and affinity chromatography on *p*-aminobenzamidine matrix. A 300-fold purification was obtained, indicating that the protein is present as a trace component in the venom. The prothrombin activators from *B. atrox* venom is thought to be a metalloprotein and is inhibited by cysteine, dithiothreitol, and *o*-phenanthroline. The action on prothrombin is identical to that of the activator from *E. carninatus* venom, in that the product is meizothrombin.	20

(Continued)

TABLE 2.3
(Continued)

Venom	Comment	Reference
Bothrops neuwiedi	The prothrombin activator from *B. neuwiedi* venom was purified (200-fold) by gel filtration (G-100), DEAE-Sephadex and affinity chromatography (zinc affinity column with iminodiacetic acid). The purified protein had a molecular weight of 60 kDa and was inhibited by EDTA, EGTA, and *o*-phenanthrolene, suggesting that it is a metalloproteinase. As with *B. atrox*, the action on prothrombin is identical to that of the *E. carinatus* prothrombin activator in that the product is meizothrombin. It is a class A (class Ia) venom prothrombin activator.	21
Bothrops cotiara	A low-molecular-weight (MW 22,931 by mass spectrometry), single-chain prothrombin activator (cotiaractivase) was isolated from *B. cotiara* venom. The purification used phenyl-Superose hydrophobic affinity chromatography followed by ion-exchange chromatography on Mono Q-Sephadex. The *B. cotiara* prothrombin activator was calcium-dependent and produced α-thrombin via cleavage of the same peptide bonds cleaved by factor Xa. The *B. cotiera* prothrombin activator is classified as a class B (class Ib) snake venom prothrombin activator.	22
Bothrops cotiara	A low-molecular-weight prothrombin activator (cotiarinase) (29 kDa on reduced SDS-PAGE; 20 kDa on nonreduced SDS-PAGE) suggesting that the protein is more compact with disulfide bonds intact. The activation of prothrombin by cotiarinase did not require the presence of calcium ions and yielded the products obtained by activation of prothrombin with prothrombinase. The lack of cofactor requirements suggests that this protein is a class A (class I) snake venom activator of prothrombin.	23
Bothrops erythromelas (jararaca-da-seca)	A class A snake venom activator (berythyactivase) can be purified (31-fold purification) from the venom of *B. erythromelas* by cation-exchange chromatography (Sulfoethyl, Resource S from Amersham).[24] The 31-fold purification suggests that the berythryactivse comprised 3%–5% of the total venom protein. The purified protein has a molecular weight of 78 kDa (reduced SDS-PAGE). A class A snake venom activator, berythryactivase produces meizothrombin from prothrombin. This protein was cloned and shown to have homology with other snake venom metalloproteinases. There is also an enzyme in *B. erythromelas* venom that activates factor X.[25] This study also reported a higher molecular weight (90 kDa) for the prothrombin-activating enzyme, as determined by gel filtration (Sephacryl S-300).	24,25
Bothrops jararaca	A prothrombin activator (bothrojaractivase) was purified from *B. jararaca* venom. As with several other snake venom proteases, bothrojaractivase is a low-molecular-weight glycoprotein (M_r = 22,829 as determined by mass spectrometry). Bothrojaractivase was purified by a combination of gel filtration (G-75 HR) followed by cation-exchange chromatography on a Mono S column FPLC system. It is a metalloproteinase that is a class A snake venom prothrombin activator.	26

Venom	Comment	Reference
Micropechnis ikaheka	A prothrombin activator (mikarin) was purified from *M. ikaheka* venom by a combination of gel filtration (G-75 Superdex and UNO S1 (monolithic cation exchange). Mikarin was shown to have a molecular weight of 47 kDa (reduced SDS-PAGE). The activation of prothrombin by mikarin requires calcium ions and thus is a class Ia (class B) snake venom prothrombin activator.	27
Daboia russelii	A small (26.8 kDa) snake venom prothrombin activator (rusviprotase) was purified from the venom of *D. russelli* by a combination of gel filtration (Sephadex G-50) and cation exchange (CM-fast flow). In addition to the ability to activate prothrombin, rusviprotease has protease activity in degrading fibrinogen (α-fibrinogenase). The activation of prothrombin by rusviprotease does not require any cofactors and thus is classified as a class A snake venom protease and considered to be a metalloprotein. There was a difference between the fold purification of protease activity and prothrombin activating activity.	28

[a] This table is not intended to be comprehensive. The reader is directed to selected review articles,[b–h] including two on the classification of snake venom prothrombin activators.[6,7]

[b] Markland, F.S., Jr., Snake venoms and the hemostatic system, *Toxicon* 36, 1749–1800, 1988.

[c] Joseph, J.S., and Kini, R.M., Snake venom prothrombin activators similar to blood coagulation factor Xa, *Curr. Drug. Targets Cardiovasc. Haemotol. Disord.* 4, 397–416, 2004.

[d] Kini, R.M., The Intriguing world of prothrombin activators from snake venom, *Toxicon* 45, 1133–1145, 2005.

[e] Lovgren, A., Recombinant snake venom prothrombin activators, *Bioengineered* 4, 143–157, 2013.

[f] Markland, F.S. Jr. and Swenson, S., Snake venom metalloproteinases, *Toxicon* 62, 3–18, 2012.

[g] Rosing, J. and Tans, G., Structural and functional properties of snake venom prothrombin activators, *Toxicon* 30, 1527, 1992.

[h] Kink, R.M., Morita, T., and Rosing, J., Classification and nomenclature of prothrombin activators isolated from snake venoms, *Thromb. Haemost.* 86, 710–711, 2001.

References

1. Slagboom, J., Kool, J., Harrison, R.A. and Casewell, N.R., Haemotoxic snake venoms: their functional activity, impact on snakebite victims and pharmaceutical promise, *Br. J. Haematol.* 177, 947–959, 2017.
2. Chen, Y.-L. and Tsai, I.-H., Functional and sequence characterization of coagulation Factor IX/Factor X-binding protein form the venom of *Echis carinatus* leucogster, *Biochemistry* 35, 5264–5271, 1996.
3. Kornalik, F. and Blombäck, B., Prothrombin activation induced by ecarin—a prothrombin converting enzyme from *Echis carniatus* venom, *Thromb. Res.* 6, 57–63, 1975,
4. Morita, T. and Iwanaga, S., Prothrombin activation from *Echis carinatus* venom, *Methods Enzymol.* 80, 303–311, 1981.
5. Franza, B.R., Jr., Aronson, D.L., and Finalyson, J.S., Activation of human prothrombin by a procoagulant from the venom of *Echis Carinatus*. Identification of a high-molecular weight intermediate with thrombin activity, *J. Biol. Chem.* 250, 7057–7068, 1975.
6. Briet, E., Noyes, C.M., Roberts, H.R. and Griffith, M.J., Cleavage and activation of human prothrombin by *Echis carinatus* venom, *Thromb. Res.* 27, 591–600, 1982.
7. Rhee, M.-J., Morris, S., and Kosow, D.P., Role of meizothrombin and meizothrombin-(des F1) in the conversion of prothrombin to thrombin by *Echis carinatus* venom coagulant, *Biochemistry* 21, 3437–3443, 1982.

(Continued)

TABLE 2.3
(Continued)

8. Bertina, R.M., van der Marel-van Nieuwkoop, W., Dubbeldam, J., *et al.*, New method for the rapid detection of vitamin K deficiency, *Clin. Chem. Acta* 105, 93–98, 1980.
9. Widdershoven, J., van Munster, P., De Abreu, R., *et al.*, Four methods compared for measuring descarboxy-prothrombin (PIVKA-II), *Clin. Chem.* 33, 2074–2078, 1987.
10. Bezeaud, A., Drouet, L., Soria, C., and Guillin, M.C., Prothrombin Sakata: an abnormal prothrombin characterized by a defect in the active site of thrombin, *Thromb. Res.* 34, 507–518, 1984.
11. Morishita, E., Saito, M., Asakura, H., *et al.*, Himi; an abnormal prothrombin characterized by a defective thrombin activity, *Thromb. Res.* 62. 697–706, 1991.
12. Collodos, M.T., Fernandez, J., Paramo, J.A., *et al.*, Purification and characterization of a variant of human prothrombin: prothrombin Segovia, *Thromb. Res.* 85, 465–477, 1997.
13. Jaffer, I.W., Chan, N., Roberts, R., *et al.*, Comparison of the ecarin chromozyme assay and diluted thrombin time for quantitation of dabigatran concentrations, *J. Thromb. Haemost.* 15, 2377–2387, 2017.
14. Comuth, W.J., Henriksen, L.Ø., van de Kerkhof, O., *et al.*, Comprehensive characteristics of the anticoagulant activity of dabigatran in relation to plasma concentration, *Thromb. Res.* 164, 32–39, 2018.
15. Choudhury, M., Senthivadivel, V., and Velmurugan, D., Inhibitory effects of ascorbic acid toward snake venom metalloproteinase (SVMP) from Indian *Echis carinatus* venom: insights from molecular modeling and binding studies, *J. Biochem. Mol. Toxicol.* 32, e22224, 2018.
16. Owen, W.G. and Jackson, C.M., Activation of prothrombin with *Oxyuranus scutellatus scutellaus* (Taipan snake) venom, *Thromb. Res.* 3, 705–714, 1973.
17. Walker, F.J., Owen, W.G., and Esmon, C.T., Characterization of the prothrombin activator from the venom of *Oxyuranus scutellatus scutellatus* (Taipan snake), *Biochemistry* 19, 1020–1023, 1980.
18. Speijer, H., Govers-Riemaslag, J.W.P., Zwall, R.F.A., and Rosing, J., Prothrombin activation by an activator from the venom of *Oxyuranus scutellatus* (Taipan snake), *J. Biol. Chem.* 261, 13258–13267, 1996.
19. Bos, M.H.A. and Camire, R.M., Procoagulant adaptation of a blood coagulation prothrombinase-like enzyme complex in Australian elapid venom, *Toxicon* 2, 1554–1567, 2010.
20. Hofmann, H. and Bon, C., Blood coagulation induced by the venom of *Bothrops atrox*. 1. Identification, purification, and properties of a prothrombin activator, *Biochimie* 26, 777–780, 1987.
21. Govers-Riemslag, J.W., Knapen, M.J., Tans, G., Zwaal, R.E., and Rosing, J., Structural and functional characterization of a prothrombin activator from *Bothrops neuweidi*, *Biochim. Biophys. Acta* 916, 388–401, 1987.
22. Senis, Y.A., Kim, P.Y., Fuller, G.L., *et al.*, Isolation and characterization of a novel prothrombin activator from the venom of *Bothrops cotiara*. *Biochim. Biophys. Acta* 1764, 863–871, 2006.
23. Kitano, E.S., Garcia, T.C., Menezes, M.C., *et al.*, Cotiarinase is a novel prothrombin activator from the venom of *Bothrops cotiara*, *Biochimie* 95, 1655–1659, 2013.
24. Silva, M.B., Schattner, M. Ramos, C.R.R. *et al.*, A prothrombin activator from *Bothrops erythromelas* (jararaca-da-seca) snake venom: characterization and molecular cloning, *Biochem. J.* 369, 129–139, 2003.
25. Maruyama, M., Kamiguti, A., Tomy, S.C., *et al.*, Prothrombin and Factor X activating properties of *Bothops erythromelas* venom, *Ann. Trop. Med. Paristol.* 86, 549–556, 1992,
26. Berger, M., PInto, A.F., and Guimarãs, J.A., Purification and functional characterization of bothrojaractivase, a prothrombin-activating metalloproteinase isolated from *Bothrops jaraaca*, *Toxicon* 51, 488–501, 2008.
27. Gao, R., Kini, M.R., and Gopalakrishnakone, P., A novel prothrombin activator from the venom of *Micropechis ikaheka*: isolation and characterization, *Arch. Biochem. Biophys.* 408, 87–92, 2002.
28. Thakur, R., Chattopadhyay, P., Ghosh, S.S., and Mukherjee, A.K., Elucidation of procoagulant mechanism and pathophysiological significance of a new prothrombin activating metalloproteinase purified from *Daboia russelii* venom, *Toxicon* 100, 1–12, 2015.

γ-carboxyglutamic acid in the amino terminal portion of the mature protein.[8–12] There are other γ-carboxyglutamic acid–containing proteins, such as osteocalcin, a bone protein.[13] Prothrombin (factor II) is present in plasma at a concentration (0.15 mg/mL)[14] well in excess of any other vitamin K–dependent protein. The congenital deficiency of prothrombin is very rare, with an estimated prevalence of 1 in 2,000,000.[15] Prothrombin is usually isolated from blood plasma as a crude material containing factors VII and X to permit "activation" to thrombin by various agents, including citrate or thromboplastin.[16] It is of interest to the protein chemist for several reasons. First, although showing sequence homology to other vitamin K–dependent proteins, it differs markedly from proteins such as factor IX and factor X on the basis of size (120 kDa, approximately twice as large) and the γ-carboxyglutamic acid domain (GLA domain) is discarded during the formation of the derivative enzyme, thrombin. It also has a substantially longer half-life (approximately 3 days)[14] than the other K-dependent proteins. This presented a problem with the use of prothrombin complex concentrates for the treatment of hemophilia B, where the half-life of the therapeutic component, factor IX, is approximately one-half that of prothrombin,[17] resulting in elevated levels of prothrombin.[18,19] Approximately 70% of prothrombin is present in the intravascular space, which is somewhat higher than some other plasma proteins such as IgG.[14] As another example, the concentration of factor IX is substantial in the interstitial space,[20] where it is bound to collagen IV on the basolateral surface of endothelial cells.[21]

It has been known for some time that there are various mechanisms for the activation of prothrombin.[22] There is the formation of biothrombin by the prothrombinase complex,[23,24] which is the prevalent mechanism in the vascular bed providing the hemostatic response. The canonical prothrombinase complex consists of factor Xa, factor Va, calcium ions, and a membrane surface. *In vitro* experiments use synthetic phospholipid vesicles for the membrane surface, while activated platelets are generally considered to provide the membrane surface *in vivo*. There is evidence to suggest that endothelial cells may provide a membrane surface for assembly of prothrombinase at the site of vascular injury.[25] It has been reported that prothrombinase production is reduced by aspirin in aspirin-sensitive individuals but enhanced in aspirin-resistant subjects.[26] It was also observed that prothrombinase production was reduced by a statin in both aspirin-sensitive and aspirin-resistant subjects.

The prothrombin molecule consists of three canonical domains.[27,28] The amino terminal 156 residue is the fragment 1 domain, which contains the γ-carboxyglutamic residues critical for incorporation into the prothrombinase complex, while the fragment 2 domain is of unknown function. The third domain is prethrombin 2, which contains the thrombin molecule, which consists of the A chain, which is disulfide bonded to the B chain, which contains the enzyme active site, two anion binding exosites, and other domains of evolving function. There are two pathways for the conversion of prothrombin to thrombin by the prothrombinase complex.[29–32] One pathway proceeds through a derivative referred to as meizothrombin, while the other pathway involves prethrombin 2. Meizothrombin is cleaved between Arg320-Ile321, forming a product with the thrombin B chain disulfide bonded to the remainder of the prothrombin molecule.[32] The B-domain in meizothrombin has the amino-terminal isoleucine residue, which is necessary for forming a salt bridge with the enzyme-active site.[33] Meizothrombin does bind fluorescent probes directed at the putative substrate binding site dansylarginine-N-(3-ethyl-1,5-pentadiyl)amide (DAPA), which binds to both thrombin and meizothrombin, with the binding to meizothrombin showing 1.5 times more fluorescence intensity.[31,34] Meizothrombin has reduced activity compared to thrombin.[17] Meizothrombin has been reported to have equivalent activity in the hydrolysis of the peptide nitroanilide substrate S-2238 (H-D-PhePipArg$p$NA) but 10% of the clotting activity with fibrinogen.[30] Meizothrombin rapidly undergoes autolysis to remove the fragment 1 domain to yield desF1-meizothrombin. DesF1-meizothrombin was observed to have the same activity in the hydrolysis of S-2238 and the clotting of fibrinogen as intact meizothrombin, and both react with dansyl-GluGlyArg-chloromethyl ketone. Meizothrombin is poorly inactivated by antithrombin in the presence of heparin.[17] This study suggested that the enzyme-active site was intact in meizothrombin, but it is likely that exosite-2 is partially blocked. Meizothrombin can be engineered (R155A) to prevent the autolytic removal of the fragment 1 moiety.[35] These investigators reported that this derivative had reduced activity (0.06×10^6 $M^{-1}sec^{-1}$; ph 7.5. 0.175 M NaCl) in the activation of factor V; the addition of phosphatidyl serine/phosphatidyl choline vesicles increased the rate to 18×10^6 $M^{-1}sec^{-1}$). The rate of activation of factor V by thrombin was reported to be 4×10^6 $M^{-1}sec^{-1}$ and was not affected by the addition of phospholipid vesicles. Later work[36] suggested that the reduced rate of factor V activation is the result of blocking anion-binding exosite-2 (reactions at pH 7.5, 0.15 M NaCl). The effect of phospholipid on factor V activation remains a point of contention.

The conformational isomerism of meizothrombin has been reported.[37] Here it is suggested that there is isomerization between a zymogen form (at low sodium concentration) and an enzyme form (at physiological concentration of sodium ions, 0.15 NaCl). In the personal opinion of the author, while the effect of sodium ions on thrombin, first noted by Phyllis Roberts in 1969,[38] has proved to be useful research tool (see Chapter 3); it is of no questionable physiological significance, since the plasma water concentration of sodium ions is 142 mOsm/L H_2O.[39] Other engineered forms of meizothrombin, such as R155A, R271A, and R284A,[40] which is described as a stable form of meizothrombin, cannot be converted to α-thrombin. This derivative had 7% of the fibrinogen clotting activity, 100% of esterase activity (tosyl-arginine methyl ester), and 35% of amidase activity (S-2238) when compared to α-thrombin. These results are similar to those obtained in an earlier study with a native form of human meizothrombin. Meizothrombin is the major intermediate in the activation of prothrombin by *Echis carinatus* venom.[41,42]

Meizothrombin is an important factor in considering the autoactivation of thrombin. A number of proteases are characterized by intermolecular or intramolecular autoactivation. Such phenomena are usually characterized by a conformational change prior to peptide bond cleavage.[43,44]

LIPIDS AND PROTHROMBIN CONVERSION

Platelets or phospholipids have an important role in prothrombin conversion. Phospholipids are an *in vitro* substitute for platelets. Platelets have been known to be important for prothrombin activation for some time,[45,46] and platelets must be activated to participate in prothrombin activation and factor X activation.[47] The activation of platelets is associated with the exposure of phosphatidyl serine on the surface via the action of an enzyme, flippase.[48] Phosphatidyl serine then provides a surface for the assembly of the prothrombinase complex.[49] The formation of the components of the prothrombinase complex is transferred from bulk solution to the surface of the platelet.[50] In a somewhat simplistic but useful analogy, the binding of components to the platelet surface serves to decrease the K_m for prothrombin,[51] although it is recognized that the interaction is more complex.[52] It is generally assumed that activated platelets provide the same component, platelet factor 3,[53] for both prothrombin activation and factor X activation (tenase). However, Peter Walsh and coworkers presented evidence to suggest that perhaps the functions are not products of the same component.[54] This issue has not been resolved. It has been reported that bile salts such as cholate can substitute for phospholipid in the conversion of prothrombin and thrombin.[55,56]

FACTOR V AND PROTHROMBIN CONVERSION

The prothrombinase complex is composed of factor Xa and activated platelets/phospholipid, calcium ions (other divalent cations can substitute for calcium ions), and factor V. Factor V is a late arrival to the scene compared to earlier studies on the identification of the various coagulation factors (see Chapter 1).[57–59] Nevertheless, the suggestion of factor V (labile factor)[58] did precede factors VII, VIII, IX, XI, XII, and XIII, but it was almost 20 years before a description of the congenital deficiency.[58] Armand Quick (famous for the Quick prothrombin time[60]) at the Marquette School of Medicine (now Medical College of Wisconsin) and Mario Stefanini presented evidence supporting the function of factor V in the conversion of prothrombin to thrombin.[61] P.A. Owren also made significant contributions to the role of factor V in the conversion of prothrombin to thrombin.[62–65] I highly recommend the reader to an excellent review in *Northwest Medicine*[66] in which Professor Owren describes his discovery of factor V and subsequent studies. His description of the acquisition of rabbit brain for his performance of the Quick prothrombin time should be an inspiration to current investigators who may feel that they face unreasonable challenges.

FACTOR X/XA AND PROTHROMBIN ACTIVATION

Factor Xa is the enzyme responsible for the conversion of prothrombin to thrombin in the vascular system. It is possible factor Xa also functions in the extravascular space, but as shown later, there are mechanisms for the generation of thrombin in the extravascular space. While factor Xa had been functionally identified as thrombokinase in the early parts of the 20th century, a functional deficiency of factor X was identified by John Graham, Emily Barrow, and Cecil Hougie at the University of North Carolina at Chapel Hill in 1957.[67–69] Factor X was the meeting point of the intrinsic clotting pathway and the extrinsic clotting pathway.[70–72] It can be converted to factor Xa by limited proteolysis by either the intrinsic pathway (intrinsic tenase)[73] or by the extrinsic pathway (extrinsic tenase).[74] Factor Xa is a two-chain tryptic-like serine protease which converts prothrombin to thrombin by the process of limited proteolysis.[75] Factor Xa in the prothrombinase complex forms α-thrombin by cleaving two peptide bonds via two pathways: one through meizothrombin and the other through the direct formation of prethrombin-2, the direct precursor of α-thrombin.

MECHANISMS OF PROTHROMBIN ACTIVATION OTHER THAN PLASMA PROTHROMBINASE

While it is critical that thrombin is formed rapidly in vascular hemostasis, the rapid formation of thrombin is not necessary in the putative regulatory functions outside of the vascular space. There are a number of possible pathways for thrombin formation in the extravascular space.

DER P 1

Der p 1 (*Dermatophagoides* peptidase *1*) is a cysteine protease allergen produced by the house dust mite (*Dermatophagoides pteronyssinus*).[76–79] Der p 1 is a protease of the papain family of proteases, and difficulties in purifying the native protein resulted in the use of papain to mimic the action of Der p 1 proteases.[80] Der p 1 is an approximately 25-kDa proteinase, occurs as a zymogen,[81] and undergoes autocatalytic activation under acidic conditions.[82] Der p 1 has been reported to activate prothrombin to thrombin, which was associated with the generation of reactive oxygen species.[83,84] Subsequent work suggests that the action of thrombin in the pulmonary bed is mediated through the toll-like receptor-4 receptor.[80,85] Other data suggests that proteases such as Der p 1 are pleiotropic in action.[86] Der p 1 is suggested to activate thrombin in the pulmonary bed, resulting disorders such as asthma. Studies in a murine model presented evidence that thrombin was harmful at high concentrations and beneficial at low concentrations.[87] It should also be noted that there is no comparison of Der 1 p with the prothrombinase complex; however, it is likely that the rate of activation of prothrombin by Der 1 p is much slower than that observed with prothrombinase. As discussed earlier, the rate of activation of prothrombin by Der 1 p is likely sufficient for the putative regulatory function(s).

MASP-1

Mannose-binding lectin (MBL)-associated serine proteases (MASPs) are associated with the complement system but are pleiotropic.[88–90] MBL is a collectin, a C-type lectin that functions in innate immunity.[91] MBL is a pattern-recognition receptor that recognizes carbohydrate arrays on the surface of pathogenic organisms resulting in a pleiotropic response, including complement activation and modulation of inflammation. The binding of MLB to targets results in the activation of MASPs, which have several functions.[89] Activation of MASPs also occurs on binding to ficolins.[89] Both MASP-1[92] and MASP-2[93,94] have been reported to form thrombin from prothrombin. There has been more work on MASP-1, which suggests that this enzyme cleaves prothrombin at sites different from those cleaved by factor Xa.[95,96] Cleavage is suggested to occur at R271 and R293 to yield a thrombin variant, which would likely possess reduced fibrinogen clotting activity. The work on the chemistry of the MASP-1 catalyzed cleavage of prothrombin was a catalytic fragment. The work was performed at pH 7.4 (tris-buffered saline) at 37°C for 120 minutes to obtain full cleavage of prothrombin in the absence of calcium ions or phospholipid. The authors[96] suggest that a wild-type protein might have regions that would result in the productive binding of calcium ions and phospholipid.

FGL -2 (FIBRINOGEN-LIKE PROTEIN 2)

A gene encoding a protein of approximate M_r of 42 kDa with homology to the β- and γ-chains of fibrinogen was identified in T-lymphocytes.[97] The term fibroleukin was advanced in 1998 by Marazzi and coworkers to describe a fibrinogen-like protein secreted by CD4+ and CD+ lymphocytes.[98] The term was developed to show the relation of the protein to the extracellular matrix and leukocytes. Fibroleukin mRNA was preferentially expressed in memory T cells. Production by peripheral blood mononuclear cells (PBMC) was lost in culture but restored by interferon-gamma (INF-γ). The protein was shown to undergo self-association to form a complex with molecular weight of 350–400 kDa. The monomer form has a M_r of 70 kDa on reducing SDS gel. Despite the disparity in the observed molecular weight, it is likely that this material is soluble fibrinogen-like protein 2 (sFLG2) and was subsequently characterized as such, a molecule with immunomodulatory activity.[99,100] There has been considerable work on the immunomodulatory effect of sFLG2.[101–103] The serum level has been suggested as a biomarker for traumatic brain injury.[104] This is a limited selection of citations describing the role of sFGL-2 in immunomodulation, as this function is not the focus of this discussion.

The identification of a macrophage membrane protein that could activate prothrombin to thrombin was suggested by a work in 1980[105] and more clearly defined in 1982 as a monocyte prothrombinase by the same group in La Jolla.[106] This later work had the advantage, at that time, of having specific factor-deficient plasma for the characterization of procoagulant activity (PCA). The PCA did clot prothrombin-deficient plasma, but did clot plasmas deficient in factors V, VII, VIII, X, and XII as well as normal plasma. The action of PCA on prothrombin did require the presence of calcium ions but no other cofactors. The PCA is suggested to be a serine protease based on inhibition by DFP. The activity was not detected in freshly isolated PBMC but increased on exposure to lipopolysaccharide or soluble antigen-antibody complexes. The expression of FGL-2 prothrombinase has been observed in epithelial cells, endothelial cells, dendritic cells, monocytes, and macrophages.[107,108] The characterization and functional expression of the *FGL-2* gene have been reported.[109] The expressed MGl-2 protein is a 439 residue type II membrane protein (type II membrane protein contains a hydrophobic transmembrane segment, which is not cleaved). mFGL-2 protein has a carboxyl-terminal fibrinogen-related domain (FRED), which does appear in other proteins.[110,111] Inhibition by DFP and mutagenesis studies[112] suggest the presence of a catalytically importantserine residue (Ser91). On the basis of these results, it was suggested that the FGL-2 prothrombinase is a clan SE serine protease (α-lytic endopeptidase).[113–116] A search for FGL-2 in the MEROPS database was unsuccessful; there may not be sufficient information to include FGL-2 in the MEROPS database.

There are several studies characterizing FLG-2 as a peptidase.[106,112,117] The most complete study[112] examined the FGL-2 protein as expressed in baculovirus and as truncated/mutant forms in CHO cells. The purified full-length protein did not have PCA activity; reconstitution of the expressed protein in a phosphatidyl choline–phosphatidyl serine vesicle (75% PC/25% PS) yielded an active procoagulant. The addition of factor Va further increased the rate of prothrombin activation. As noted earlier, FLG-2 is inactivated by DFP. It was observed that phenylmethylsulfonyl fluoride (PMSF) did not inactivate FLG-2. This is a bit unusual, since both reagents react with the active site serine of trypsin and thrombin at similar rates[118] (see also Table 4.1). The active site triad (asp-his-ser) present in canonical serine proteases is absent in FLG-2, perhaps providing an explanation. I would note that there are regulatory serine proteases such as factor IXa that are not inhibited by DFP.[119,120] It is of interest that a monoclonal antibody elicited from a segment close to Ser91 (residues 76–87, EEVFKEVQNLKE) inhibited FLG-2 prothrombinase activity.[121] There have been no further studies on the characterization of this interesting enzyme. There have been a number of studies on the potential role of FGL-2 in various pathologies.[122–129] The observation on the possible role of FGL-2 in pulmonary hypertension[129] deserves further comment. Pulmonary hypertension is suggested to be of importance in the development of idiopathic pulmonary fibrosis,[130] and there is data to suggest that a thrombin inhibitor, dabigatran, can prevent the development of this condition.[131,132]

STAPHYLOCOAGULASE

Staphylocoagulase is a protein formed by *Streptococci* sp. which forms a product, staphylothrombin, on interaction with human prothrombin, which can clot fibrinogen.[133] Staphylocoagulase does not form thrombin from prothrombin, but rather forms a complex where the amino terminal Ile-Val sequence of staphylocoagulase is inserted into prothrombin to form the salt bridge with aspartic acid residue, leading to a formation of a competent active site.[134] The ability of staphylococci to form a material that clots blood plasma has been known since early in the last century[135,136] and was established as a protein by Tager in 1948.[137] Staphylocoagulase is an approximately 60-kDa protein that rapidly associates with prothrombin (3.3 × 10^6 M^{-1}sec^{-1} at pH 7.5, 37°C), forming an equimolar complex.[138] Staphylocoagulase interacts poorly with bovine prothrombin; although the binding is as tight as the human form, the catalytic site is incompletely formed, reflecting differences in the sequences of human and bovine prothrombin.[139] Bovine prothrombin bound to a matrix is used for the affinity purification of staphylocoagulase.[138,140] The fibrinogen clotting activity of staphylothrombin is approximately 50% that of free human α-thrombin with somewhat reduced amidase activity (S-2238).[141] Staphylothrombin does react with DFP and the active site titrant, 4-nitrophenyl-4-guanidinobenzoate (NPGB).[142] The reactivity of residues at the enzyme-active site of staphylothrombin permitted the development of fluorescent derivatives of prothrombin where a fluorescent peptide chloromethyl ketone reacted with the active-site histidine.[143] This labeled prothrombin derivative could be used for the *in vivo* detection of *S. aureus* endocarditis.[144]

von Willebrand Binding Protein

von Willebrand binding protein (vWBP) is an approximately 60-kDa protein secreted by *S. aureus* that is characterized by binding to von Willebrand factor.[145] The vWBP protein was identified by panning a phage-display library from *S. aureus* against recombinant von Willebrand factor. The mature vWBP protein was purified from culture supernatant. The same research group subsequently reported that vWBP is a procoagulant,[146] which, as with staphylocoagulase, reacts with prothrombin to form a complex with fibrinogen-clotting (thrombin) activity. The catalytic activity is based, again as with staphylocoagulase, on the insertion of the amino terminal segment Val-Val into prothrombin, forming a salt bridge leading to a competent active site.[147] Complete activation is accomplished by binding to the substrate fibrinogen.[147,148] The latter study[148] demonstrated that PPACK reacted with the thrombin active site histidine in the prethrombin-2 domain in the vWBP-prothrombin complex.

MISCELLANEOUS PROTEASES

The activation of prothrombin requires the cleavage of two peptide bonds, R273-T274 and R332-I333, to form α-thrombin. Cleavage of one of the two peptide bonds, R332-I333, yields meizothrombin, which has reduced fibrinogen clotting activity. The point here is that there are likely a number of peptide bonds in prothrombin which can be cleaved to provide a derivative with fibrinogen clotting activity. This is likely the situation with proteases derived from bacteria and fungi that are suggested to activate prothrombin. Furthermore, many of the studies rely on peptide nitroanilide substrates for measurement of activity. All of that said, it does not require much thrombin activity to cause a physiological effect, and there is good evidence that there are regions of the thrombin molecule not involved in catalysis that have biological activity.

Early work was reviewed in 1994[149] showing that partially purified proteases from *Candida albicans*, *Pseudomonas aeruginosa*, and *Serratia marcescens* could form thrombin-like activity. Some examples of bacterial enzymes that form thrombin-like activity from prothrombin are shown in Table 2.4.

TABLE 2.4
Some Bacteria and Yeast that Secrete Enzymes that Can Form Thrombin from Prothrombin

Species	Comment	Reference
Candida albicans	Purified protein from culture supernatant fraction, M_r 43 kDa. Activation of prothrombin was best at pH 4.5, as determined with 7-amino-4-methyl-coumarin (AMC).	1
Serratia marcescens	A partially purified protein was obtained from culture supernatant, M_r 56 kDa. The reaction was performed at pH 7.5 and activity measured with AMC.	1
Pseudomonas aeruginosa	Commercially obtained protein, M_r 48 kDa. The reaction was performed at pH 7.5 and activity measured with AMC.	1

(Continued)

TABLE 2.4
(Continued)

Species	Comment	Reference
Porphyromonas gingivalis	A 95-kDa enzyme (HRgpA) was more potent than a smaller form (RgpA) in forming thrombin from prothrombin at pH 7.6. Activation of prothrombin did occur at the canonical cleavage sites. The thrombin formed was rapidly degraded by either enzyme. The rate of formation is modest compared to the prothrombinase complex but sufficient to form fibrin in the gingival crevice.	2
Aeromonas hydrophilia	The enzyme was partially purified from culture supernatant. An M_r of 35 kDa was obtained by SDS electrophoresis and gel filtration, and evidence was presented to show that the enzyme was a metalloprotein. There was modest activation of prothrombin as measured with S-2238. There was a curious effect of salt (NaCl, KCl, LiCl) with two optima: one at no salt and the other at 1.5 M.	3
Bacillus megaterium	A metalloprotein of approximately M_r 35 kDa (designated bacillolysin MA) as obtained from the culture supernatant of *B. megaterium* A95442. The purified protein was observed to activate prothrombin but at much higher enzyme concentrations than those useful for fragmentation of plasminogen or activation of protein C. Factor IX was refractory to activation by bacillolysin MA.	4
Vibrio vulnificus	A protein was purified from the culture supernatant of *V. vulnificus* ATCC 29307. Mass spectrometry provided a weight of 34,077.37 Da. Proteolytic activity was assessed with azocasein as a substrate. Activity was inhibited by metal ion chelators such as EDTA and 1,10-phenanthroline. The enzyme digested prothrombin, fibrinogen, bovine serum albumin, and other proteins. Manganese ions had no effect on the digestion pattern (SDS gel) of prothrombin, while zinc ions restricted digestion. Fibrinogen clotting activity and amidase (Boc-VPR-pNA) were associated with the action of the *Vibrio* protease on prothrombin. The enzyme is pleiotropic.	5–7
Acremonium sp.	Subtilisin-like enzyme produces a meizothrombin-like species by cleavage at Y316-I317. The enzymes had a monomer weight of approximately 35 kDa but appeared to function as a dimer.	8
Bacillus subtilis	A subtilisin-like enzyme was obtained from *B. subtilis*, which can cleave prothrombin at A470-N471, yielding a meizothrombin-like molecule with greatly reduced fibrinogen clotting activity (0.5%–1%) compared to α-thrombin.	9

References

1. Kaminishi, H., Hamatake, H., Cho, T., *et al.*, Activation of blood clotting factors by microbial proteinases, *FEMS Microbiol. Lett.* 121, 327–332, 1994.
2. Imanmura, T., Banbula, A., Pereira, P.J., Travis, J., and Potempa, J., Activation of human prothrombin by arginine-specific cysteine proteinases (Gingipains R) form porphyromonas gingivalis, *J. Biol. Chem.* 276, 18984–18991, 2001.

3. Keller, T., Seitz, R., Dobt, J., and Konig, H., A secreted metallo protease from *Aeromonas hydrophilia* exhibits prothrombin activator activity, *Blood Coagul. Fibrin.* 15, 169–178, 2004.

4. Narasaki, R., Kuribayashi, H., Shimigu, K., *et al.*, Bacillolysin MA, a novel bacterial metalloproteinase that produces angiostatin-like fragments from plasminogen and activates protease zymogens in the coagulation and fibrinolysis systems, *J. Biol. Chem.* 280, 14278–14287, 2005.

5. Chang, A.K., Kim, H.Y., Park, J.E., *et al.*, *Vibrio vulnificus* secretes a broad-specificity metalloprotease capable of interfering with blood homeostasis through prothrombin activation and fibrinolysis, *J. Bacteriol.* 187, 6909–6916, 2005.

6. Kim, H.Y., Chang, A.K., Park, J.E., *et al.*, Procaspase-3 activation by a metalloprotease secreted by *Vibrio vulfnicus*, *Int. J. Mol. Med.* 20, 591–595, 2007.

7. Park, J.E., Park, J.W., Lee, W., and Lee, J.S., Pleiotropic effects of a vibrio extracellular protease on the activation of contact system, *Biochem. Biophys. Res. Commun.* 450, 1099–1103, 2014.

8. Liu, C., Matsushita, Y., Smimizu, K., Makimura, K., and Hasumi, K., Activation of prothrombin by two subtilisin-like serine proteases from *Acremonium* sp, *Biochem. Biophys. Res. Commun.* 358, 356–362, 2007.

9. Pontarollo, G. Acquasaliente, L., Peterle, D., *et al.*, Non-canonical proteolytic activation of human prothrombin by subtilisin from *Bacillus subtilis* may shift the procoagulant-anticoagulant equilibrium toward thrombosis, *J. Biol. Chem.* 292, 15161–15179, 2017

ACTIVATION OF PROTHROMBIN TO THROMBIN IN 25% TRISODIUM CITRATE AND OTHER CONCENTRATED SALT SOLUTIONS

The conversion of prothrombin to thrombin in concentrated solutions of trisodium citrate was first reported in detail by Seegers in 1949.[150] The rate and extent of activation of prothrombin were dependent on the concentration of trisodium citrate, with most reactions being performed in 25% trisodium citrate (approximately 0.85 M, assuming the dihydrate, MW 294.099; unadjusted pH 8.0–8.5). As a note, do not confuse the trisodium salt with the disodium salt (sodium citrate, acid), as there will be no activation with the disodium salt. The activation of prothrombin in sodium citrate is pH dependent, with optimal activation at alkaline pH.[151] As part of their comprehensive study, Schultze and Schwick[151] showed that the activation of their prothrombin (which likely contained factor X and factor VII) proceeded best at pH 7.10–9.95 with a distinct decrease at 6.65 and more so at 5.85. They also showed that the reaction proceeded best at 37°C and 45°C with a distinct drop at 28°C; the difference was in rate (extent of lag phase) and not yield. The concentration of pro-thrombin in the reaction mixture was also critical

The effect of citrate was noted by Seegers as a result of adding the salt to prevent premature activation of his prothrombin preparations.[152,153] Activation of some pro-thrombin preparation was also observed with other polyvalent inorganic anions such as ammonium sulfate, potassium oxalate, and potassium hydrogen phosphate.[154] As noted earlier, the rate and extent of thrombin formation from prothrombin in triso-dium citrate is dependent on the concentration of sodium citrate with little, if any, activation at 5% sodium citrate.[154] Seegers also observed that diphenyl sulfones had a profound effect on the rate and extent of prothrombin activation in 25% trisodium citrate; both acceleration (2-methyl,4-amino, 4′,6′-diamino diphenyl sulfone and

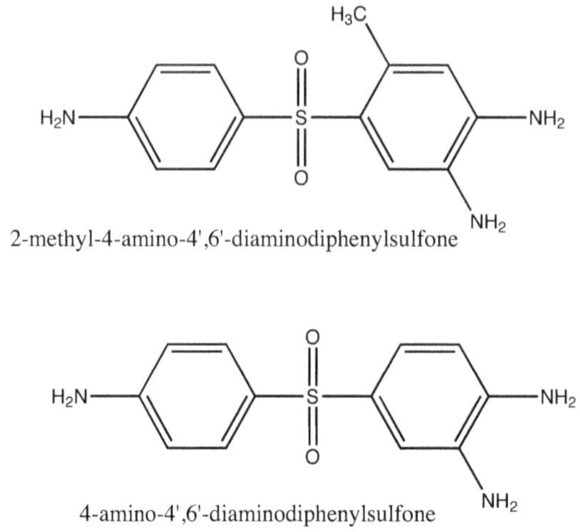

2-methyl-4-amino-4',6'-diaminodiphenylsulfone

4-amino-4',6'-diaminodiphenylsulfone

FIGURE 2.1   The structures of diphenylsulfones.

inhibition (4-amino, 4',6'diamino diphenyl sulfone, Figure 2.1) were observed. As with citrate activation, this was a fortuitous finding by Seegers, as he had added the diphenyl sulfones to inhibit possible bacterial growth.[153] Certain sulfones reduced the lag phase and increased the yield of thrombin. It is reasonable to suggest the inhibitory effect of diphenyl sulfones would be directed at factor Xa (autoprothrombin C).[153] Rationalization of the enhancement of the rate of formation of and yield of thrombin is more difficult. This would be an interesting project but unlikely to gain support in the current age, which is dominated by cell biology. There is one paper (that I could find) on the effect of diphenyl sulfones on proteolytic enzymes which suggest a modest inhibition of "leucoprotease."[155] Diphenyl sulfone has a log P value of 2.4, suggesting that the derivatives would be poorly soluble in water, and indeed Seegers commenst that the solubility of the various diphenyl sulfones limited the use to the addition of only several milligrams.[150] Inhibition of citrate activation of prothrombin is also inhibited by DFP and *p*-toluenesulfonyl-L-arginine methyl ester (TAMe).[156,157] The citrate activation of prothrombin is also inhibited by the addition of diluted serum,[158] likely due to antithrombin.

The majority of work on the citrate activation of prothrombin was performed by Professor Seegers and his associates. He interpreted the results as the autocatalytic activation of prothrombin in trisodium citrate. Although this was later shown to be an incorrect interpretation of the data, it was not unreasonable at the time, considering the state of the art in protein purification and analysis. It is clear now that the early prothrombin preparations contained other coagulation factors, as it was possible for other investigators to obtain prothrombin preparations that did not form thrombin in 25% trisodium citrate.[159,160] Lechner and Deutsch[159] were able to demonstrate the presence of factors VII and X in the preparations of prothrombin obtained by the method of Seegers and associates. Somewhat earlier, Seegers and Landaburu[161] had obtained a prothrombin preparation which did not undergo citrate activation but attributed the difference to changes in the structure of the prothrombin. Somewhat later, Teng and Seegers[162] described the activation of factor X in 25% trisodium citrate and ascribed the activation to thrombin, which they described as reciprocal proenzyme activation, where factor Xa (autoprothrombin C) activated prothrombin in 25% trisodium citrate and thrombin activated factor X in trisodium citrate. I don't know of any other work describing the activation of factor X by thrombin, but I don't think anyone has tried to perform the reaction in 25% trisodium citrate. I should add that Seegers, in what might have been his final opus, maintained that thrombin did activate factor X in 25% trisodium citrate.[153] Inhibition of thrombin formation from prothrombin in 25% trisodium citrate by soybean trypsin inhibitor[163,164] provided support for the role of factor Xa in the process of citrate activation. Factor Xa is one of the few activated coagulation factors, all of which can be considered to be trypsin-like serine proteases, that is inhibited by soybean trypsin inhibitor.[163] Although there was no thrombin formation, the chemical changes[165] (electrophoretic mobility in a Tiselius apparatus[166] and material soluble in 7% trichloroacetic acid) were unchanged when compared to a control activation without the inhibitor. A similar effect is observed by inclusion of 3,4,4′-triaminodiphenyl sulfone, the inhibitor described earlier. Final clarification of the mechanism of citrate activation came from the work of the late Dave Aronson and Doris Menaché, who showed that factor VII was also required for the citrate activation of prothrombin.[167] Current work on the citrate activation of prothrombin either uses partially purified prothrombin,[168] such as a barium sulfate eluate, or a Cohn fraction III/IV, or added a source of factors VII and X, such as defibrinated plasma.[169] Phyllis Roberts, at MCV in Richmond, observed that the rate of prothrombin activation was slower in potassium citrate than in sodium citrate[170] and suggested that the decreased rate of thrombin formation was due to the inhibition of factor Xa activity by potassium ions and enhancement by sodium ions. While most of the work cited earlier suggests that purified prothrombin will not activate in concentrated trisodium citrate, a patent[171] described the activation of recombinant prothrombin to thrombin in 35% trisodium citrate (added as a solid) for 32 hours at 37°C. The activation of purified plasma prothrombin is also claimed with 35% trisodium citrate (24–28 hours, 37°C). The patent also claims pH 6.5–7.5, which is more acidic than the more alkaline pH; Schultze and Schwick[151] note that citrate activation is more rapid at alkaline pH than at pH 6.6. I know of no further work on this particular technical approach. It does have considerable commercial value for the manufacture of thrombin as only a pure chemical, not another biologic such as thromboplastin or factor Xa or a reptile venom product. I would note that there has been renewed interest in the autoactivation of prothrombin,[172] and it is possible that high concentrations of citrate may favor the open conformation, which is suggested to be susceptible to autoactivation.[173,174] Citrate activation has been useful for preparing multiple forms of thrombin (e.g., β-thrombin, γ-thrombin).[168,169,175–177]

AUTOACTIVATION OF PROTHROMBIN

The autoactivation of prothrombin has been of interest since the identification of the protein (see Chapter 1). While subsequent work on the phenomenon of citrate activation supported a requirement of factors VII and X,[186] another group[171] suggested that both recombinant prothrombin and highly purified plasma prothrombin could form thrombin via autoactivation in the presence of 35%–40% trisodium citrate. More recently, Di Cera and coworkers have presented evidence supporting the autoactivation of prothrombin.[178] They reported the incorporation of PPACK into prethrombin 2 and prothrombin. The autoactivation of prethrombin 2 was not observed with the S205A mutant. Further work suggested that an "open" conformation of prothrombin would be more susceptible to autoactivation.[179,180] Histone H4 has been reported to promote the autoactivation of prothrombin, as assessed by electrophoretic analysis and hydrolysis of tripeptide nitroanilide substrate.[181] The reaction proceeded in the presence of 5 mM $CaCl_2$, 150 mM NaCl at 20 mM Tris, pH 7.4. Histone H4 was observed to bind to prothrombin. Subsequent work[182] showed that histone H4 enhanced the rate of prothrombin activation by factor Xa in the presence of 5 mM $CaCl_2$, 145 mM NaCl, 20 mM Tris, pH 7.4. The effect of histone in terms of fold stimulation is reduced in the presence of phospholipid, with inhibition observed in the presence of factor Va.

Staffan Magnusson reported that prothrombin could be activated to thrombin in the presence of EDTA.[183,184] Specifically, 2.6–2.9 gm of bovine prothrombin was dissolved in 40–100 mM Tris, pH 7.5, containing 20–50 mM EDTA to a concentration of 70–200 mg/mL. Thrombin was formed after 20–30 minutes incubation at room temperature (usually 23°C).

ACTIVATION OF PROTHROMBIN IN THE PRESENCE OF POLYMERIC POLYAMINES/BASIC PROTEINS AND OTHER POLYAMINES

The effect of histone H4 on the activation of prothrombin has been discussed earlier. These studies, however, are not the first observations on the effect of polymeric polyamines/basic proteins. Ken Miller. at the time at the New York State Department of Health and later at the University of Miami, reported the activation of prothrombin in the presence of polylysine.[185] Activation occurred with a 10-fold molar excess of polylysine (MW 5,000) at alkaline pH (8.0–8.5). The enhancing effect of polylysine was sensitive to ionic strength but dependent somewhat on the cation. Inhibition was observed at an ionic strength of 0.1 with calcium nitrate or magnesium nitrate, while inhibition by sodium nitrate occurred at a higher (0.2–0.3) ionic strength. This early study, while very useful, was performed with preparations of prothrombin that also contained amounts of factor X and factor VII. David Aronson and Doris Ménaché showed that factor Xa (thrombokinase) was required for the activation of prothrombin by polylysine.[186] This was one of several observations in a study which focused on the activation of prothrombin by factor Xa. Factor Xa was observed to slowly activate prothrombin in the absence of any cofactors at pH 7.5, with slight stimulation observed with 5 mM $CaCl_2$. Stimulation of factor Xa–catalyzed activation of prothrombin was observed with NaCl up to a concentration of approximately 0.3 M. Activation of prothrombin by factor Xa was twice as fast in 1.0 M sodium acetate than that observed in the presence of either 1.0 M potassium acetate or 1.0 M lithium acetate. These latter studies were part of a study evaluating the effect of anions on the factor Xa–catalyzed activation of prothrombin. Subsequent work by Vogel and coworkers[187] observed that polylysine could facilitate the activation of prothrombin and prethrombin-1 but not prethrombin-2. As with Miller's original study, relatively high concentrations (10-fold molar excess) of polylysine were required. There was modest inhibition of the reaction with 5 mM $CaCl_2$. The effect of polylysine was independent of molecular weight in the range of 6–72 kDa. Protamine sulfate was somewhat effective in the factor Xa–catalyzed activation of prothrombin. No activation of prothrombin or prethrombin 1 was observed with only polylysine. In later work,[188] Miller observed that a number of basic compounds could facilitate the activation of prothrombin, including neomycin, polymyxin, colistin, polybrene, and polyvinylamine.

As might be expected, polylysine was observed to facilitate the activation of factor X by factor IXa.[189] This observation provided for the development of an assay for factor IXa.[190–192] The interaction of prothrombin complex concentrates with DEAE-cellulose[193] is likely responsible for the observed thrombogenicity of certain prothrombin complex concentrates.[194] The activation of prothrombin complex concentrates is likely mediated through factor VII.[195,196]

LABORATORY PRODUCTION OF THROMBIN

While all of these approaches can yield thrombin, only activation of a partially purified prothrombin preparation such as a barium citrate eluate from plasma (usually bovine) or for human thrombin, a Cohn fraction, or derivative thereof was found useful for the large-scale production of thrombin.[197–200] Purification is usually accomplished with a cation-exchange resin such as sulfopropyl-sephadex.[178–181] As noted elsewhere, this approach was an extension of Rasmussen's early work with IRC-50 ion-exchange resin.[201,202] Magnussen[203] noted that on occasion, preparation of bovine thrombin with exceptionally high activity was obtained. I have had the same experience[204] with preparations with specific activity exceeding 6,000 NIH units/mg protein as assayed directly from the sulfopropyl-sephadex (SP-Sephadex) column. Our normal procedure was to freeze the purified thrombin in the 0.25 sodium phosphate buffer, pH 6.5, or dialyzed into 10 mM NaCl or another solvent. Regardless of the solvent, the specific activity of the purified material was between 2,500 and 3,000 units/mg of protein. This material was stable when stored at −20°C for at least 3 months. Human thrombin purified by this procedure usually had higher specific activity (3,000–3,500 units/mg).

REFERENCES

1. Aronson, D.L., Comparison of the actions of thrombin and the thrombin-like venom enzymes Ancrod and Batroxobin, *Thromb. Haemost.* 36, 9–13, 1976.
2. Macfarlane, R.G. and Barnett, B., The haemostatic possibilities of snake-venom, *Lancet* 227(2, July–December), 985–987, 1934.
3. Anon., Snake venom for haemophilia, *Br. Med. J.* 2, 867, 1934.
4. Marsh, N. and Williams, V., Practical applications of snake venom toxins in haemostasis, *Toxicon* 45, 1171–1181, 2005.

5. Kumar, L.S.V., Potential use of snake venom derivatives as hemostatic agents in dentistry, *Gen. Dent.* 64, e10–e11, 2016.

6. Seon, G.M., Lee, M.H., Kwon, B.J., *et al.*, Functional improvement of hemostatic dressing by addition of recombinant batroxobin, *Acta Biomater.* 48, 175–185, 2017.

7. Schmitmeirer, S., Markland, F.S., Ritter, M.R., Sawer, D.E., and Chen, T.C., Functional effect of cortorristatin, a snake venom disintegrins, on human glioma cell invasion, *In Vitro, Cell Commun. Adhes.* 10, 1–16, 2003.

8. Stenflo, J., Fernlund, P., Egan, W., and Roepstorff, P., Vitamin K dependent modifications of glutamic acid residues in prothrombin, *Proc. Natl. Acad. Sci. USA* 71, 2730–2733, 1974.

9. Fernlund, P., Stenflo, J., Roepstoff, P., and Thomsen, J., Vitamin K and the biosynthesis of prothrombin V. γ-Carboxyglutamic acids, the vitamin K-dependent structures in prothrombin, *J. Biol. Chem.* 250, 6125–6133, 1975.

10. Morris, H.R. and Dell, A., Mass spectrometric identification and sequence location of the ten residues of the new amino acid (γ-carboxyglutamic acid) in the *N*-terminal region of prothrombin, *Biochem. J.* 153, 663–679, 1976.

11. Nelsestuen, G.L., Role of γ-carboxyglutamic acid. An unusual protein transition required for the calcium-dependent binding of prothrombin to phospholipid, *J. Biol. Chem.* 251, 5648–5656, 1976.

12. Burnier, J.P., Borowski, M., Furie, B.C., and Furie, B., Gamma-carboxylic acid, *Mol. Cell. Biochem.* 39, 191–207, 1981.

13. Hauschka, P.V., Osteocalcin: the vitamin K-dependent Ca^{2+}-binding protein of bone matrix, *Haemostasis* 16, 258–272, 1986.

14. Shapiro, S.S. and Martinez, J., Human prothrombin metabolism in normal man and in hypocoagulable subjects, *J. Clin. Invest.* 48, 1292–1298, 1969.

15. Lancellotti, S. and De Cristofaro, R., Congenital prothrombin deficiency, *Semin. Thromb. Hemost.* 35, 367–381, 2009.

16. Lundblad, R.L., Kingdon, H.S., and Mann, K.G., Thrombin, *Methods Enzymol.* 45, 156–176, 1976.

17. Björkman, S. and Berntorp, E.E., Pharmacokinetics, in *Textbook of Hemophilia*, ed. C.A. Lee, E.E. Berntrop, and W.K. Hoots, Chapter 19, pp. 106–111, Blackwell Publishing, Malden, MA, USA, 2005.

18. Mannuci, P.M., Bauer, K.A., Gringeri, A., *et al.*, Thrombin generation is not increased in the blood of hemophilia B patients after the infusion of a purified factor IX concentrate, *Blood* 76, 2540–2545, 1990.

19. Sørensen, B., Spahn, D.R., Innerhofer, P., *et al.*, Clinical review: prothrombin complex concentrates—evaluation of safety and thrombogenicity, *Crit. Care.* 15, 201, 2011.

20. Björkman, S., Carlsson, M., and Berntorp, E., Pharmacokinetics of factor IX in patients with haemophilia B, *Eur. J. Clin. Pharmacol.* 41, 325–332, 1994.

21. Cheung, W.E., van der Born, J., Kühn, K., *et al.*, Identification of the endothelial binding site for factor IX, *Proc. Natl. Acad. Sci. USA* 93, 11068–11073, 1996.

22. Scheraga, H.A. and Laskowski, M., Jr., The fibrinogen-fibrin conversion, *Adv. Prot. Chem.* 12, 1–131, 1957.

23. Mann, K.G., Nesheim, M.E., Hibbard, L.S., and Tracy, P.B., The role of factor V in the assembly of the prothrombinase complex, *Ann. N. Y. Acad. Sci.* 370, 378–388, 1981.

24. Crook, M., Platelet prothrombinase in health and disease, *Blood Coagul. Fibrinolysis* 1, 167–174, 1990.

25. Ivaneiu, L., Krishnaswamy, S., and Camire, R.M., New insights into the spatiotemporal localization of prothrombinase *in vivo*, *Blood* 124, 1705–1714, 2014.

26. Undas, A., Siudak, Z., Brummel-Ziedins, K., Mann, K.G., and Tracz, W., Prothrombinase formation at the site of microvascular injury and aspirin resistance: the effect of simvastatin, *Thromb. Res.* 125, 283–285, 2010.

27. Mann, K.G., Prothrombin, *Methods Enzymol.* 45, 123–156, 1976.

28. Mann, K.G., Elion, J., Butkowski, R.J., Downing, M., and Nesheim, M.E., Prothrombin, *Methods Enzymol.* 80(C), 286–302, 1981.

29. Nesheim, M.E. and Mann, K.G., The kinetics and cofactor dependence of two cleavages involved prothrombin conversion, *J. Biol. Chem.* 258, 5386–5396, 1983.

30. Rosing, J., Zwaal, R.F.A., and Tans, G., Formation of meizothrombin as an intermediate in the Factor Xa-catalyzed prothrombin activation, *J. Biol. Chem.* 261, 4224–4228, 1986.

31. Krishnaswamy, S., Mann, K.G., and Neisheim, M.E., The prothrombinase-catalyzed activation of prothrombin proceeds through the intermediate meizothrombin in an ordered, sequential reaction, *J. Biol. Chem.* 261, 8977–8984, 1986.

32. Doyle, M.F. and Haley, P.E., Meizothrombin: active intermediate formed during prothrombinase-catalyzed activation of prothrombin, *Methods Enzymol.* 222, 299–312, 1993.

33. Fersht, A.R., Conformational equilibria in α- and δ-chymotrypsin. The energetics and importance of the salt bridge, *J. Mol. Biol.* 64, 497–509, 1972.

34. Nesheim, M.E. and Mann, K.G., The kinetics and cofactor dependence of the two cleavages involved in prothrombin activation, *J. Biol. Chem.* 258, 5386–5391, 1983.

35. Tans, G., Nicolaes, G.A.F., Thomassen, M.C.L.G.D., *et al.*, Activation of human factor V by meizothrombin, *J. Biol. Chem.* 269, 15969–15972, 1994.

36. Bradford, H.N. and Krishnaswamy, S., Occlusion of anion-binding exosite 2 in meizothrombin explains its impaired ability to activate factor V, *J. Biol. Chem.* 294, 2422–2435, 2019.

37. Bradford, H.N. and Krishnaswamy, S., Meizothrombin is an unexpectedly zymogen-like variant of thrombin, *J. Biol. Chem.* 287, 30414–30425, 2012.

38. Roberts, P.S. and Burkat, R.K., Inhibition of the esterase activity of thrombin by Na⁺, Inhibition of the esterase activity of thrombin by Na⁺, *Proc. Soc. Exp. Biol. Med.* 127, 447–450, 1969.

39. Kurtz, I. and Nguyen, M.K., Evolving concepts in the quantitative analysis of the determinants of the plasma water sodium concentration and the pathophysiology and treatment of the dysnatremias, *Kidney Int.* 68, 1982–1993, 2005.

40. Côté, H.C.F., Stevens, W.K., Bajzar, L., *et al.*, Characterization of a stable form of meizothrombin derived from recombinant prothrombin (R155A, R271A, R284A), *J. Biol. Chem.* 269, 11374–11380, 1994.

41. Rhee, M.-J., Morris, S., and Kosow, D.P., Role of meizothrombin and meizothrombin-(des-F1) in the conversion of prothrombin to thrombin by the *Echis carinatus* venom coagulant, *Biochemistry* 21, 3437–3443, 1982.

42. Briet, E., Noyes, C.M., Roberts, H.R., and Griffith, M.J., Cleavage and activation of human prothrombin by *Echis carinatus* venom, *Thromb. Res.* 27, 591–600, 1982.

43. Fischer, B.E., Schlokat, U., Mitterer, A., *et al.*, Differentiation between proteolytic activation and autocatalytic conversion of human prothrombin. Activation of recombinant human prothrombin and recombinant D49N-prothrombin by snake venoms from *Echis carinatus* and *Oxyranus scutellatus*, *Protein. Eng.* 9, 921–926, 1996.

44. Shibata, T., Kobayashi, Y., Ikeda, Y., and Kawabata, S.-I., Intermolecular autocatalytic activation of serine proteases zymogen factor C through an active transition state responding to lipopolysaccharide, *J. Biol. Chem.* 293, 11589–11599, 2018.

45. Eagle, H., Studies on blood coagulation: IV. The nature of the clotting deficiency in hemophilia, *J. Gen. Physiol.* 18, 813–819, 1935.

46. Walsh, P.N. and Lipscomb, M.S., Comparison of the coagulant activities of platelets and phospholipids, *Br. J. Haematol.* 33, 9–16, 1976.

47. Rosing, J., van Rijn, J.L., Bevers, E.M., *et al.*, The role of activated human platelets in prothrombin and factor X activation, *Blood* 65, 319–332, 1985.

48. Nagata, S., Suzuki, J., Segawa, K., and Fujii, T., Exposure of phosphatidyl serine on the cell surface, *Cell Death Differ.* 23, 952–961, 2016.

49. Mann, K.G., Nesheim, M.E., Church, W.R., Haley, P., and Krishnaswamy, S., Surface-dependent reactions of the vitamin K-dependent enzyme complexes, *Blood* 76, 1–16, 1990.

50. Willems, G.M., Giesen, P.L., and Hermens, W.T., Adsorption and conversion of prothrombin on a rotating disc, *Blood* 82, 497–504, 1993.

51. Rosing, J., Tans, G., Govers-Riemslag, J.W., Zwaal, R.F., and Hemker, H.C., The role of phospholipids an factor Va in the prothrombinase complex, *J. Biol. Chem.* 255, 274–283, 1980.

52. Bradford, H.N., Orcutt, S.J., and Krishnaswamy, S., Membrane binding by prothrombin mediates its constrained presentation to prothrombinase for cleavage, *J. Biol. Chem.* 288, 27789–27800, 2013.

53. Henry, R.L., Platelet function, *Semin. Thromb. Haemost.* 4, 93–122, 1977.

54. Walsh, P.N., Camp, E., and Dende, D., Different requirements for intrinsic factor-Xa forming activity and platelet factor C activity and their relationship to platelet aggregation and secretion, *Br. J. Haematol.* 40, 311–331, 1978.

55. Bathels, M. and Seegers, W.H., Substitution of lipid with bile salts in the formation of thrombin, *Thromb. Diath. Haemorrh.* 22, 13–27, 1969.

56. Bajar, S.S. and Hanahan, D.J., Interaction of short chain and long chain fatty acid phosphoglycerides and bile salts with prothrombin, *Biochim. Biophys. Acta* 444, 118–130, 1976.

57. Quick, A.J., On the composition of prothrombin, *Amer. J. Physiol.* 140, 212–220, 1943.

58. Quick, A.J. and Stefanini, M., The concentration of component A in blood, its assay and relation to the labile factor, *J. Lab. Clin. Med.* 34, 973–982, 1949.

59. Friedman, L.A., Quick, A.J., Higgins, F., Hussey, C.V., and Hickey, M.E., Hereditary labile factor (factor V) deficiency, *JAMA* 175, 370–374, 1961.

60. Horsti, J., Has the Quick or the Owren prothrombin time method the advantage in harmonization for the International Normalized Ratio system?, *Blood Coagul. Fibrinolysis* 13, 641–646, 2002.

61. Quick, A.J. and Stafanini, M., Nature of action of the labile factor in formation of thrombin, *Am. J. Physiol.* 160, 572–575, 1950.

62. Owren, P.A., Parahemophilia: haemorrhagic diathesis due to absence of a previously unknown clotting factor, *Lancet* 249(6449), 446–448, 1947.

63. Owren, P.A., The fifth coagulation factor ('factor V'). Preparation and properties, *Biochem. J.* 43, 136–139, 1948.

64. Borchgrevink, C.F. and Owren, P.A., Surgery in a patient with factor V deficiency (proaccelerin deficiency, *Acta Med. Scand.* 170, 743–746, 1961.

65. Owren, P.A., Prothrombin and accessory factors: clinical significance, *Am. J. Med.* 14, 201–215, 1953.

66. Owren, P.A., Coagulation of blood, *Northwest Med.* 56, 31–39, 1957.

67. Hougie, C., Barrow, E.M., and Graham, J.B., Stuart clotting defect. I. Segregation of an hereditary hemorrhagic state from the heterogeneous group heretofore called stable factor (SPCA, procovertin, Factor VII) deficiency, *J. Clin. Invest.* 36, 485–496, 1957.

68. Hougie, C., Barrow, E.M., and Graham, J.B., The Stuart factor: a hitherto unrecognized blood coagulation factor, *Bibl. Haematol.* 7, 336–340, 1958.
69. Graham, J.B., Stuart factor: discovery and designation as factor X, *J. Thromb. Haemost.* 1, 871–877, 2000.
70. Macfarlane, R.G., An enzyme cascade in the blood clotting mechanism, and it function as a biochemical amplifier, *Nature* 202. 498–499, 1964.
71. Davie, E.W. and Ratnoff, O.D., Waterfall sequence for intrinsic blood clotting, *Science* 145, 1310–1312, 1964.
72. Davie, E.W., A brief historical review of the waterfall/cascade of blood coagulation, *J. Biol. Chem.* 278, 50819–50832, 2003.
73. Blostein, M.D., Furie, B.C., Rajotte, I., and Furie, B., The Gla domain of factor IXa binds to factor VIIIa in the tenase complex, *J. Biol. Chem.* 278, 31297–31302, 2003.
74. Kovalenko, T.A., Pauteleev, M.A., and Sveshnikova, A.N., The mechanism and kinetics of initiation of blood coagulation by the extrinsic tenase complex, *Biophysics* (Oxford) 62, 291–300, 2017.
75. Krishnaswamy, S., Church, W.R., Nesheim, M.E., and Mann, K.G., Activation of human prothrombin by human prothrombinase. Influence of factor Va on the reaction mechanism, *J. Biol. Chem.* 262, 3291–3299, 1987.
76. Schulz, O., Sewell, H.F., and Shakih, E. Proteolytic cleavage of CD25, the α subunit of the human T cell interleukin 2 receptor, by Der p 1, a major mite allergen with cysteine protease activity, *J. Expt. Med.* 187, 271–275, 1998.
77. López-Rodríguez, J.C., Manosalva, J., Cabrera-García, J.D., *et al.*, Human glutathione-S-transferase pi potentiates the cysteine redox mechanism, *Redox Biol.* 26, 101256, 2019.
78. Reithofer, M. and Jahn-Schmidt, B., Allergens with protease activity from house dust mites, *Int. J. Mol. Sci.* 18(7), 1386, 2017.
79. Meno, K., Thorsted, P.B., Ipsen, H., *et al.*, The crystal structure of recombinant prodder p 1, a major house dust mite proteolytic allergen, *J. Immunol.* 175, 3835–3845, 2005.
80. Chevigné, A. and Jacquet, A., Emerging roles of the protease allergen Der p 1 in house dust mite-induced airway inflammation, *J. Allergy Clin. Immunol.* 142, 398–400, 2018.
81. Chua, K.Y., Steward, G.A., Thomas, W.R., *et al.*, Sequence analysis of cDNA coding for a major house dust mite allergen, Der p 1. Homology with cysteine proteases, *J. Exp. Med.* 167, 175–182, 1988.
82. Chevigné, A., Campizi, V., Szpakowska, M., *et al.*, The Lys-Asp-Tyr triad within the mite allergen Der p 1 propeptide is a critical structural element for the pH-dependent initiation of the protease maturation, *Int. J. Mol. Sci.* 18, E1087, 2017.
83. Zhang, J., Chen, J., and Allen-Philbey, K., Innate generation of thrombin and intracellular oxidants in airway epithelia by allergen Der 1 p, *J. Allergy Clin. Immunol.* 138(4), 1224–1227, 2016.
84. Zhang, J., Chen, J., Mangat, S.C., *et al.*, Pathways of airway oxidant formation by house dust mite allergens and viral RNA converge through myosin motors, pannexons, and Toll-like receptor 4, *Immun. Inflamm. Dis.* 6, 276–296, 2018.
85. Zhang, J., Chen, J., and Robinson, C., Cellular and molecular events in the airway epithelium defining the interaction between house dust mite group 1 allegens and innate defences, *Int. J. Mol. Sci.* 19(11), E3549, 2018.
86. Kubo, M., Innate and adaptive type-2 immunity in lung allergic inflammation, *Immunol. Rev.* 278, 162–172, 2017.
87. Mikaye, Y., Alessandro-Gabazza, D.N.D., *et al.*, Dose-dependent differential effect of thrombin in allergic bronchial asthma, *J. Thromb. Haemost.* 11, 1903–1915, 2013.

88. Yongging, T., Drentin, N., Duncan, R.C., Wijeyewickrema, L.C., and Pike, R.N., Mannose-binding lectin serine proteases and associated proteins of the lectin pathway of complement: two genes, five proteins and many functions?, *Biochim. Biophys. Acta* 1824, 253–262, 2012.

89. Garred, P., Genster, N., Pilely, K., *et al.*, A journey through the lectin pathway of complement-MBL and beyond, *Immunol. Rev.* 274, 74–97, 2016.

90. Dobó, J., Pál, G., Cervenak, I., and Gál, P., The emerging role of mannose-binding lectin-associated serine proteases (MASPs) in the lectin pathway of complement and beyond, *Immunol. Rev.* 274, 98–111, 2016.

91. de Messias-Reason, I., and Boldt, A.B.W., Introduction, in *Mannose-Binding Lectin in the Innate Immune System*, Chapter 1, pp. 1–23 (references on pp. 45–76), Nova Science Publications, Happauge, NY, USA, 2009.

92. Hess, K., Ajjan, R., Phoenix, F., *et al.*, Effects of MASP-1 of the complement system on activation of coagulation factors and plasma clot formation, *PLoS One* 7(4), e35690, 2012.

93. Krarup, A., Wallis, R., Presanis, J.S., Gál, P., and Sim, R.B., Simulataneous activation of complement and coagulation by MBL-associated serine protease 2, *PLoS One* 2(7), e623, 2007.

94. Gulla, K.C., Gupta, K., Krarup, A., *et al.*, Activation of mannan-binding lectin-associated serine protease leads to generation of a fibrin clot, *Immunology* 129, 482–495, 2010.

95. Jenny, L., Dobó, J., Gál, P., and Schroeder, V., MASP-1 of the complement system promotes clotting via prothrombin activation, *Mol. Immunol.* 66, 398–405, 2015.

96. Jenny, L., Dobó, J., Gál, P., *et al.*, MASP-1 induced clotting-the first model of prothrombin activation by MASP-1, *PLoS One* 10(12), e0144633, 2015.

97. Koyama, T., Hall, L.R., Haser, W.G., Tonegawa, S., and Saito, H., Structure of a cytotoxic T-lymphocyte-specific gene shows a strong homology to fibrinogen β and γ chains, *Proc. Natl. Acad. Sci. USA* 84, 1609–1613, 1987.

98. Marazzi, S., Blum, S., Hartmann, R., *et al.*, Characterization of human fibroleukin, a fibrinogen-like protein secreted by T lymphocytes, *J. Immunol.* 161, 138–147, 1998.

99. Yang, G. and Hooper, W.C., Physiological functions and clinical implications of fibrinogen-like protein 2: a review, *World J. Clin. Infect. Dis.* 3, 37–46, 2013.

100. Liu, X.-g., Liu, Y., and Chen, F., Soluble fibrinogen-like protein 2, the novel effector molecule for immnoregulation, *Oncotarget* 8, 3711–3723, 2017.

101. Shrivastavi, S., Moley, J.H., and Darling, A., The interface between coagulation and immunity, *Am. J. Transplant.* 7, 499–506, 2007.

102. Liu, H., Shalev, I., Manuel, J., *et al.*, The FGL-2 Fcγ RIIB pathway: a novel mechanism leading to immunosuppression, *Eur. J. Immunol.* 38, 3114–3126, 2008.

103. Shalev, I., Liu, H., Koscik, C., *et al.*, Targeted deletion of *fgl2* leads to imparted regulatory T cell activity and development of autoimmune glomerulonephritis, *J. Immunol.* 180, 249–260, 2008.

104. Chen, T.-J., Ji, M.-X., Tao, Z.-Q., *et al.*, The relationship between serum fibrinogen-like protein 2 concentrations and 30-day mortality of patients with traumatic brain injury, *Clin. Chim. Acta* 489, 53–57, 2019.

105. Levy, G.A. and Edginton, T.S., Lymphocyte cooperation is required for amplification of macrophage procoagulant activity, *J. Expt. Med.* 151, 1232–1244, 1980.

106. Schwartz, B.S., Levy, G.A., Fair, D.S., and Edgington, T.S., Murine lymphoid procoagulant activity induced by bacterial lipopolysaccharide and immune complexes is a monocyte prothrombinase, *J. Expt. Med.* 155, 1464–1479, 1982.

107. Parr, R.L., Fung, L., Reneker, J., *et al.*, Association of mouse fibrinogen-like protein with murine fibrinogen-like protein with murine hepatitis virus-induced prothrombinase activity, *J. Virol.* 69, 5033–5038, 1995.

108. Yang, Y. and Hooper, W.C., Physiological function and clinical implications of fibrinogen-like 2: a review, *World J. Clin. Infect. Dis.* 3, 37–46, 2013.
109. Yuwaraj, S., Ding, J., Liu, M., Marsden, P.A., and Levy, G.A., Genomic characterization, localization, and functional expression of *FGL2*, the human gene encoding fibroleukin: a novel human procoagulant, *Genomics* 71, 330–338, 2001.
110. Gordy, M.A., Pila, E.A., and Harrington, P.C., The role of fibrinogen-related protein in the gastropod immune response, *Fish Shellfish Immunol.* 46, 39–49, 2015.
111. Thomsen, T., Schlosser, A., Holmskov, V., and Sorenson, G.L., Ficolins and FIBCD1: soluble and membrane bound pattern recognition molecules with acetyl group selectivity, *Mol. Immunol.* 48, 369–381, 2011.
112. Chan, C.W.Y., Chan, M.W.C., Liu, M., *et al.*, Kinetic analysis of a unique direct prothrombinase, *fgl2*, and identification of a serine residue critical for the prothrombinase activity, *J. Immunol.* 168, 5170–5117, 2002.
113. Barrett, A.J. and Rawlings, N.D., Families and clans of serine peptidases, *Arch. Biochem. Biophys.* 318, 247–250, 1995.
114. Barrett, A.J., Rawlings, N.D., and O'Briens, E.A., the MEROPS database of proteolytic enzymes as a protease information system, *J. Struct. Biol.* 134, 95–102, 2001.
115. www.ebi.ac.uk/merops
116. Rawlings, N.D., Barrett, A.J., Thomas, P.D., *et al.*, The database of proteolytic enzymes, their subunits and inhibitors in 2017 and a comparison with the PANTHER database, *Nucleic Acids Res.* 46, D624–D632, 2018.
117. Fung, L.S., Neil, G., Leibowitz, J., *et al.*, Monoclonal antibody analysis of a unique macrophage procoagulase activity induced by murine hepatitis virus strain 3 infection, *J. Biol. Chem.* 286, 1789–1795, 1990.
118. Fahrney, D.E. and Gold, A.M., Sulfonyl fluorides as inhibitor of esterases I. Rates of reaction with acetylcholinesterase, α-chymotrypsin, and trypsin, *J. Amer. Chem. Soc.* 85, 997–2006, 1965.
119. Lundblad, R.L. and Davie, E.W., The activation of antihemophilic factor (Factor 8) by activated Christmas factor (activated factor 9), *Biochemistry* 3, 1720–1725, 1964.
120. Di Scipio, R.G., Kurachi, K., and Davie, E.W., Activation of human factor IX (Christmas factor), *J. Clin. Invest.* 61, 1528–1538, 1978.
121. Li, W.-Z., Wang, J., Long, R., *et al.*, Novel antibody against a glutamic acid-rich human fibrinogen-like protein 2-derived peptide near Ser91 inhibits hfgl2 prothrombinase activity, *PLoS One* 9(4), e94551, 2014.
122. O'Brien, M., Morrison, J.J., and Smith, T.J., Expression of prothrombin and protease activated receptors in human myometrium during pregnancy and labor, *Biol. Reprod.* 78, 20–26, 2008.
123. Su, K., Chen, F., Yan, W.-H., *et al.*, Fibrinogen-like protein 2/fibroleukin prothrombinase contributes to tumor hypercoagulability *via* IL-2 and IFN-γ, *World J. Gastroenterol.* 14, 5980–5989, 2008.
124. Melnyk, M.C., Shalev, J., Zhang, J., *et al.*, The prothrombinase of FLG2 contributes to the pathogenesis of experimental arthritis, *Scand. J. Rheumatol.* 40, 269–278, 2011.
125. Liu, Y., Li, X., Zeng, Q., *et al.*, Downregulation of FGL2/prothrombinase delays HCCLM6 xenograft tumour growth and decreases tumour angiogenesis, *Liver Int.* 32, 1585–1595, 2012.
126. Rabizadeh, E., Cherny, I., Wolach, O., *et al.*, Increased activity of cell membrane-associated prothrombinase, fibrinogen-like protein 2, in peripheral blood mononuclear cells of B-cell lymphoma patients, *PLoS One* 9(10), e101371, 2014.
127. Liu, J., Tan, Y., Zhang, J., *et al.*, C5aR, TNF-α, and FGL2 contribute to coagulant and complement activation in virus-induced fulminant hepatitis, *J. Hepatol.* 62, 354–362, 2015.

128. Rabizadeh, E., Cherny, I., Lederfein, D., *et al.*, The cell-membrane prothrombinase, fibrinogen-like protein 2, promotes angiogenesis and tumor development, *Thromb. Res.* 136, 118–124, 2015.

129. Fan, C., Wang, J., Mao, C., *et al.*, The Fgl2 prothrombinase contributes to the pathological process of experimental hypertension, *J. Appl. Physiol.*, 127(6), 1677–1687, 2019, doi: 1152/japplphysiol.00396.2019.

130. Elia, D., Caminati, A., Zompatori, M., *et al.*, Pulmonary hypertension and chronic lung disease: where are we headed?, *Eur. Respir. Rev.* 28(153), 190065, 2019.

131. Bogatkevich, G.S., Ludwicka-Bradley, A., and Silver, R.M., Dabigatran, a direct thrombin inhibitor, demonstrates antifibrotic effects on lung fibroblasts, *Arthritis Rheum.* 60, 3455–3464, 2009.

132. Shea, B.S., Probst, C.K., Brazee, P.L., *et al.*, Uncoupling of the profibrotic and hemostatic effects of thrombin in lung fibrosis, *JCI Insight* 2(9), 86608, 2017.

133. Morita, T., Inagami, H., and Iwanaga, S., Staphylocoagulase, *Methods Enzymol.* 80, 311–319, 1980.

134. Panizzi, P., Friedrich, R., Fuenes-Prior, P., Bode, W., and Bock, P.E., The Staphylocoagulase family of zymogen activator and adhesion proteins, *Cell. Mol. Life. Sci.* 61m, 2793–2798, 2004.

135. Fisher, A.M., The plasma coagulating properties of staphylococci, *Bull. Johns Hopkins Hospital* 59, 415–426, 1936.

136. Neter, E., Fibrinolytic, anticoagulating and plasma clotting properties of Streptococci, *J. Bacterol.* 34, 243–254, 1937.

137. Tager, M., Concentration, partial purification, properties, and nature of Staphylocoagulase, *Yale J. Biol. Med.* 20, 487–501, 1948.

138. Hendrix, H., Lindhout, T., Mertens, K., Engels, W., and Hemker, H.C., Activation of human prothrombin by stoichiometric levels of Staphylocoagulase, *J. Biol. Chem.* 258, 3637–3644, 1983.

139. Friedrich, R., Panizzi, P., Kawabata, E.I., *et al.*, Structural basis for reduced Staphylocoagulase-mediated prothrombin activation, *J. Biol. Chem.* 281, 1128–1197, 2006.

140. Igarashi, H., Morita, T., and Iwanaga, S., A new method for purification of staphylocoagulase by a bovine prothrombin-Sepharose column, *J. Biochem.* 86, 1615–1618, 1979.

141. Kawabata, S., Morita, T., Iwanaga, S., and Igarashi, H., Difference in enzymatic properties between α-thrombin-staphylocoagulase complex and free α-thrombin, *J. Biochem.* 97, 1073–1078, 1985.

142. Kawabata, S., Morita, T., Iwanaga, S., and Igarashi, H., Enzymatic properties of staphylothrombin, an active molecular complex formed between Staphylocoagulase and human prothrombin, *J. Biochem.* 98, 1603–1614, 1985.

143. Panizzi, P., Friedrich, R., Fuentes-Prior, P., *et al.*, Novel fluorescent prothrombin analogs as probes of Staphylocoagulase-prothrombin interactions, *J. Biol. Chem.* 281, 1169–1178, 2006.

144. Panizzi, P., Nahredorf, M., Figueiredo, J.-L., *et al.*, *In vivo* detection of *Staphylococcus aureus* endocarditis by targeting pathogen-specific prothrombin activation, *Nat. Med.* 17, 1142–1146, 2011.

145. Bjerketorp, J., Nilsson, M., Ljungh, A., *et al.*, A novel von Willebrand factor binding protein expressed by *Staphylococcus aureus*, *Microbiology* 148, 2037–2044, 2002.

146. Bjerketorp, J., Jacobsson, K., and Frykberg, L., The von Willebrand factor binding protein (vWBP) of *Straphylococcus aureus* is a coagulase, *FEMS Microbiol. Lett.* 234, 319–324, 2004.

147. Kron, H.K., Panizzi, P., and Bock, P.E., Von Willebrand factor-binding protein is a hysteric conformational activator of prothrombin, *Proc. Natl. Acad. Sci. USA* 106, 7786–7791, 2009.

148. Kroh, H.K. and Bock, P.E., Effect of zymogen domains and active site occupation on activation of prothrombin by von Willebrand factor-binding protein, *J. Biol. Chem.* 287, 39149–39157, 2012.

149. Kaminishi, H., Hamatake, H., Cho, T., *et al.*, Activation of blood clotting factors by microbial proteinases, *FEMS Microbiol. Lett.* 121, 327–332, 1994.

150. Seegers, W.H., Activation of purified prothrombin, *Proc. Soc. Exptl. Biol. Med.* 77, 677–680, 1949.

151. Schultze, H.E. and Schwick, G., Uber den mechanisimus der Thrombinbildung im isölieren System, *Hoppe-Seyler Zeitsch. fur Phyiol. Chem.* 289, 26–43, 1951.

152. Seegers, W.H., Development of thrombin activity in concentrated salt solutions, in *Prothrombin in Enzymology, Thrombosis, and Hemophilia*, Chapter 4, pp. 88–98, C.C. Thomas, Springfield, IL, USA, 1967.

153. Seegers, W.H., IV., Activation of prothrombin and factor X in 25% sodium citrate, *Sem. Thromb. Hemost.* 17, 213–220, 1981.

154. Seegers, W.H., McClaughry, R.I., and Fahey, J.L., Some properties of purified prothrombin and its activation with sodium citrate, *Blood* 5, 421–433, 1950.

155. Grob, D., Proteolytic enzymes. I. Control of their activity, *J. Gen. Physiol.* 29, 219–247, 1946.

156. Seegers, W.H., Marciniak, E., and Cole, E.R., Autocatalysis in prothrombin activation, *Am. J. Physiol.* 203, 347–400, 1962.

157. Miller, K.D. and Van Vunakis, H., The effect of diisopropyl fluorophosphate on the proteinase and esterase activities of thrombin and on prothrombin and on its activation, *J. Biol. Chem.* 223, 227–237, 1956.

158. Seegers, W.H. and Landaburu, R.H., Inhibition of prothrombin activation in sodium citrate solutions, *Proc. Soc. Exptl. Biol. Med.* 95, 710–713, 1957.

159. Lechner, K. and Deutsch, E., Activation of factor X, *Thromb. Diath. Haemorrh.* 13, 314–329, 1965.

160. Esnouf, M.P., Prothrombin, in *Human Blood Coagulation, Haemostasis and Thrombosis*, ed. R. Biggs, Chapter 3, pp. 42–45, Blackwell Scientific, Oxford, UK, 1972.

161. Seegers, W.H. and Landaburu, R.H., Purification of prothrombin and thrombin by chromatography on cellulose, *Can. J. Biochem. Physiol.* 38, 1405–1416, 1960.

162. Teng, C.-M. and Seegers, W.H., Activation of factor X and thrombin zymogens in 25% sodium citrate solution, *Thromb. Res.* 22, 203–212, 1981.

163. Milstone, J.H., Effect of blood thrombokinase, as influenced by soy bean trypsin inhibitor, ultracentrifugation, and accessory factors, *J. Gen. Physiol.* 38, 757–769, 1955.

164. Alkjaersig, N., Deutsch, E., and Seegers, W.H., Prothrombin derivatives and the inhibition of thrombin formation with soy bean trypsin inhibitor, *Am. J. Physiol.* 180, 367–370, 1955.

165. Lorand, L., Alkjaersig, N., and Seegers, W.H., Carbohydrate and nitrogen distribution during the activation of purified prothrombin in sodium citrate solution, *Arch. Biochem. Biophys.* 45, 312–318, 1953.

166. Tiselius, A., Electrophoresis past, present, and future, *Clin. Chim. Acta* 3, 1–9, 1958.

167. Aronson, D.L. and Mustafa, A.J., The activation of human factor X in sodium citrate: the role of factor VII, *Thromb. Haemost.* 36, 104–113, 1976.

168. Lundblad, R.L., Nesheim, M.E., Straight, D.L., *et al.*, Bovine α- and β-thrombin. Reduced fibrinogen-clotting activity of β-thrombin is not a consequence of reduced affinity for fibrinogen, *J. Biol. Chem.* 259, 6991–6995, 1984.

169. Mann, K.G., Heldebrant, C.M., and Fass, D.N., Multiple active forms of thrombin II. Mechanism of production from prothrombin, *J. Biol. Chem.* 246, 6106–6114, 1971.
170. Roberts, P.S., The effects of Na⁺ and K⁺ on the citrate activation of prothrombin and on the esterase activities of thrombokinase, *Biochim. Biophys. Acta* 201, 340–349, 1970.
171. Ralston, A.H. and Drohan, W.N., Activation of pure prothrombin to thrombin with about 30% to 40% sodium citrate, US Patent 6,245,548, 2001.
172. Pozzi, N., Chen, Z., Zapata, F., *et al.*, Autoactivation of thrombin precursors, *J. Biol. Chem.* 288, 11601–11610, 2013.
173. Pozzi, N., Chen, Z., Gohara, D.W., *et al.*, Crystal structure of prothrombin reveals conformational flexibility and mechanism of activation, *J. Biol. Chem.* 288, 22734–22744, 2013.
174. Pozzi, N., Bystranowska, D., Zhu, X., and Di Cera, E., Structural architecture of prothrombin in solution revealed by single molecule spectroscopy, *J. Biol. Chem.* 291, 18107–18116, 2016.
175. Lanchantin, G.F., Friedmann, J.A., and Hart, D.W., Esterase and clotting activities derived from citrate activation of human prothrombin, *J. Biol. Chem.* 242, 2491–2501, 1967.
176. Mann, K.G., Heldebrant, C.M., and Fass, D.N., Multiple active forms of thrombin I. Partial resolution, differential reactivities, and sequential formation, *J. Biol. Chem.* 246, 5994–6001, 1971.
177. Fass, D.N., and Mann, K.G., Activation of fluorescent-labeled prothrombin, *J. Biol. Chem.* 248, 3280–3287, 1973.
178. Pozzi, N., Chen, Z., Zanata, F., *et al.*, Autoactivation of thrombin precursors, *J. Biol. Chem.* 288, 11601–11610, 2016.
179. Pozzi, N., Chen, Z., Gehara, D.W., *et al.*, Crystal structure of Prothrombin reveals conformational flexibility and mechanism of activation, *J. Biol. Chem.* 288, 22734–22744, 2013.
180. Pozzi, N., Bystnamowska, D., Zuo, X.X., and Di Cera, E., Structural architecture of prothrombin in solution revealed by single molecule spectroscopy, *J. Biol. Chem.* 291, 18107–18116, 2016.
181. Barranco-Medina, S., Pozzi, N., Vogt, A.D., and Di Cera, E., Histone H4 promotes prothrombin activation, *J. Biol. Chem.* 288, 35749–35757, 2013.
182. Pozzi, N., and Di Cera, E., Dual effect of histone H4 on prothrombin activation, *J. Thromb. Haemost.* 14, 1814–1818, 2016.
183. Magnusson, S., Sottrup-Jensen, L., Petersen, T.E., and Claeys, H., The primary structure of prothrombin. The role of vitamin K in blood coagulation and a thrombin catalyzed 'negative feedback' control mechanism for limiting the activation of prothrombin, in *Prothrombin and Related Coagulation Factors*, ed. H.C. Hemker and J.J. Veltkamp, pp. 25–46, Leiden University Press, Leiden, Netherlands, 1975.
184. Magnusson, S., Petersen, T.E., Sottrup-Jensen, L., and Claeys, H., Complete primary structure of prothrombin. Isolation, structure and identification of the carboxylated glutamic acid residues and regulation of prothrombin activation to thrombin, in *Proteases and Biological Control*, ed E. Reich, D.B. Rifkin, and E. Shaw, pp. 123–149, Cold Spring Harbon Press, Cold Spring Harbor, NY, USA, 1975.
185. Miller, K.D., The nonenzymic activation of prothrombin by polylysine, *J. Biol. Chem.* 235, PC63–PC64, 1960.
186. Aronson, D.L. and Ménaché, D., Action of human thrombokinase on human prothrombin and *p*-tosyl-l-arginine methyl ester, *Biochim. Biophys. Acta* 167, 378–387, 1968.
187. Vogel, C.N., Bukowski, R.J., Mann, K.G., and Lundblad, R.L., Effect of polylysine on the activation of prothrombin. Pollylysine substitutes for calcium ions and factor V in the factor Xa catalyzed activation of prothrombin, *Biochemistry* 15, 3265–3269, 1976.

188. Miller, K.D., Nonenzymic control of prothrombin activation, *Ann. N. Y. Acad. Sci.* 370, 336–347, 1981.
189. Lundblad, R.L. and Roberts, H.R., The acceleration by polylysine of the activation of factor X by factor IXa, *Thromb. Res.* 25, 319–329, 1982.
190. McCord, D.M., Monroe, D.M., Smith, K.J., and Roberts, H.R., Characterization of the functional defect in factor IX Alabama. Evidence for a conformational change due to high affinity calcium binding in the first epidermal growth factor domain, *J. Biol. Chem.* 265, 10250–10254, 1990.
191. Nishimura, H., Takeya, H., Miyata, T., *et al.*, Factor IX Fukuoka. Substitution of ASN92 by his in the second epidermal growth factor-like domain results in defective interaction with factors VIIa/X, *J. Biol. Chem.* 268, 24041–24046, 1993.
192. Hertzberg, M.S., Facey, S.L., and Hogg, P.J., An Arg/Ser substitution in the second epidermal growth factor-like module of factor IX introduces an *O*-linked carbohydrate and markedly impairs activation by factor XIa and factor VIIa/tissue factor and catalytic efficiency of factor IXa, *Blood* 94, 156–163, 1999.
193. White, G.C., II., Roberts, H.R., Kingdon, H.S., and Lundblad, R.L., Prothrombin complex concentrates: potentially thrombogenic materials and clues to the mechanism of thrombosis in vivo, *Blood* 49, 159–170, 1977.
194. Blatt, P.M., Lundblad, R.L., Kingdon, H.S., McLean, G., and Roberts, H.R., Thrombogenic materials in prothrombin complex concentrates, *Ann. Intern. Med.* 81, 766–770, 1974.
195. Pedersen, A.H., Lund-Hansen, T., Bisgaard-Frantzen, H., Olsen, F., and Petersen, L.C., Autoactivation of human recombinant coagulation factor VII, *Biochemistry* 28, 9331–9336, 1989.
196. Casey, B.J., Behrens, A.M., Hess, J.R., *et al.*, FVII dependent coagulation activation in citrated plasma by polymeric hydrogels, *Biomacromolecules* 11, 3248–3255, 2010.
197. Lundblad, R.L., A rapid method for the purification of bovine thrombin and the inhibition of the purified enzyme with phenylmethylsulfonyl fluoride, *Biochemistry* 10, 2501–2506, 1971.
198. Lundblad, R.L., Kingdon, H.S., and Mann, K.G., Thrombin, *Methods Enzymol.* 45, 156–176, 1976.
199. Workman, E.F., Jr. and Lundblad, R.L., On the preparation of bovine α-thrombin, *Thromb. Haemost.* 39, 193–200, 1978.
200. Fenton, J.W., Fasco, M.J., Stackrow, A.B., *et al.*, Human thrombins production, evaluation, and properties of α-thrombin, *J. Biol. Chem.* 252, 3587–3588, 1977.
201. Rasmussen, P.A., Purification of thrombin by chromatography, *Biochim. Biophys. Acta* 16, 157–158, 1955.
202. Magnussen, S., Preparation of highly purified thrombin (E.C.3.4.4.13) and determination of its N-terminal amino acid residues, *Arkiv Kemi* 24, 349–358, 1965.
203. Magnussen, S., Thrombin and prothrombin, in *The Enzymes*, Vol. III, ed. P.D. Boyer, Chapter 9, pp. 277–321, Academic Press, New York, USA, 1971.
204. Lundblad, R.L., unpublished observations. 1968–1990.

3 Structural Biology of Thrombin

INTRODUCTION

Thrombin is a regulatory tryptic-like serine protease functioning in hemostasis,[1] wound healing,[2,3] and other regulatory functions. Thus, thrombin is pleiotropic in function,[4,5] and it is clear that various domains of thrombin have different functions contributing to the observed pleiotropy. Indeed, some peptides derived from limited proteolysis of thrombin appear to have specific functions. A regulatory protease differs from a digestive protease in that a regulatory protease will cleave only a limited number of the potentially susceptible peptide bonds in a protein, while a digestive protease would cleave most, if not all, peptide bonds, an exception being a peptide bond containing proline.[6,7]

As a tryptic-like serine protease, thrombin is specific for the cleavage of arginine and lysine peptide bonds in proteins; however, lysine peptide bonds are rarely cleaved by thrombin in proteins. However, thrombin does readily cleave lysine ester bonds.[8,9] Some of these studies were performed with partially purified preparations of thrombin. As a regulatory protease, thrombin cleaves only four peptide bonds in fibrinogen, while trypsin, a digestive protease, cleaves 26 peptide bonds in fibrinogen in a "fast" reaction and approximately 80 more peptide bonds in a slow reaction.[10] The crystal structure of various forms of thrombin have been solved, as have the structures of various thrombin-inhibitor complexes.[11,12] In addition to the site-specific mutant forms of thrombin described later, there are several "naturally occurring" mutant thrombins, which have been described by the late Ruth Ann Henriksen and coworkers.[13–19]

CLASSIFICATION OF THROMBIN AS A PROTEOLYTIC ENZYME

There is a continuing discussion on the use of the terms proteinase, protease, and peptidases. All three are hydrolases (Category 3 IUBMB nomenclature). The term proteinase and protease are, in my opinion, interchangeable and refer to an activity that hydrolyzes proteins yielding peptides. The term peptidase refers to an enzyme that hydrolyzes a peptide bond but is usually used to refer to an enzyme acting on peptides. Category 3.4 describes an enzyme that is a peptidase. Category 3.4.21 refers to serine endopeptidases. There has been a continuing effort to classify proteases.[20–30] The International Union of Biochemistry and Molecular Biology (IUBMB) Committee on Nomenclature (www.qmul.ac.uk/sbcs/iubmb/enzyme/) is the successor to the Enzyme Commission. The Enzyme Commission (IUBMB) considers thrombin to be a serine protease (3.4.21); thrombin is 3.4.21.5. IUBMB considers a protease to be a peptidase, which is

DOI: 10.1201/b22204-3

classified as a hydrolase (Category 3 in the Enzyme Commission classification process). MEROPS (www.ebi.ac.uk/merops/)[28-30] is a classification system specific for peptidases (per IUBMB nomenclature). MEROPS classifies peptidases into clans (clan members evolved from a common ancestor). Clans in turn are divided into families (members are homologous). Clans and families can be further divided into subclans and subfamilies. In the specific case of thrombin, the designation is S01.271, where S is the subclass, 01 the chymotrypsin family, and 217 is thrombin.[30]

PRIMARY STRUCTURE AND FUNCTIONAL DOMAINS

The primary structure of thrombin is shown in Table 3.1, including the designation of functional residues. I was a graduate student in the early days of primary structure analysis. The information provided great excitement to those interested in homology. The sequence of chymotrypsin provided great help to those interested in other serine proteases. Thrombin is a classic serine protease, having aspartic acid, histidine, and serine at the active site.[30-33] This combination of residues is referred to as the catalytic triad.[31,32,34] As noted by Carter and Wells,[35] the catalytic triad, together with the oxyanion hole,[36-38] comprise the basic machinery for the catalytic step of bond-making and bond-breaking for serine proteases with other factors such as the S_1 cation binding site, the apolar binding sites, and exosites, providing specificity for the cleavage site. The serine hydroxyl functions as a nucleophile attacking the carbonyl carbon of the scissile peptide bond, forming an acyl-enzyme intermediate, while histidine functions as a catalytic base. The aspartic acid stabilizes the necessary tautomeric form of histidine.[34,39] The aspartic acid is stabilized in its ionized form to maintain the tautomeric form of histidine as the catalytic base. Prethrombin-2 is essentially inactive because of the lack of the salt bridge between the amino-terminal isoleucine of the B chain and aspartic acid 199; prethrombin-2 can, however, bind substrates and inhibitors.[40] While prethrombin-2 is considered one of two intermediates in the canonical activation of prothrombin, prethrombin-2 can also be considered to be the zymogen form for thrombin, similar to that for other proteins such as profactor D.[41] Jing and coworkers[41] emphasize the flexibility of these zymogen forms with the conversion to a active enzyme associated with increased structural order. Staphylocoagulase has been known to "activate" prothrombin without peptide bond cleavage by the formation of a complex,[42] which is active in clotting fibrinogen; the formation of fibrin has been suggested to be important i establishing bacterial infection.[43] Bock and coworkers were able to modify the thrombin-active site histidine in a prothrombin-staphylocoagulase complex.[44] While at the Rockefeller University in New York, I tried to identify intrinsic activity in pepsinogen. I was not successful in those efforts, but later work by Bustin and Conway-Jacobs[45] clearly showed that pepsinogen had catalytic activity upon acidification prior to peptide bond cleavage, resulting in the formation of pepsin. The point here is that the zymogen forms of thrombin can have biological activity.[46] The reader is directed to a discussion of the concept of zymogenicity.[47]

TABLE 3.1
Primary Structure of Thrombin

#	H	B	CHTG	#	H	B	CHTG	#I	H	B	CHTG
1	I	I	16	87	H	H	91	173	C	C	168
2	V	V	17	88	P	R	92	174	K	K	169
3	E	E	18	89	R	R	93	175	D	A	170
4	G	G	19	90	Y	Y	94	176	S	S	171
5	S	Q	20	91	N	N	95	177	T	T	172
6	D	D	21	92	W	W	96	178	R	R	173
7	A	A	22	93	R	K	97	179	I	I	174
8	E	E	23	94	E	E	97a	180	R	R	175
9	I	V	24	95	N	N	98	181	I	I	176
10	G	G	25	96	L	L	99	182	T	T	177
11	M	L	26	97	D	D	100	183	D	D	178
12	S	S	27	98	R	I	101	184	N	N	179
13	P	P	28	99	D	A	102	185	M	M	180
14	W	W	29	100	I	I	103	186	F	F	181
15	Q	Q	30	101	A	A	104	187	C	C	182
16	V	V	31	102	L	L	105	188	A	A	183
17	M	M	32	103	M	L	106	189	G	G	184
18	L	L	33	104	K	K	107	190	Y	Y	184A
19	F	F	34	105	L	L	108	191	K	K	185
20	R	R	35	106	K	K	109	192	P	P	186
21	K	K	36	107	K	R	110	193	D	G	186A
22	S	S	36A	108	P	P	111	194	E	E	186B
23	P	P	37	109	V	I	112	195	G	G	186C
24	Q	Q	38	110	A	A	113	196	K	K	186D
25	E	E	39	111	F	L	114	197	R	R	187
26	L	L	40	112	S	S	115	198	A	G	188
27	L	L	41	113	D	D	116	199	D	D	189
28	C	C	42	114	Y	Y	117	200	A	A	190
29	G	G	43	115	I	I	118	201	C	C	191
30	A	A	44	116	H	H	119	202	E	E	192
31	D	S	45	117	P	P	120	203	G	G	193
32	L	L	46	118	V	V	121	204	D	D	194
33	I	I	47	119	C	C	122	205	S	S	195
34	S	S	48	120	R	L	123	206	G	G	196
35	D	D	4949	121	P	P	124	207	P	G	197
36	R	R	50	122	D	D	125	208	P	P	198
37	W	W	51	123	R	P	126	209	F	F	199
38	V	V	52	124	E	Q	127	210	V	V	200
39	L	L	53	125	T	T	128	211	H	M	201
40	T	T	54	126	A	A	129	212	K	K	202
41	A	A	55	127	A	A	129A	213	S	S	203
42	A	A	56	128	S	K	129B	214	P	P	204
43	H	H	5	129	L	L	129C	215	F	Y	204A

(Continued)

TABLE 3.1
(Continued)

#	H	B	CHTG	#	H	B	CHTG	#I	H	B	CHTG
44	C	C	58	130	L	L	130	216	N	N	204b204B
45	L	L	59	131	Q	H	131	217	N	N	205
46	L	L	60	132	A	A	132	218	R	R	206
47	Y	V	60A	133	G	G	133	219	W	W	207
48	P	P	60B	134	Y	F	134	220	Y	Y	208
49	P	P	60C	135	K	K	135	221	Q	G	209
50	W	W	60D	136	G	G	136	222	M	M	210
51	D	D	60E	137	R	R	137	223	G	G	211
52	K	K	60F	138	V	V	138	224	I	I	212
53	N	N	60G	139	T	T	139	225	V	V	213
54	F	F	60H	140	G	G	140	226	S	S	2'4
55	T	T	60I	141	W	W	141	227	W	W	215
56	E	V	61	142	G	G	142	228	G	G	216
57	N	D	62	143	N	N	143	229	E	E	217
58	D	D	63	144	L	R	144	230	G	G	219
59	L	L	64	145	K	R	145	231	C	C	220
60	L	L	65	146	E	E	146	232	D	D	221
61	V	V	66	147	T	T	147	233	R	R	221A
62	R	R	67	148	W	W	148	234	D	R	222
63	I	I	68	149	T	T	149	235	G	G	223
64	G	G	69	150	A	T	149A	236	K	K	224
65	K	K	70	151	N	S	149B	237	Y	Y	225
66	H	H	71	152	V	V	149C	238	G	G	226
67	S	S	72	153	G	A	149D	239	F	I	227
68	R	R	73	154	K	E	149E	240	Y	Y	228
69	T	T	74	155	Q	V	150	241	T	T	229
70	R	R	75	156	Q	Q	151	242	H	H	230
71	Y	Y	76	157	P	P	152	243	V	V	231
72	E	E	77	158	S	S	153	244	F	F	232
73	R	R	77a	159	V	V	154	245	R	R	233
74	N	K	78	160	L	L	155	246	I	L	234
75	I	V	79	161	Q	Q	156	247	K	K	235
76	E	E	80	162	V	V	157	248	K	K	236
77	K	K	81	163	V	V	158	249	W	W	237
78	I	I	82	164	N	N	159	250	I	I	238
79	S	S	83	165	L	L	160	251	Q	Q	239
80	M	M	84	166	P	P	161	252	K	K	240
81	L	L	85	167	I	L	162	253	V	V	241
82	E	D	86	168	V	V	163	254	I	I	242
83	K	K	87	169	E	E	164	255	D	D	243
84	I	I	88	170	R	R	165	256	Q	R	244
85	Y	Y	89	171	P	P	166	257	F	L	245
86	I	I	90	172	V	V	167	258		G	246
								259		S	247

The term zymogen dates to the 1800s, where the term defines a substance from which a ferment (enzyme) is derived.[48] Most zymogens require processing by a separate factor,[49,50] but there are a number of examples of autocatalytic activation, such as pepsinogen as noted earlier. Irrespective of process, the conversion of zymogen to enzyme frequently involves the process of limited proteolysis, resulting (usually) in the formation of a separate molecular species. The zymogen forms of classical, chymotrypsin-like serine proteases have been demonstrated by Neurath, Walsh, and others to have low levels of catalytic activity (10^{-4}–10^{-7} compared to the active enzymes).[51–55] This early work was reviewed by Neurath in an excellent but seldom-quoted article on limited proteolysis and zymogen activation.[56]

THROMBIN-ACTIVE SITE

As noted earlier, thrombin is a regulatory protease as opposed to a digestive protease such as trypsin.[57,58] Both thrombin and trypsin cleave ester bonds, where the carboxyl group is contributed by arginine and lysine. However, the enzymes have a far different specificity in proteins. There are multiple structural factors driving the catalytic efficiency of thrombin, which are discussed later. A specific substrate may drive concerted proton transfer, which may increase catalytic efficiency.[59,60] In the cleavage of peptide bonds, it is likely that the process of acylation of the active serine residue via the formation of a tetrahedral intermediate is the rate-limiting step and would depend, in part, on the nucleophilicity of the serine residue. However, there are other approaches to enhancing the nucleophilicity of the active-site serine residue,[61] and it could be argued that, as with concerted proton transfer, it would depend on the binding of a specific substrate to an extended binding site. Case and Stein[62] also reported on the importance of subsite interactions in driving the efficiency of the acylation step for serine proteases.

The active-site serine residue S205 in thrombin can be modified with DFP, resulting in a total loss of enzymatic activity.[63] This work was important in defining thrombin as an enzyme and the conversion of fibrinogen to fibrin as an enzymatic process (see Chapter 1). Much later, PMSF[64] was shown to react with thrombin also, with the loss of activity toward all substrates. The site of modification of thrombin with PMSF was not established but assumed to be the active-site serine residue. Thrombin has also been shown to be inactivated by related compounds such as sarin (isopropyl methylfluorophosphate; isopropyl methylphosphonofluoridate).[65] DFP inactivates proteases such as trypsin and chymotrypsin. It also inactivates acetylcholinesterase, which was the basis for its development as war gas.[66,67] The specificity of the alkylfluorophosphate compounds was greatly improved with the development of amidine fluorophosphates.[68]

At the time of these early studies, it was argued that the steric bulk of the substituting groups were responsible for the loss of activity at the enzyme-active site.[69–72] Subsequent work from a number of laboratories showed that other approaches, such as the conversion of the active-site serine to dehydroalanine, established the importance of a specific residue in catalysis; in this case, it was the active-site serine in catalysis by chymotrypsin.[73] This paper from the late Dan Koshland's laboratory at UC-Berkeley contains a nice discussion of this specific question at the time regarding the role of serine in chymotrypsin. Ashton and Scheraga[74] subsequently prepared anhydrothrombin via the base-catalyzed elimination of the phenylmethyl-sulfonic acid from thrombin, which had been inactivated with PMSF. The result of the overall reaction is the elimination of one molecule of water from the active serine residue. Anhydrothrombin is catalytically inactive but still binds to protein ligands such as hirudin[75] and factor VIII.[76] Thrombin where the active-site serine residue has been mutated to alanine (S195A) was described by Pei *et al.*[77] and was catalytically inactive toward fibrinogen and a peptide nitroanilide substrate. Gan and coworkers[78] examined an important (arginine-glycine-aspartic) acid sequence in human thrombin (Arg197-Gly198-Asp199). Asp199 is a component of the S_1 binding site. The effect of mutations on protein structure was evaluated by circular dichroism (CD); a difference in conformation was observed with the D199E mutant. The researchers observed that the S205A mutant had 0.1% of the activity of wild-type thrombin in the hydrolysis of Chromozym TH (Tos-Gly-Pro-Arg-pNA), the D199E had 0.03% the activity of wild-type thrombin, and R197E had 26% of the activity of the wild type. S205A thrombin was inactivated by PPACK, but there was some question about contamination with the wild type. The D199E mutant reacted with PPACK much more slowly than wild-type thrombin. The D199E mutant also had reduced activity in activating platelets and clotting fibrinogen. While these observations are important, it would be useful if studies such as that Wells and coworkers performed with active-site residue replacements in subtilisin BPN[79,80] were performed with thrombin, with rigorous kinetic analysis with specific protein substrates.[62,81] The S205A thrombin has been useful in the study of the interaction of thrombin and heparin cofactor II[82] and for the binding of substrate.[83]

The participation of a histidine residue in the catalytic function of thrombin can be inferred by homology to chymotrypsin and trypsin, where early studies suggested participation of a histidine residue.[84–87] The histidine at the active site of thrombin was identified through the use of active site–directed reagents. It may be surprising to current biochemists, but the concept of an enzyme-active site was still evolving 70 years ago.[78,88,89] The development of active site–directed reagents is frequently traced to work by Baker and colleagues, who developed a specific reagent, 4-(acetoacetamido)-salicyclic acid, which was developed to alkylate residues outside of the enzyme-active site in what they described as exo-alkylation.[90] The concept of exo-alkylation was used in the development of reagents to label exosites in thrombin.[91,92] The work from the Baker laboratory was based on earlier work to develop reagents for the modification of specific sites in proteins.[93,94] Singer and coworkers also contributed to the early development of affinity labeling, with work on the labeling of binding sites in antibodies.[95] In this work, *p*-(arsenic acid)-benzenediazonium fluoroborate was used to label antibodies directed against benzenearsonic acid. This early work provided the basis for Elliott Shaw and coworkers to develop the peptide chloromethyl ketones as active site–specific reagents for serine proteases.[96] Their first work was the development of tosyl-phenylalanine chloromethyl ketone (TPCK) for identification of a histidine residue at the active site of chymotrypsin.[97] Shaw and coworkers subsequently developed a reagent, tosyl-lysyl chloromethyl ketone (TLCK; chloro-3-tosylamido-7-amino-2-heptanone) for identifying the active-site histidine in trypsin.[98] The lysine derivative was developed in lieu of the arginine derivative; synthesis of the arginine derivative was not possible. In his early reviews,[96,98] Shaw discussed the development of these and other active site–directed reagents. These reactions react with the protein in two steps. The first step is the binding of the reagent to the enzyme (referred to as the k1 step; this reaction is similar to the formation of an enzyme-substrate complex). The second step is the formation of a covalent bond with the protein (referred to as the k2 step). In the case of the peptide chloromethyl-ketone, this is an alkylation step, and the rate of the k2 step can be considered an index of the nucleophilicity of the modified residue, in this specific situation, a histidine residue. The reader is referred to a review by Bryce Plapp on affinity labeling.[99] Plapp notes that active-site affinity labeling is different from suicide of k_3 inhibitors, such as dabigatran, which is described as a prodrug.[100–102]

It was work by Glover and Shaw[103] that established the presence of histidine at the thrombin-active site by isolation of the peptide modified by TLCK (1-chloro-3-tosylamido-7-amino-2-hepatone). The reaction of TLCK with thrombin is specific but quite slow. The development of tripeptide lysylchloromethyl ketones, which provide P_2 and P_3 interactions, proved to be efficient inhibitors of thrombin.[104,105] Kettner and Shaw[106,107] developed D-Phe-Pro-Arg-chloromethylketone (PPACK, FPR-CH$_2$Cl), which is a more rapid, specific inhibitor of thrombin. My laboratory reported that the rate of inactivation of bovine β-thrombin by TLCK was substantially less than that of bovine α-thrombin. These results were interpreted as reduced nucleophilicity of the active-site histidine.[108] In later work from another laboratory[109] on the reaction of PPACK with a human β-thrombin derived from limited proteolysis of α-thrombin, it was suggested that the nucleophilicity of histidine was not reduced in the human β-thrombin. Work from other laboratories with bovine thrombin showing enhanced esterase activity also showed enhanced reactivity with TLCK.[110,111] Orthner and Kosow[112] showed that sodium ions increased the rate of hydrolysis of tripeptide nitroanilide substrates by human α-thrombin; the rate of inactivation of human α-thrombin by TLCK is also increased by sodium ions. These investigators suggested that the enhanced reactivity was due to an increase in the k_2 step. In the study from my laboratory cited earlier,[108] diethylpyrocarbonate (DEP, ethoxyformic anhydride) also reacted with bovine α-thrombin more rapidly than with β-thrombin, which supported the hypothesis that the active-site histidine is less reactive (reduced nucleophilicity) in β-thrombin. The reduced nucleophilicity (basicity) of histidine in β-thrombin would have less of an effect in increasing the basicity of the active-site serine. The effect of a histidine residue on catalytic activity is not limited to serine proteins. Work on papain[113] and factor XIIIa[114] where there is an active-site thiol shows that the histidine residue increases the nucleophilicity of the serine by forming a histidinium cation. This work supports a role for histidine as a general acid/base catalyst by thrombin serving first as a protein acceptor (base) from the active-site serine and subsequently as a proton donor (acid) to the leaving group.

The importance of aspartic acid in the active site of thrombin was suggested from the study of other serine proteases. It is generally accepted that the work of David Blow established the importance of aspartic acid in chymotrypsin. Blow presented a great discussion of the aspartate story in 1997.[115] This aspartic acid residue (D99) is critical for the maintenance of the active-site histidine in a tautomeric form, which drives the nucleophilicity of the active-site histidine residue.[31,32,116]

There has been significantly more interest in the elucidation of the structural factors in thrombin which are responsible for the degree of specificity as compared to trypsin. The concept of an extended binding site for proteolytic enzymes can be traced to a seminal paper of Schechter and Berger.[117] These investigators proposed that there are binding sites (designated S_1–S_n) on the protein, which recognize complementary amino acids in the sequence of substrate on the carboxyl function of the scissile peptide (P_1-P_n) bond and S'–S_n' on the amino side of the peptide bond (P_1'–P_n'). It would be expected that the efficiency of catalysis would be improved with the length of substrate, providing for a larger number of interactions between enzyme (S) and substrate (P). Gallwitz and coworkers[118] examined the sequence of amino acids around several peptide bonds cleaved by thrombin in a number of substrates. This information allowed these investigators to advance a consensus sequence, LTPRGVRL, for cleavage by thrombin. It was suggested that the proline residue at P_2 and the arginine at P_3' are critical, although it is recognized that effective cleavages occur at other residues, such as isoleucine at P_2 in PAR-3 and factor V or threonine in PAR-1. These investigators developed a model system consisting of a linker peptide between two thioredoxin molecules to evaluate changes in the consensus sequence. Mutating the P_2 proline resulted in a 20-fold decrease in the rate of cleavage, while a change in the P_3' arginine resulted in a 14-fold decrease in the rate of cleavage; a combination of the two mutations causes a 200- to 400-fold decrease in the rate of cleavage. Other work[119] suggests more tolerance in the P_3, P_4, or P_4' positions. The importance of the primary sequence surrounding the primary binding site is supported by the development of tripeptide nitroanilide substrate.[120–122] Svendsen and coworkers[120] compared the rate of hydrolysis of several nitroanilide substrates (Table 3.2). These studies were performed with partially purified bovine thrombin, and it would appear that substrate concentrations in excess of K_M were used. A similar pattern was observed with trypsin, with the exception

---

TABLE 3.2
The Hydrolysis of Some Peptide Nitroanilide Substrates by Bovine Thrombin[a]

Substrate	Relative Rate of Hydrolysis	[Thrombin][b]	Rate[c]
Bz-DL-arginine pNA (BAPNA)	1	50	1.67
H-Val-Arg-pNA	2.12	10	0.71
Bz-Val-Arg-pNA	4.49	5	0.75
H-Phe-Arg-pNA	3.89	5	0.65
H-Phe-Val-Arg-pNA	15.58	5	2.6
Bz-Phe-Val-Arg-pNA	958.08	0.5	16
Bz-D-PheVal-Arg-pNA	5.99	5	1

[a] Data derived from Svendsen, L., Blombäck, B., Blombäck, M., and Olsson, P.I., Synthetic chromogenic substrates for determination of trypsin, thrombin, and thrombin-like enzymes, *Thromb. Res.* 1, 267–278, 1972.

[b] NIH units/mL

[c] μmole *p*-Nitroanline/mL/min × 10^{-3}

of the difference between benzoyl-Phe-Val-Arg-pNA and benzoyl-D-Phe-Val-Arg-pNA; with bovine thrombin, the rate of hydrolysis of the D-isomer is less than 1% of that of the L-Phe peptide, while with bovine trypsin, the rate of hydrolysis of the peptide nitroanilide containing D-phenylalanine is 30% that of the L-phenylalanine peptide nitroanilide. The importance of substrate length and binding to an apolar site in thrombin catalysis is dramatically demonstrated by the difference in the rate of hydrolysis of benzoyl-Phe-Val-Arg-pNA and Phe-Val-Arg-pNA. Le Bonniec and coworkers[123] showed that sequence C-terminal from the scissile peptide bond (P′ positions) was also important. This study showed that Glu25 is important in binding amino acid residues in the P'_3 or P'_4 positions. Pozgay and coworkers[124] published a more extensive study of the thrombic hydrolysis of tripeptide nitroanilide substrates. These results are discussed later in the section on thrombin specificity and substrates. Notable is that they did not observe an effect of D-Phe in the P_3 position with Bz-D-Phe-Gly-Arg-pNA ($k_{cat}/k_m = 6.48 \times 10^5$) vs. Bz-Phe-Gly-Arg-pNA ($k_{cat}/K_M = 1.34 \times 10^5$).

The importance of the S_1 binding site (Asp199), which is thought to bind arginine and lysine, is supported by studies of Tadashi Inagami, Stanford Moore Professor of Biochemistry at Vanderbilt University, which were published almost 50 years ago. I have always been interested in these studies but never got to running the same experiments with thrombin. I should also add that I have a special relationship to Stanford Moore, having been a research associate in his laboratory at the Rockefeller University, also some 50 years ago. However, I digress! Inagami and York[125] first showed that methyl guanidine could stimulate the rate of hydrolysis of N-acetylglycine ethyl ester; it was of interest that greater stimulation was observed with ethylamine, while inhibition was observed with ethylguanidine. The pH optima shifted to a lower pH. In subsequent work, Inagami and Hatano[126] observed that methyl guanidine enhanced the rate of inactivation of trypsin by iodoacetamide (6-fold but still slow, 12 M⁻¹hr⁻¹, compared to 1.9 M⁻¹hr⁻¹ in the absence of methylguanidine). The effect of methyl guanidine also requires a high concentration (0.4 M); earlier work[125] has suggested a K_a of 18 mM for methylguanidine activation of N-glycine ethyl ester hydrolysis and a K_i of 13 mM for inhibition of the hydrolysis of N^α-benzyl-L-arginine ethyl ester. The observations of the enhancement of the hydrolysis of a neutral substrate, N-acetylglycine ethyl ester, and the rate of inactivation with iodoacetamide support the enhancement of the nucleophilicity of the active-site histidine by productive S_1 occupancy. Ascenzi and coworkers[127] reported on the effect of ethylamine (ammonium, lysine) or acetamidine (guanidinium, arginine) on the hydrolysis of a neutral ester, N^α-carbobenzoxy alanine-p-nitrophenyl ester (CbzAlaONP), as shown in Table 3.3. It would appear that the major effect of acetamide and ethyl amine on the hydrolysis of ZalaONP is on the formation of the acyl-enzyme intermediate, which is the rate-limiting step in the hydrolysis of ester substrates. However, there is a significant effect of k_3, which also involves histidine function as a general acid. An examination of the data in Table 3.3 suggests that occupancy of the S_1 site enhances the catalytic activity of thrombin to a far greater extent than trypsin. The data is also consistent with the hypothesis that productive substrate binding enhances the nucleophilicity character of the histidine residue, enabling its function as a base. An examination of the pH dependence of the k_2 step

TABLE 3.3
Effect of Alkyl Amines and Amidines on the Hydrolysis of a Neutral Substrate by Trypsin, Thrombin, and Chymotrypsin[a]

Enzyme	Substrate	$k_2(s^{-1})$	k_2/K_s^e	k_{cat}/K_m^f	K_m (μM)	k_2/k_2^g	$k_2/K_s/$ k_2/K_s^f	$k_{cat}/K_m/$ k_{cat}/K_m^f
Trypsin	ZAlaONP[b]	0.69	0.007	0.0065	72	1	1	1
Trypsin	ZAlaONP+EA[c]	4.0	0.033	0.037	80	5.8	4.16	5.7
Trypsin	ZAlaONP+A[d]	3.8	0.030	0.026	100	5.5	4.21	4
Thrombin	ZAlaONP	0.19	0.000013	0.0013	120	1	1	1
Thrombin	ZAlaONP+EA	0.65	0.0049	0.004	100	3.4	377	3.1
Thrombin	ZAlaONP+A	2.0	0.013	0.012	125	10.5	1000	9.2
Chymotrypsin	ZAlaONP	0.50	0.004	0.0033	90	1	1	1
Chymotrypsin	ZAlaONP+EA	0.50	0.0042	0.003	85	1	1.05	0.9
Chymotrypsin	ZAlaONP+A	0.60	0.0048	0.0026	90	1.2	1.2	0.8

[a] Data adapted from Ascenzi, P., Menegatti, E., Guareri, M., Bortolloti, F., and Antonini, E., Catalytic properties of serine proteases. 2. Comparison between human urinary kallikrein and human urokinase, bovine β-trypsin, bovine thrombin, and bovine α-chymotrypsin, *Biochemistry* 21, 2483–2490, 1982.
[b] Carbobenzoxy-Ala-nitrophenyl ester
[c] Ethylamine
[d] Acetamide
[e] $μM^{-1}s^{-1f}$ $μM^{-1}s^{-1}$
[g] Experimental (either + EA or +A)/control

(transition from an enzyme-substrate complex to an acyl enzyme intermediate) demonstrated the importance of two ionizable groups: one with a pKa of 3.7 and a second with a pKa of 7.1. The binding of acetamide was dependent on a group with pKa 4.1, similar to that observed for the binding of benzamidine.

THROMBIN AS A PROTEIN

Some vital statistics of thrombin as a protein are given in Table 3.4. The various data in this table suggest that thrombin is an asymmetric spheroid in solution. Not shown in Table 3.4, an axial ratio of 2.8 was given for bovine thrombin by Harmison and Seegers.[128] Other investigators describe a more symmetrical spheroid with an axial ratio of approximately 1.0.[129] Crystallographic studies suggest a degree of asymmetry.

TABLE 3.4
Physical Characteristics of Thrombin

Protein	Parameter
Bovine thrombin	Isoelectric point 5.6 by paper electrophoresis[1] 5.3 (0.1 ionic strength), 5.75 (0.2 ionic strength) by free boundary electrophoresis.[2,3]
Human or bovine thrombin	Isoelectric point 8–9 by isoelectric focusing.[4]

Protein	Parameter
Human α-thrombin	Thrombin has a positive charge below pH 9.0, suggesting an isoelectric point above pH 9.0. Charge density determined by polyelectrolyte titration. Number of surface charges is 7 at pH 7.4. Data said to imply even distribution of charge density over the surface of thrombin and show that thrombin is highly charged at physiological pH.[5]
Human α-thrombin	Isoelectric point 7.0; 7.3; 7.6 isoionic forms determined for DIP-thrombin by isoelectric focusing on 5% acrylamide gel, ampholyte gradient pH 3.5–10.[6] The heterogeneity is suggested to be due to differences in glycosylation.
Human α-thrombin	Extinction coefficient (absorption coefficient) was shown to vary from 1.68 to 2.33 mL mg^{-1}cm^1, dependent on experimental conditions.[a,6] Careful measurements provided the value of 1.75 mL mg^{-1}cm^{-1} in 1.0 M HOAc; 1.74 mL mg^{-1}ml^{-1} in 0.21 M NaCl-0.02 M sodium phosphate, pH 7.2; and 1.83 mL mg^{-1}mL^{-1} in 0.1 M NaOH.[b]
Human α-thrombin	Molecular weight 38,000 nonreduced SDS; 31,500–40,000 by sedimentation equilibrium in 0.21 M NaCl-0.02 M sodium phosphate, pH 7.2, 36,500–36,000 by sedimentation equilibrium in 5.0 M GuCl.[6]
Human α-thrombin	Partial specific volume 0.732 mL/gm[6]
Bovine α-thrombin	Partial specific volume 0.736 mL/gm[7]
Bovine α-thrombin (citrated activated)[c]	Molecular weight 30,000 by sedimentation equilibrium in sodium phosphate–potassium chloride, pH 7.0 (I = 0.16). Aggregation of the protein was noted.[8] Partial specific volume 0.735.
Bovine α-thrombin	$S_{w,20}$ = 3.76; partial specific volume, 0.69 mL/gm; $D_{w,20}$ = 8.41 × 10^{-7} cm^2; viscosity 0.0376 mL/gm; molecular weight 33,700.[9]
Bovine α-thrombin	Molecular weight 36,700, sedimentation equilibrium in 6.0 M GuCl.[10]
Equine α-thrombin (citrated activation)[c]	Molecular weight 34,600 by analytical ultracentrifugation ($S_{w,20}$ = 3.87); diffusion constant, $D_{w,20}$ = 8.66 × 10^{-7} cm^2 s^{-1}; partial specific volume, 0.686; f/f_o = 1.16.[11]
Bovine α-thrombin	$S_{w,20}$ = 3.76, molecular weight, 33,700 (33,700 by amino acid composition); $D_{w,2}$ = 8.70 × 10^{-7}cm^{2}s^{-1}; partial specific volume, 0.69 mL/gm; intrinsic viscosity, 0.0376 mL g^{-1}: f/f_o = 1.16.[12]
Human α-thrombin (citrate activation)[c]	Molecular weight, 35,000 by gel filtration on G-100 Sephadex (0.097 M sodium citrate, pH 7.8)[13]
Bovine α-thrombin	Extinction coefficient 2.1 mL mg^{-1}cm^{-1}; molecular weight 38,000 ultracentrifugation in 2.0 mM EDTA; f/f_o= 1.1; $S_{w,20}$ = 3.3[14]
Bovine α-thrombin	Molecular weight 37,000 by sedimentation equilibrium in 6.0 M GuCl, 35,000 by SDS-PAGE, 36,961 by chemical composition. Partial specific volume, 0.726 mg gm^{-1}.[15]

[a] Aggregation of a protein sample can affect absorption by light scattering. This requires correction of the value obtained at 280 nm to be corrected. There also may be differences in values obtained with a double-beam spectrophotometer when compared to a single-beam instrument.

[b] Ionization of tyrosine in base increased the absorbance at 280 nm, although the increase in absorbance is generally measured at 295–297 nm (Donovan, J.W., The spectrophotometric titration of the sulfhydryl and phenolic groups of aldolase, *Biochemistry* 3, 67–74, 1964; Melo, E.P., Aires-Barrow, M.P., Coasta, S.M., and Cabral, J.M., Thermal unfolding of proteins at high pH range studied by UV absorbance, *J. Biochem. Biophys. Methods* 34, 45–59, 1997: Su, D., Aquillan, C. and Gadda, G., Characterization of conversed active site residues in class I nitronate, *Arch. Biochem. Biophys.* 677, 108058, 2019).

[c] It is the author's experience that citrated activated bovine thrombin can contain quantities of β-thrombin, which would have a molecular mass similar to bovine α-thrombin but might have different hydrodynamic properties.

(Continued)

TABLE 3.4
(Continued)

References

1. Levine, W.G. and Neuhaus, O.W., *Proc. Soc. Exptl. Biol. Med.* 106, 54, 1959.
2. Seegers, W.H., Harmison, O.R., Ivanovic, N., and Heene, D.L., *Thromb. Diath. Haemorh.* 15, 343, 1966.
3. Magnusson, S., Thrombin and prothrombin, in *The Enzymes*, Vol. III, ed. P.D. Boyer, Chapter 3, pp. 227–321, Academic Press, New York, USA, 1971.
4. Berg, W., Hillvärn, B., Arwin, H., Stenberg, M., and Lundstrom, I., The isoelectric point of thrombin and its behavior compared to prothrombin at some solid surfaces, *Thromb. Haemost.* 42, 972–982, 1979.
5. Heuck, C.C., Schiele, U., Horn, D., Fronda, D., and Ritz, E., The role of surface charge on the acceleration action of heparin on the antithrombin III-inhibited activity of α-thrombin, *J. Biol. Chem.* 260, 4598–4803, 1985
6. Fenton, J.W., II., Fasco, M.J., Stackrow, A.B., *et al.*, Human thrombin. Production, evaluation, and properties of α-thrombin, *J. Biol. Chem.* 257, 3587–3598, 1977.
7. Magnusson, S., *Thromb. Diath. Haemrrh.* 51(suppl), 257–261, 1972.
8. Conly, M. and Scheraga, H.A., Aggregation of thrombin, *Arch. Biochem. Biophys.* 95, 428–434, 1961.
9. Harmison, C.R., Landaburu, R.H., and Seegers, W.H., Some physicochemical properties of bovine thrombin, *J. Biol. Chem.* 236, 1693–1696, 1961.
10. Mann, K.G. and Batt, C.W., The molecular weights of bovine thrombin and its primary autolysis products, *J. Biol. Chem.* 244, 6555–6557, 1969.
11. Inada, Y., Matsushima, A., Kotaku, I., Hussain, S.P., and Shibata, K., Molecular weight and amino acid composition of equine thrombin, *J. Biochem.* 68, 193–198, 1970.
12. Harmison, C.R. and Seegers, W.H., Some physicochemical properties of bovine autoprothrombin II, *J. Biol. Chem.* 237, 3074–3076, 1962.
13. Lanchantin, G.F., Friedman, J.A., and Hart, D.N., The conversion of human prothrombin to thrombin by sodium citrate Analysis of the activation mixture, *J. Biol. Chem.* 240, 3276–3281, 1965.
14. Olsen, P.H., Esmon, N.L., Esmon, C.T., and Laue, T.M., Ca2+ dependence of the interaction between protein C, Thrombin and the elastase fragment of thrombomodulin. Analysis by ultracentrifugation, *Biochemistry* 31, 746–754, 1992.
15. Owen, W.G., Esmon, C.T., and Jackson, C.M., The conversion of prothrombin to thrombin. I. Characterization of the reaction product formed during the activation of bovine prothrombin, *J. Biol. Chem.* 249, 594–605, 1975.

EXOSITES

Most proteins have a heterogeneous surface containing areas or domains of unique composition that can bind to surfaces, proteins, synthetic substrates, and low molecular-weight substances, which can include inorganic ions such as sodium or calcium. These various sites are separate from the active site or the primary substrate binding site, but, as will be apparent, occupancy of these sites does influence active-site function. There are a variety of such proteins, including hydrophobic patches,[130–132] which can have a role in enzyme function[130] and protein purification;[131,132]cationic patches,[133,134] which can target a protein to a membrane surface; and anionic patches.[135] Exosites are a protein surface patch that is some distance from the enzyme-active site that bind substrates or regulatory molecules, while an endosite is a primary binding domain close to the active site. The concept of exosites and endosites was developed by Baker in 1967.[136] Endo-alkylation described the covalent modification of a residue or residues at the enzyme-active site. This reaction was later described as "active-site-directed irreversible inhibition," or affinity labeling. The term endo-alkylation refers to a reaction with a residue or residues outside of the enzyme-active site. In the case of matrix metalloproteins such as collagenase or rADAMTS enzymes such as aggrecanases, exosites serve to bind extracellular matrix substates,[137] and development of a monoclonal antibody against the exosite is suggested as a therapeutic approach to the prevention of osteoarthritis.[138,139] Matrix metalloproteinases (MMPs) also have exosites important for function.[140,141] Robichaud and coworkers[142] showed that exosites located at the N-terminal or C-terminal from the MMP1-MMP3 consensus collagen cleavage site (residues 769–783) influenced the catalytic efficiency (k_{cat}/K_m) for several MMPs. The k_{cat}/K_m for MMP-1 cleavage of the Glu-Ile peptide bond increased from 4,100 s$^-$M^{-1} to 37,000 s^{-1}M^{-1} with coupling of a putative exosite (residue 784 to residue 792) C-terminal to the consensus sequence. The k_{cat}/K_m for MMP-8 increased from 9,000 s^{-1}M^{-1} to 22,000 s^{-1}M^{-1}. The increase in MMP-1 activity was due to a large increase in k_{cat} (0.0071^{-1} increasing to 0.105 s^{-1}), as was the increase for MMP-8 (0.088 s^{-1} increasing to 0.30 s^{-1}); there was only a small change in K_m with either enzyme.

It has been difficult to develop active site–directed inhibitors of MMPs,[143] so the exosite present in MMP, which provides enzyme specificity toward a substrate, has emerged as an attractive therapeutic target.[144–146] The same specificity issues exist for ADAM and ADAMTS proteases,[147] where an antibody against the exosite of ADAMTS5(aggrecanase-2) has been advanced as a therapeutic option.[148] Exosites have also been demonstrated in complement factor D,[149] botulinum neurotoxin,[150–152] cathepsin K,[153] caspase,[154] subtilisin-like protease AprV2,[155] and insulin-degrading enzyme[156,157] and MMP.[149] In the case of factor D,[151] a monoclonal antibody was developed that could block the activation of C3bB but had no effect on the hydrolysis of a synthetic substrate. A low-molecular-weight natural product, D-chicoric acid, was shown to bind to an exosite with partial inhibition of botulinum neurotoxin A protease and was synergistic with synthetic hydroxamate in the inactivation of the protease.[152] It is of interest that it was earlier shown that substrate binding to botulinum neurotoxin A protease made the catalytic site competent.[150] While the binding of D-chicoric acid inhibits the activity of botulinum neurotoxin A protease, the binding to the exosite by D-chicoric acid might make the active site competent to promote interaction with the synthetic hydroxamate. The earlier work shows that there is a rich literature on the function of exosites in a variety of proteases, which are rarely referenced in the thrombin exosite literature; conversely, there is rarely mention in the other literature of the seminal work on the concept of exosites present in the thrombin literature.

Thrombin has been shown to have a number of regulatory sites. Binding of substrates or specific ligands to these sites can modulate catalytic efficiency. Thrombin has two exosites—clusters of basic amino acids—which have been shown to be of significance in the regulation of activity. To the best of my knowledge, David Bing, John Fenton, and coworkers[158,159] were the first to demonstrate an exosite in human thrombin with a reagent (exo affinity labeling reagent[160]) that bound at the active site and covalently modified a residue or residues in the cationic binding site (anion binding site). John Fenton worked with S.J. Singer at the University of California at Santa Barbara.[161] The work of Bing and colleagues clearly, as acknowledged, built on the prior work of Bernard Baker and associates.[162–164] While Bing and coworkers built on prior work, it is clear that they saw the opportunity to identify a functional exosite in human thrombin with m-[o-(2-chloro-5-fluorosulfonylphenylureido) phenoxybutoxy]benzamidine (Figure 4).[158,159] This reagent binds to the S_1 site, providing specificity via the benzamidine moiety,[165–167] and the covalent modifies a residue in the anion binding site via reaction with the sulfonyl fluoride. I could not find work at that time identifying the site(s) of modification; however, based on some recent work on chemical crosslinking, a number of different residues might be modified, including histidine, lysine, serine, threonine, tyrosine, or serine.[168] Bing and coworkers[158,159] also observed modification of the A chain to thrombin, suggesting proximity to the active site. It is of interest that much later work[169,170] showed that the A chain is located close to the active site. Bing and coworkers subsequently used this reagent to modify an exosite in C1s, a participant in the complement system,[19,171] as well as bovine factor Xa and plasmin.[171]

It was a surprise to me that there was little early interest in exosites in thrombin following the initial work. There were several papers showing the importance of exosites in the interaction with cells.[172,173] In the same time period, Sonder and Fenton[174] compared the reaction of mCP(PBA)-F and analogues with bovine thrombin and human thrombin. The rate of reaction, as judged by loss of fibrinogen clotting activity on reaction with mCP(PBA)-F, was similar for human thrombin ($t_{1/2}$ = 200 seconds) and bovine thrombin ($t_{1/2}$ = 133 seconds). There was one derivative, m-[o-(2-chloro-5-fluorosulfonylphenylureido)phenoxypropoxy]benzamidine, which reacted with bovine thrombin much faster ($t_{1/2}$ = 417 seconds) than human thrombin ($t_{1/2}$ = 2,105 seconds). While it has been known for some time that the specific fibrinogen clotting activity of bovine thrombin is lower than that of human thrombin, to the best of my knowledge, this was the first demonstration of a difference in chemical reactivity between bovine and human thrombin.

THROMBIN BINDING TO FIBRIN

It has been known for some time that thrombin bound to fibrin and could be eluted by increased ionic strength (see Chapter 1); in fact, adsorption and elution from fibrin were the early methods for thrombin (ferment) purification. Kaminski and McDonagh[175,176] studied the interaction of thrombi with a fibrin monomer-agarose column and with a fibrin gel. They observed that thrombin was bound to a fibrin-Sepharose column prepared by the action of thrombin on fibrinogen coupled to the matrix by cyanogen bromide technology.[177] The thrombin could be eluted with calcium ions; other divalent cations were effective, as was added thrombin. Binding of thrombin to the fibrin-Sepharose column was sensitive to ionic strength, with little binding occurring at an ionic strength of 0.25. These studies used ^{125}I-labeled thrombin, which was affinity-purified on a fibrin-Sepharose column. Kaminski and McDonagh later studied the binding of inhibited thrombins to fibrinogen and fibrin. The binding of PPACK-thrombin and PMSF-inhibited thrombin was identical to native thrombin, as well as thrombin in the presence of DAPA. Thrombin modified with pyridoxal-5-phosphate (PLP) did not bind to fibrin. PLP modifies lysine residues in proteins, so it is likely that exosite-1 was modified to prevent interaction with fibrin. Work from Frank Church's laboratory at the University of North Carolina at Chapel Hill showed that PLP modified two lysine residues (K21 and K65) in exosite-1.[178] These investigators also reported that hirudin prevented binding to fibrin consistent with subsequent work showing the hirudin bound to exosite-1, the binding site on thrombin for fibrinogen or fibrin (see later). Larry Berliner and coworkers at Ohio State University studied factors influencing the binding of thrombin to fibrin monomer bound to an agarose column.[179] This matrix was prepared by covalently linking fibrinogen to a cyanogen bromide–activated agarose and converting the fibrinogen-agarose to fibrin-agarose by the addition of thrombin. The residual thrombin was removed by washing with 3.0 M NaCl. This paper contains an excellent description of the preparation of fibrin-agarose. Berliner and coworkers[179] showed that thrombin bound to the fibrin-agarose column and was eluted with approximately 170 mM NaCl. The addition of adenosine triphosphate (ATP) decreased the salt concentration required for elution (80 mM NaCl at 10 mM ATP); pyrophosphate was somewhat less effective (95 mM at 10 MM pyrophosphate). Other polyanions (phosphate, citrate, oxalate) had a modest effect. Serotonin, tryptophan, and benzamidine have little effect on the NaCl concentration required for elution. The lack of effect of serotonin and tryptophan suggests that the fibrin-binding domain did not overlap the apolar binding site; the lack of effect of benzamidine suggests a lack of overlap of the fibrin binding site with the S_1 binding site. The data suggests the importance of an anion binding site in interaction with fibrin.

Subsequent work by Fenton and coworkers[180] used adsorption and elution of human thrombin from a cation-exchange resin (Amberlite CG-50) to further demonstrate the importance of an anion-binding site in the interaction with fibrinogen. γ-Thrombin, a degradation product of α-thrombin with lower fibrinogen clotting activity, likely resulting from peptide bond clearage in the anion-binding site (exosite-1),[181] required a lower salt concentration (0.24 M NaCl) for elution than did α-thrombin (0.35 M NaCl). PPACK-thrombin was eluted at only a slight lower salt concentration (0.325 M) than that required for native human α-thrombin. A hirudin-α-thrombin complex was not retained by the ion-exchange column. Nitrated human α-thrombin (modified with tetranitromethane) was also not retained by the ion-exchange column; it was suggested that this might represent a chemical modification in the anion-binding site or aggregation of the protein. It is also possible that the formation of 3-nitrotyrosine, which lowers the pKa of the phenolic hydroxyl,[182] decreases the isoelectric point of the protein,[183] which in turn would reduce affinity for a negatively charged matrix such as Amberlite CG-50. There was good correlation between the elution behavior of the various thrombin derivatives on the ion-exchange resin column and on a fibrin-agarose column. It was noted that TLCK-modified human α-thrombin bound to the fibrin-agarose column somewhat more tightly than native protein; the elution behavior of PPACK-thrombin on fibrin-agarose was similar to that of native α-thrombin. There are differences in the product obtained from the reaction of PPACK or TLCK with thrombin,[184–186] which are discussed in greater detail in the section on the chemical modification of thrombin.

    This early work on exosites in thrombin referred to this functional cluster of basic amino acids as the anion binding site or cation binding site, with the former based on the charge nature of the site, the latter on the charge nature of the ligand. Subsequent work refers to this site as exosite-1. Another site was identified as exosite-2. Exosite-2 is another cluster of basic amino acids located on the other side of the thrombin molecule important in binding heparin for acceleration of the rate of reaction with antithrombin.[178,187] Exosite-1 is located proximal to the active site and consists of Arg20, Lys21, Lys65, Arg68, Arg70, Arg73, Lys77, Lys106, Lys107, and Lys154, while exosite-2 consists of Arg89, Arg95, Arg98, Arg129, Arg170, Lys174, Arg178, Arg245, Lys267, and Lys272.[188,189] It is my sense that the composition of the two sites is not exact and may depend on the ligand. The fact that the sites are discontinuous rather than continuous does present a challenge. It would be of interest to determine the extent to which exosite function would be lost on reduction and carboxymethylation, such as is the case for discontinuous epitopes.[190] It is recognized that proteolysis of human α-thrombin to yield γ-thrombin involves cleavage at positions in exosite-1 and exosite-2. While John Fenton, Dave Bind, and coworkers were the first to suggest the importance of domains distant from the active site in thrombin, to the best of my knowledge, Guillin and coworkers[191] were the first to introduce the term exosite to describe the two anion-binding sites in thrombin. This is a rather nice short review article that covers a variety of topics, including thrombin variants and antibodies to thrombin. This work is as timely today some 25 years later as it was in 1995.

    After what in retrospect appears to be a slow start, more work on the exosite structure and function in thrombin appeared in the 1990s. It is m sense that the absence of three-dimensional structural information made progress difficult prior to the publication of the crystal structure in 1989.[192] Chang and Detwiler[193] showed that the anion-binding site of human α-thrombin was necessary for a reaction with platelet-derived nexin. γ-Thrombin, which lacks intact exosites, did not block the reaction between human α-thrombin and platelet-derived nexin; hirugen, a peptide analog of hirudin, did block the reaction between human α-thrombin and platelet-derived nexin. While I can't find definitive information, it is my sense that the γ-thrombin obtained by "autolysis" is structurally different from that obtained by limited proteolysis by trypsin. I would also note that I spent a considerable amount of time trying to show that my purified bovine α-thrombin underwent autolysis with lack of success. Prolonged storage at 23°C at neutral to alkaline pH resulted in a loss of activity but no change in covalent structure. Platelet-derived nexin is known as protease nexin-1 (Serpin E2) and is 43-kDa protease inhibitor expressed by a variety of cells in addition to platelets,[194,195] which, despite low levels of expression by platelets, is suggested to have an important role in hemostasis and thrombosis.[195,196]

    Exosite-1 was shown to be important for the interaction of thrombin with heparin cofactor II.[197,198] Hortin and coworkers[199,200] showed that a peptide derived from heparin cofactor II (residues 54–75, Gly-Glu-Glu-Asp-Asp-Asp-Tyr-Leu-Asn-Leu-Glu-Lys-Ile-Phe-Ala-Glu-Asp-Asp-Asp-Tyr-Ile-Asp). This is an acidic peptide that was shown to inhibit the cleavage of fibrinogen by thrombin; the peptide also blocked the binding of thrombin to an immobilized peptide derived from hirudin [hirudin PA(54–66)]. Both the heparin cofactor II peptide and hirudin peptide blocked the reaction of thrombin with intact heparin cofactor II. The heparin cofactor II 54–75 peptide was observed to enhance the hydrolysis of S-2366 (pyro-L-Glu-Pro-Arg-pNA). While data for this enhancement was not presented in this paper, a more complete study of this effect on the hydrolysis of tripeptide nitroanilide substrates was subsequently published[200] and will be discussed in greater detail later. One of the coauthors of this paper was Doug Tollefsen who is generally given credit for the discovery of heparin cofactor II.[201]

Hirulog-1 is a synthetic peptide derived from hirudin [(D-Phe)-Pro-Arg-Pro-(Gly)$_4$-Asn-Phe-Glu-Glu-Ile-Pro-Glu-Tyr-Leu], which was shown to inhibit both fibrinogen clotting activity and hydrolysis of a tripeptide nitroanilide substrate (Spectrozyme TH) with a K_i of 2.3 nM.[202] A shorter peptide from hirudin (S-Hir$_{53-64}$) inhibited fibrinogen clotting with a K_i of 44 nM but did not inhibit the hydrolysis of a tripeptide nitroanilide substrate. These results suggest that hirulog-1 binds to both exosite-1 and the catalytic site of thrombin, while S-Hir$_{53-64}$ binds only to exosite-1. Hirulog-1 was a weaker inhibitor of tripeptide nitroanilide hydrolysis by γ-thrombin (K_i = 1,400 nM) and bovine α-thrombin (K_i = 42.4 nM). The difference with γ-thrombin is attributed to disruption of exosite-1, while the difference observed with bovine thrombin reflects the absence of Lys154. There was a small amount of inhibition of human factor Xa by hirulog-1, but while factor Xa has a functional exosite,[203] these investigators attributed the inhibition of factor Xa by hirulog-1 to contamination with thrombin. Hirulog-1 had no effect on the activity of trypsin or plasmin at concentrations of the inhibitor three orders of magnitude. John Maraganore and coworkers,[204] while still at Biogen, were the first to identify a specific residue in the anion-binding site of thrombin. These investigators used a fluorodinitrobenzene (FDNB) derivative of a peptide derived from hirudin to covalently label Lys154. Lysine154 is considered to be located in exosite-1. Another group studied the effect of fragments of hirudin obtained by cyanogen bromide cleavage.[205] Native hirudin does not contain any methionine, so these investigators inserted a methionine residue to replace asparagine at position 53. The enabled the specific cleavage by cyanogen bromide at the inserted methionine residue to yield a C-terminal fragment of residues 53–64 and an N-terminal fragment of residues 1–52. The N-terminal fragment was a competitive inhibitor of thrombin, while the C-terminal fragment lowered the K_m for H-D-Phe-Pip-Arg-pNA, resulting in stimulation of the rate of hydrolysis; this effect is discussed in greater detail later. It was suggested that the C-terminal fragment bound to exosite-1, as this peptide protected thrombin from cleavage by trypsin.

Sheehan and coworkers[206] showed that exosite-1 was critical for the reaction with heparin cofactor II but not for the reaction with antithrombin. These investigators prepared four mutants of human thrombin, with changes in residues considered to be in exosite-1: K52E, R68E, R70E, and K154A. The reaction of these derivative forms was studied with heparin cofactor II and antithrombin in the presence and absence of heparin. The rate of inactivation of thrombin was measured by the hydrolysis of S-2238 (H-D-Phe-Pip-Arg-pNA). It was reported that the mutation of the four residues did not affect the hydrolysis of the tripeptide nitroanilide substrates. With one exception, there was no difference in the reaction of any of the mutant thrombins with antithrombin in the presence or absence of heparin. The rate of reaction of mutant K52E with antithrombin (1.8×10^5 M^{-1}min^{-1}) was 50% that of recombinant thrombin (3.7×10^5 M^{-1}min^{-1}) in the absence of heparin; the rate of reaction of the K52E mutant with antithrombin in the presence of heparin was slightly greater (5.4×10^9 M^{-1}min^{-1}) than the recombinant enzyme (4.5×10^8 M^{-1}min^{-1}) or the plasma-derived enzyme (3.9×10^8 M^{-1}min^{-1}). The authors note that K52 is close to the S′ substrate binding site. The authors suggest that there is interaction of the S′ site with the P′ site in antithrombin, which would be less important in the interaction in the presence of heparin. Mutations in exosite-1 had a far larger effect on the reaction with heparin cofactor II in either the presence or absence of heparin or dermatan sulfate (Table 3.5). Mutagenesis of two arginine residues in exosite-1 (R68E, R70E) decreased the rate of reaction of heparin cofactor II with the thrombin in the presence or absence of heparin, as did the mutagenesis of one lysine residue, K52E. Mutant K154A showed little effect on the reaction with heparin cofactor II in the absence of heparin or dermatan sulfate, but there was a 50% reduction in the rate of inactivation in the presence of dermatan sulfate, with a lesser effect in the presence of heparin. The K154A mutant differs from the other mutant in that this mutation represents charge neutralization, while the other mutants are charge reversal, with a change from a positive charge to a negative charge. The results suggest that exosite-1 is important for the reaction of thrombin with heparin cofactor II. Another group[207] evaluated the effect of mutations in the anion-binding site where glutamine was substituted for basic amino acids in thrombin exosite-1. Glutamine is close to isosteric for arginine[208] and lysine.[209] The results from this study are shown in Table 3.6. Changes in exosite-1 had some effect on the reaction with antithrombin and protease nexin-1 in the presence or absence of heparin. With some mutations, there was a decrease in the rate of reaction with the serpins but acceleration in the rate of reaction of the other serpins with some other mutants. It is clear that the mutations in exosite-1 consistently had a large effect on the rate of reaction of heparin cofactor with thrombin. Subsequent work from the Washington University laboratory[210] showed that mutations in exosite-2 (R89E, R345E, K248E, K252E) affected the affinity of the mutant thrombin for dermatan sulfate agarose. While native thrombin (plasma-derived or recombinant) bound to a dermatan sulfate column and was eluted with 0.24 M NaCl, the mutant thrombins did not bind to the dermatan sulfate column. The rate of inactivation of the mutant thrombins with heparin cofactor II was not significantly different from native thrombin in the presence or absence of dermatan sulfate or heparin. It was concluded that binding of glycosaminoglycans to exosite-2 was not significant for the reaction with heparin cofactor II but was important for the reaction with antithrombin. The work did suggest that the accelerating effect of heparin or dermatan sulfate on the reaction of heparin cofactor II with thrombin was due to an effect of the glycosaminoglycans on heparin cofactor II, not thrombin. Another group[211] used a different set of exosite-2 mutants (R93A, R97A, R101A) to study the importance of exosite-2 on the reaction of thrombin with several serpins. Heparin accelerated the rate of reaction of antithrombin or protein C inhibitor (PCI) with native thrombin, but not with the exosite-2 mutants. Hirugen did not block the acceleration of the inactivation of thrombin by PCI, suggesting that heparin and hirugen bound at different sites—in this case, heparin binds to exosite-2 and hirugen binds to exosite-1. Heparin or dermatan sulfate did enhance the rate of reaction with the exosite-2 mutants (R93A, R97A, R101A) with heparin cofactor II to the same extent as observed with native thrombin. These results, combined with those described earlier and later structural work[212] support a mechanism where a conformational change occurs in the presence of heparin cofactor II with glycosaminoglycans such as heparin or dermatan sulfate, providing the basis for the rate enhancement differing from the mechanism proposed for antithrombin or protein C inhibitor. I was privileged to have both Mike Griffith and Frank Church in my laboratory in Chapel Hill. They both brought a unique experience from their doctoral work, which they applied to their subsequent work which far exceeded my efforts. One of my mentors, Stanford Moore, told me once that as my career progressed, the time would arrive when my greatest contribution was to create an atmosphere where other individuals could do good work. I think that he also told me once "the work of your students should exceed that of yourself." I am not sure about this one, since he was a Nobel Laureate and it is unlikely that I will achieve that distinction. Frank stayed at UNC and worked with Harold Roberts for a time before striking out on his own, becoming a distinguished researcher and an even more distinguished educator. Mike went in a similar direction and worked with Harold Roberts at UNC but left UNC to work for the Hyland Division of Baxter Healthcare and was responsible for the development of a viral-safe ultra-high-purity factor VIII product. Somewhat later I left UNC to work for Hyland and ended up reporting to Mike who was vice president for research and development, fulfilling the prophecy of what goes around comes around. I understand that Harold Roberts did not take kindly to the loss of Mike Griffith and subsequently attempted to push the president of Hyland into a swimming pool. I would be remiss if I did not mention Christine Vogel, later Christine Vogel Sapan, who also made major contributions to our work on thrombin; her work on the *in vivo* reaction of thrombin with antithrombin has been very useful.[213] I should also comment on the work by John Sheehan and coworkers. It is rare that you have three truly distinguished authors on the same paper. Dr. Sheehan was a fellow at Washington University at the time and is now a distinguished professor at the University of Wisconsin, while Dr. Tollefsen remains at Washington University as a professor of medicine. Unfortunately J. Evan Sadler passed too early in 2018.

TABLE 3.5
Effect of Mutagenesis in Exosite-1 on Reaction with Heparin Cofactor II[a]

Enzyme	Heparin Cofactor II	Heparin Cofactor II + Heparin	Heparin Cofactor II + Dermatan Sulfate
Plasma thrombin	93[b]	71	70
Recombinant thrombin	100	100	100
K23E	9	7	11
R68E	55	2	1
R70E	37	17	12
K154A	91	76	50

[a] Data from Sheehan, J.P., Wu, Q., Tollefsen, D.M., and Sadler, J.E., Mutagenesis of thrombin selectively modulates inhibition by serpins heparin cofactor II and antithrombin III. Interaction with the anion-binding exosite determines heparin cofactor II specificity, *J. Biol. Chem.* 268, 3639–3645, 1993.

[b] Percentage of rate observed with recombinant thrombin

TABLE 3.6
Effect of Specific Site Mutations in Exosite-1 on Interaction with Antithrombin or Heparin Cofactor II in the Presence or Absence of Heparin[a]

Thrombin Species	ATT[b]	ATT + Heparin[b]	PN-1[b,c]	PN-1 + Heparin[b]	HCII[b,d]	HCII + Heparin[b]
pIIa[e]	0.75	0.63	1.15	1.02	0.99	0.60
rIIa[f]	1.00	1.00	1.00	1.00	1.00	1.00
R20Q[g]	0.89	1.08	1.00	0.27	0.50	0.40
K21Q	0.97	1.13	1.09	0.69	0.43	0.40
R62Q	0.94	1.48	0.81	0.45	0.44	<0.01
R68Q	0.67	0.77	0.73	0.49	0.31	0.07
R70Q	1.00	0.99	1.02	0.56	0.47	0.47
R73Q	0.89	1.23	0.85	0.31	0.60	0.57
K77Q	0.92	0.92	0.76	0.81	0.50	0.34
K106Q	0.98	1.24	1.30	1.08	0.38	0.34
K107Q	1.06	1.31	1.24	0.85	0.42	0.35
K154Q[hhhh]	1/48	1.32	1.50	0.84	0.84	0.65

[a] The data in this table is taken from Myles, T., Church, F.C., Whinna, H.C., Monard, D., and Stone, S.R., Role of thrombin anion-binding exosite-1 in the function of thrombin-serpin complexes, *J. Biol. Chem.* 273, 31203–31208, 1988 and Myles, T., Le Bonniec, B.F., and Stone, S.R., The dual role of thrombin's anion-binding exosite-1 in the recognition and cleavage of the protease-activated receptor 1, *Eur. J. Biochem.* 268, 70–77, 2001.

[b] Ratio of second-order rate constants for formation of thrombin-serpin complex, k_{on}mutant/k_{on}rIIa

[c] PN-1, protease nexin-1

[d] HCII, heparin cofactor II

[e] Plasma-derived thrombin

[f] Recombinant thrombin

[g] Thrombin mutant with number based on B-chain of human thrombin

It is clear from this work that heparin and other sulfated mucopolysaccharides accelerate the reaction of thrombin with heparin cofactor II and antithrombin. However, the mechanism of action of the sulfated mucopolysaccharides is quite different. The sulfated mucopolysaccharide binds to heparin cofactor II, forming a bimolecular complex, which then reacts with thrombin. In the case of antithrombin, heparin binds to both thrombin and antithrombin and results in rate enhancement.

An autoantibody was isolated from the plasma of a patient with recurrent arterial thrombosis.[214] The isolated IgG inhibited human α-thrombin action on human umbilical vein endothelial cells (HUVEC), but had little effect on γ-thrombin; a similar effect was observed with platelets where the isolated antibody blocked aggregation by human α-thrombin but not with human γ-thrombin; there was no effect of the antibody on platelet aggregation by SFLLR. The antibody competed with thrombomodulin for the site responsible for protein C activation. In another study, an IgA paraprotein was isolated from the plasma of a patient with a prolonged prothrombin time (PT) and activated partial thromboplastin time (aPTT).[215] A mix of patient plasma and normal plasma did not correct the prolonged clotting times, suggesting the absence of a deficiency. The purified IgA paraprotein was a potent inhibitor of the thrombin clotting time but enhanced the hydrolysis of a tripeptide nitroanilide substrate (S-2238). Surface plasmon resonance analysis provided a Kd of 1.1 nM for the binding of the intact antibody to thrombin. An Fab fragment was prepared by digestion with papain followed by anion-exchange chromatography. Crystallographic analysis of the Fab-thrombin complex showed Fab binding to the exosite domain.

Triabin, a protein binding to the anion-binding site of thrombin, was isolated from the saliva of *Triatoma pallidipennis*, a triatomine bug (also known as a kissing bug or vampire bug).[216] It was found that two proteins influenced platelet aggregation: one blocked collagen-aggregation of platelets and another component inhibited thrombin-induced platelet aggregation. The two proteins were isolated by chromatography on Superose 12. The separation is not based on size difference, but likely on ion exchange. Superose is an agarose-based material, and the different grades are based on the extent of crosslinking. The agarose matrix does have a diffuse negative charge, permitting it to act as a very weak cation exchange column. The fraction inhibiting thrombin was further purified by affinity chromatography on thrombin-agarose with elution at pH 2.5. The purified material has a molecular weight of 18 kDa on nonreduced SDS-PAGE and 21 kDa on reduced SDS-PAGE. The purified protein, triabin, inhibited thrombin activation of platelets, fibrinogen clotting, and the aPTT. There was a modest (20%–30%) inhibition of thrombic hydrolysis of H-D-Phe-Pip-Arg-pNA (S-2238). A complex between triabin and thrombin could be observed on nondenaturing PAGE. Triabin also inhibited the activation of protein C in the presence of thrombomodulin and blocked the proteolysis of thrombin by trypsin to form γ-thrombin The data is consistent with triabin binding to exosite-1; a K_i of 3 pM was determined with a fibrinogen-clotting assay. Bothrojaracin is a thrombin inhibitor isolated from the venom of *Borthrops jararaca*.[217] Bothrojaracin was shown to have a molecular weight of 27 kDa and an isoelectric point of 4.3. Bothrojaracin inhibits both fibrinogen clotting activity and platelet aggregation but did not affect the hydrolysis of S-2238 (H-D-Phe-Pip-Arg-pNA) at substrate concentrations below K_M. Native PAGE demonstrated that bothrojaracin formed a complex with thrombin; the complex was not stable on SDS-PAGE. Subsequent work suggested that bothrojaracin also formed a complex with prothrombin in murine or rat plasma, which could be demonstrated on native (nondenaturing) PAGE,[218] confirming earlier work[219,220] that suggested interaction with an incomplete exosite-1 in human prothrombin. The data does suggest that bothrojaracin could interact with both exosite-1 and exosite-2.[219–222] Bothrojaracin has been shown to block factor V activation by thrombin.[222] There has been interest in the development of therapeutics based on bothrojaracin.[218,223] Avathrin is a peptide isolated from the salivary gland of a tick, *Amblyomma variegatum*, which was shown to be a potent inhibitor of thrombin.[224] Another peptide, variegin, was also shown to be an inhibitor. Both peptides inhibit fibrinogen clotting and hydrolysis of S-2238, suggesting binding to both exosite-1 and the enzyme active site, similar to hirudin. Both peptides are also substrates but cleaved at a slow rate. Crystallographic analysis of a complex of avathrin-thrombin complex showed binding to exosite-1. Both peptides are derived from a precursor, and the authors suggest that multiple peptides could inhibit thrombin.

While there were previous studies suggesting the importance of exosite-1 of thrombin in the activation of platelets,[214,225–227] Myles and coworkers[228] were the first to show that exosite-1 was crucial in the direct activation of platelets as opposed to binding. These investigators used the various exosite-1 mutant thrombins prepared earlier,[207] where glutamine is substituted for arginine or lysine to evaluate the cleavage of a peptide, PAR-1 38–60, derived from the PAR-1 sequence and platelet aggregation. These investigators also evaluated the action of these mutants on fibrinogen and on binding of hirudin. This data is collected in Table 3.7. The combined data for the interaction of the mutant thrombins with antithrombin, protease nexin 1, heparin cofactor II, and hirudin from two studies by this group[207,228] is collected in Table 3.8. Mutations in exosite-1 had little effect on the reaction of antithrombin with α-thrombin, either in the presence or absence of heparin; an exception was the R68Q mutant. Mutation at this residue had a modest effect on the rate of reaction of thrombin with antithrombin in the presence (0.77/1.00) or absence (0.67/1.00) of heparin. Mutation at this residue had a more marked effect on reaction with protease nexin-1, heparin cofactor II, the cleavage of PAR-1 peptide, and reaction with hirudin. The R62A mutant had reduced reactivity with protease nexin-1 in the presence of heparin; activity in the cleavage of the PAR-1 peptide and reactivity with hirudin are essentially absent in this mutant. Cleavage of the thrombin B-chain in this region produces β-thrombin and γ-thrombin, both of which demonstrate defective exosite-1

TABLE 3.7
Effect of Exosite-1 Mutations on PAR-1 Cleavage, Platelet Aggregation, Fibrinogen-Clotting, and Binding of Hirudin[a]

Enzyme	PAR(38–60) Cleavage[b]	Platelet Aggregation[c]	Fibrinogen-Clotting[d]	Hirudin Binding[e,r]
pIIa[f]	1.00	1.00	0.81	0.88
rIIa	1.00	1.00	1.00	1.00
R20Q	0.45	1.71	0.49	0.19
R62Q	0.02	7.20	<0.01	<0.01
R68Q	0.05	5.65	0.22	0.02
R70Q	0.53	2.00	0.51	0.31
R73Q	0.38	2.00	0.19	0.29
K77Q	0.52	0.94	0.2	0.83

[a] Data taken from Myles, T., Le Bonniec, B.F., and Stone, S.R., The dual role of thrombin's anion-binding exosite-I in the recognition and cleavage of the protease-activated receptor 1, *Eur. J. Biochem.* 268, 70–77, 2001.

[b] Relative rate of reaction of cleavage of the PAR(38–60) peptide by mutant thrombin to that of recombinant thrombin

[c] ([II_a] × $t_{1/2}$ platelet aggregation)^{mutant}/([II_a] × $t_{1/2}$ platelet aggregation)^{rIIa}

[d] Relative rate of fibrinogen clotting

[e] Relative rate of hirudin binding (k_{on} × 10^8(M^{-1}s^{-1}) with rII_a = 1.00

[f] Human plasma thrombin

[g] Recombinant human thrombin

TABLE 3.8
Effect of Specific Site Mutations in Exosite-1 on Interaction with Serpins, PAR-1, and Hirudin

Thrombin Species	ATT[b]	ATT + Heparin[b]	PN-1[a,b]	PN-1 + Heparin[b]	HCII[b,c]	HCII + Heparin[b]	k_{cat}/K_M[d] × 10^{-7}	Hirudin[e]
pIIa[f]	0.75	0.63	1.15	1.02	0.99	0.60	9.1	88
rIIa[g]	1.00	1.00	1.00	1.00	1.00	1.00	9.2	100
R20Q[h]	0.89	1.08	1.00	0.27	0.50	0.40	4.1	19
K21Q[h]	0.97	1.13	1.09	0.69	0.43	0.40	3.6	21
R62Q[h]	0.94	1.48	0.81	0.45	0.44	<0.01	0.18	<1
R68Q[h]	0.67	0.77	0.73	0.49	0.31	0.07	0.46	2
R70Q[h]	1.00	0.99	1.02	0.56	0.47	0.47	4.8	31'
R73Q[h]	0.89	1.23	0.85	0.31	0.60	0.57	3.3	29
K77Q[h]	0.92	0.92	0.76	0.81	0.50	0.34	4.7	83
K106Q[h]	0.98	1.24	1,30	1.08	0.38	0.34	5.8	57
K107Q[h]	1.06	1.31	1.24	0.85	0.42	0.35	5.6	78
K154Q[hhhhh]	1/48	1.32	1.50	0.84	0.84	0.65	15.5	64

Source: Myles, T., Church, F.C., Whinna, H.C., Monard, D., and Stone, S.R., J. Biol. Chem. 273, 31203–31208, 1988; Myles, T., Le Bonniec, B.F., and Stone, S.R., Eur. J. Biochem. 268, 70–77, 2001.

[a] PN-1, protease nexin-1
[b] Ratio of second-order rate constants for formation of thrombin-serpin complex, k_{on}mutant/k_{on}rIIa
[c] HCII, heparin cofactor II
[d] Specificity constant for cleavage of PAR-38–60 peptide derived from platelet PAR-1 receptor
[e] Percentage of activity based on the second-order rate constant for reaction with hirudin
[f] Plasma-derived thrombin
[g] Recombinant thrombin
[h] Thrombin mutant with number based on B-chain of human thrombin

function. It would appear that the binding of hirudin is more affected by residues across the exosite-1 domain. Residues Arg62, Lys65, Glu70, and Glu76 in exosite-1 were identified as being important in the binding of thrombomodulin using alanine-scanning mutzgeneis.[229] These investigators also found that mutation of two aspartic acid residues (Asp99, Asp178) to alanine increased thrombomodulin binding This is one of several studies that suggests interaction between exosite-1 and exosite-2.

Work cited earlier[222] showed that bothajaraciin, a protein obtained from B. jararaca, inhibited the activation of factor V by thrombin. It was suggested that bothrojaracin bound to both exosite-1 and exosite-2 in thrombin. Other investigators have also shown that exosite-1 is important in the activation of factor V and factor VIII.[230] These studies used an acidic pentapeptide, Asp-TyrSO$_4$-Asp-TyrSO$_4$-Gln; β-thrombin, a derivative of α-thrombin obtained by limited proteolysis;[231] and a form of meizothrombin that has only exosite-1 exposed.[232] The human β-thrombin is obtained by cleavage at Arg62 and Arg73.[231] The data obtained support a role of exosite-1 in the activation of factor V and factor VIII in blood coagulation. As noted by these investigators, the effectiveness of meizothrombin suggests an important role for this form of thrombin in the hemostatic response. Subsequent work suggests a more complex relationship of thrombin with factor V, which would involve interaction of factor V with exosite-2 during activation by thrombin.[233] Exosite-1 is incomplete in prothrombin and unavailable to most ligands, but is, as shown earlier, available in meizothrombin, with both sites available in α-thrombin. Maurer and coworkers at the University of Louisville have studied the development (maturation) of exosite-1.[234,235] In the 2017 study,[234] they used nuclear magnetic resonance (NMR) to show the change in exosite-1 in the transition of prothrombin to thrombin. They showed that several peptides derived from PAR-3 bound to a proexosite in prothrombin and that the binding became stronger with the conversion of prothrombin to thrombin. A specific interaction between a glutamic acid residue in a PAR-3 peptide and Arg73 in exosite-1 was important. There were several hydrophobic residues on the PAR-3 peptide important in stabilizing the interaction, and it was suggested that a pocket formed by Phe19, Leu60, and Ile85 in exosite-1 was important. In subsequent work[235] this group extended these observations to peptides derived from the PAR-1 receptors, showing binding to the immature exosite in prothrombin, which become stronger on the conversion to thrombin. The interaction with the PAR-1 peptide involved Arg68, and again the hydrophobic cluster of Phe19, Leu60, and Ile85 was important in stabilizing the interaction. They also showed that a peptide from GpIbα, which bound to exosite-2, increased the avidity of binding of PAR-1 peptides and PAR-3 peptides to thrombin. Previous work[228] showed that the R68Q mutant of human α-thrombin showed reduced activity in the cleavage of a peptide derived from PAR-1, reduced activity in platelet aggregation, and reduced inhibition with heparin cofactor II or hirudin.

Most of the work discussed here has focused on electrostatic interactions, interactions between positively charged residues in exosite-1 and negatively charged residues in the ligand.[236] There is one study which did suggest the importance of hydrophobic interactions between hidrudin and exosite-1.[237] The importance of a hydrophobic core in stabilizing the complex of hirudin with exosite-1 has also been suggested by other investigators.[236]

THERAPEUTIC APPROACHES BASED ON EXOSITE-1

The importance of exosite-1 in the action of thrombin on physiological substrates suggests that it might be therapeutic target. Some therapeutic approaches to exosite-1 have been mentioned earlier. In addition, two recent studies are worthy of mention at the time of this writing (March 2020). One group[238] isolated a peptide from the hydrolysis of casein (trypsin), YQEPVLGPVR (designated PICA) as a ligand to thrombin exosite-1. This peptide was slightly acidic (pI 6.0) and inhibited the aPTT, the PT, and the thrombin clotting time (TCT). PICA did not inhibit the hydrolysis of S-2238, demonstrating that the active site is not blocked. PICA did influence the conformation of thrombin, and *in silico* studies (molecular docking) demonstrated interaction of PICA with exosite-1 in thrombin. Jansen is developing a monoclonal antibody, JNJ-64179375, that has high affinity ($K_D = 0.8$ nM) toward exosite-1 as a therapeutic.[238,239]

Some of the studies noted earlier suggested that the binding of ligand to exosite-1 enhances the catalytic activity of thrombin. I have discussed the effect of exosite occupancy on the active MMPs previously. This was done in part to allow comparison with the thrombin studies described later. It was also done to show that thrombin and other blood coagulation proteins are legitimate biochemicals. I started work with blood coagulation proteins in 1962 with Earl Davie at the University of Washington in Seattle. I should note that I have just been informed of Earl's passing in early June 2020. He was a major force in blood clotting, bringing a major conceptual change in our understanding of the interaction of the various proteins.[240] The Department of Biochemistry at the University of Washington in the 1960s was a great department with two future Nobel Laureates, Ed Fisher and Ed Krebs; Don Hanahan; Phil Wilcox; Milton Gordon; and Joe Kraut, among others. The chair was Hans Neurath. It was a great collection of talent. The department was small enough that you knew everybody. The Davie lab in 1962 consisted of Henry Kingdon and Antero So, who followed Earl from Western Reserve in Cleveland to Seattle, and Kathy Downey and myself, who waited for Earl's arrival. Kathy and Antero worked on protein biosynthesis and went on to become professors at the University of Miami. Henry Kingdon went to the National Institutes of Health (NIH), where he worked with Earl Stadtman with seminal work on the control of glutamine synthetase, then to the University of Chicago and later to the University of North Carolina at Chapel Hill. He left UNC to join the Hyland Division of Baxter Healthcare. I later worked with Henry at the Hayward facility. It was a good group, later joined by Art Thompson and Fran Pitlick. While blood coagulation proteins had respect in the clinical community, there was some concern among the biochemistry community, who looked askance at a protein called pro-serum prothrombin conversion accelerator; reminding them of proteins as the old yellow enzyme[241] fell on deaf ears.

EFFECT OF LIGAND BINDING TO EXOSITE-1 ON THE ACTIVITY OF THROMBIN

The following section will discuss the effect of ligand binding at exosite-1 on the activity of α-thrombin. While it should be intuitive, the binding of ligand to exosite-1 can result in a conformational change in thrombin.[242–245] There are also studies showing that modification at the enzyme-active site can modify the conformation of thrombin. PPACK reacts with the active-site histidine residue in thrombin, and there is a difference in conformation between PPACK-thrombin and thrombin.[246] Another study showed that addition of dabigatran (Pradaxa) to thrombin inhibited binding to fibrin and factor Va, while argatroban enhanced binding.[247] Dabigatran had no effect on the binding of a peptide from a GpIbα (269–286 ppp; the "ppp" indicates three phosphotyrosine residues), while argatroban enhanced binding. This peptide is considered to be specific for binding to exosite-2.[248] DAPA, an active-site probe, also enhanced binding to fibrin and factor Va, as well as the GpIbα peptide. It is possible that ligand binding at exosite-1 modulates binding of ligands at exosite-2. This discussion establishes that occupancy of either exosite-1 or exosite-2, or both, can influence the conformation of thrombin. It is also clear that the activity of thrombin is dependent on conformation, with the effect of thrombomodulin on thrombin driving the activation of protein C being a prime example.[249] The studies described next show that a variety of ligands can bind exosite-1 with an effect on catalytic activity.

Hortin and coworkers may have been the first to observe that peptides derived from proteins known to bind to exosite-1 also have an effect on the catalytic activity of thrombin.[199] Peptides derived from heparin cofactor II (51–75) or hirudin (54–66) were studied for their effect on thrombin activity. Both peptides inhibited fibrinogen clotting activity but enhanced the hydrolysis of two tripeptide nitroanilides: Tos-Gly-Pro-Arg-pNA and S-2366 (pyroGlu-Pro-Arg-pN) (Table 3.9). It was noted that the enhancement was greater at lower substrate concentrations, suggesting that the enhancement is due to a change in K_M rather than k_{cat}. Hortin and Benutto[250] subsequently published a more extensive study of the effect of several peptides derived from biological ligands of exosite-1 [heparin cofactor II (54–75), fibrinogen (γB, 410–427), thrombomodulin (426–444), and hirudin (54–65), as well as hirudinSO$_4$ (54–75)]. The several peptides were evaluated for their effect on fibrinogen clotting activity and S-2366. I could not find a K_m, for S-2366 for human α-thrombin, but there is a value of 39 μM for bovine thrombin.[251] Hortin and Benutto used a concentration of 1 mM S-2366 (in 10 mM HEPES-140 mM NaCl, pH 7.4 at 37°C), which is likely considerably above K_M for the human enzyme. Shortly after the observation of Hortin and coworkers[199,200] showing that a peptide derived from heparin cofactor II stimulated the hydrolysis of peptide nitroanilide substrates by thrombin, Dennis and coworkers[205] showed that the C-terminal cyanogen bromide fragment from hirudin (53–65) resulted in an increase in the specificity constant for the hydrolysis of D-Phe-Pip-Arg-pNA, primarily from a decrease in K_M. These investigators subsequently published a more extensive study on the effect of these peptides on the catalytic activity of human α-thrombin.[252] Hirudin (54–65), hirudin SO$_4$ (54–65), heparin cofactor II (54–75), and fibrinogen γB chain (410–427) inhibited the activation of protein C in the absence of calcium ions and thrombomodulin. The observed inhibition with the hirudin peptides and heparin cofactor II peptide was 40%–60% and did not increase at higher peptide concentrations. Inhibition by the fibrinogen peptide was different than that observed with the other peptides, with an initial simulation at lower (10 μM) concentration and progressive inhibition at higher peptide concentrations. There was a different pattern of inhibition in the presence of calcium ions and thrombomodulin with a greater extent of inhibition observed; in particular, the effective concentration of hirudin (54–65) was 10-fold less. The response in the hydrolysis of S-2366 was similar for the four peptides, with a decrease in K_M from 0.25 mM to 0.12 mM; there was a difference in the concentration of peptide used, with low concentrations for hirudin peptides and higher for the heparin cofactor II peptide and fibrinogen peptide. Thrombin had a similar response in the hydrolysis of other peptide nitroanilide substrates; the K_m for Chromozym TH (Tos-gly-pro-arg-pNA) decreased from 0.025 mM to 0.015 mM, while that for S-2251 (D-Val-Leu-Lys-pNA) decreased from 2 mM to approximately 1.5 mM. Since the active-site heparin cofactor II is a leucine residue, a peptide nitroanilide substrate (succinyl-ala-ala-pro-leu-pNA) was used to see if occupancy of exosite-1 would cause thrombin to cleave this substrate; the substrate was not hydrolyzed by thrombin in the presence or absence of several synthetic peptides. Thus, occupancy by the test peptides did not change the specificity of thrombin, such as observed with the reaction with intact heparin cofactor II.

TABLE 3.9
Effect of Peptides from Heparin Cofactor II and Hirudin on Thrombin Activity[a]

Peptide	Inhibition of Fibrinogen Clotting Activity[a]	S-2366 Hydrolysis[b]
HirudinSO$_4$ (54–65) 2 μM	0.17	113
Hirudin (54–65) 2 μM	1.3	112
Fibrinogen γ-B-chain (410–427) 90 μM	130	100
HCII (49–75) 20 μM	28	112

[a] Adapted from Hortin, G.L. and Benutto, B.M., Inhibition of thrombin clotting activity by synthetic peptide segments of its inhibitors and subsratrates, *Biochem. Biophys. Res. Commun.* 169, 437–443, 1990.

[a] IC$_{50}$ (μM)

[b] % of control in the absence of peptide

Fibrinogen contains several binding sites for thrombin, which are also found in fibrin.[253] As noted earlier, thrombin can bind to fibrin and does retain catalytic activity. One of the binding sites in fibrin is located in the E domain. Fibrinogen fragment E is obtained by the digestion of fibrinogen by plasmin and contains regions of the α, β, and γ chains linked by disulfide bonds and has a molecular weight of approximately 50 kDa.[254] Fragment E obtained from human fibrinogen was shown to inhibit fibrinogen clotting, serotonin release from platelets, binding to platelets (GpIb), and binding to thrombomodulin.[255] The binding to GpIb was assessed by crosslinking thrombin bound to platelets. The effect of fragment E on GpIb binding implies either a direct effect on exosite-2 or an effect of binding at exosite-1 on exosite-2, a so-called allosteric effect. These investigators evaluated the effect of fragment E on the hydrolysis of several synthetic substrates by thrombin. Fragment E increased the rate of hydrolysis of S-2238 (H-D-Phe-Pip-Arg-pNA), Chromozym TH (Tos-Gly-Pro-Arg-pNA), S-2288 (H-D-Ile-Pro-Arg-pNA), and CBS-6525 (N^ε-Cbz-D-Lys-Pro-Arg-pNA), all at 0.4 nM concentration, considerably below K$_M$. Inhibition of hydrolysis was observed for S-2765 (N^α-Cbz-D-Arg-Gly-Arg-pNA) and CBS-4625 (Ethoxy-N^ε-Lys-Gly-Arg-pNA) at 80 nM. This is considerably below the K$_m$ for S-2765 as of this writing (March 2020). A more rigorous analysis of the effect on the hydrolysis of peptide nitroanilide substrates with S-2765 (inhibited) and CBS-6525 (enhanced) is shown in Table 3.10. The effect of inhibition or enhancement is observed in K$_M$ and not in k$_{cat}$.

Recombinant rabbit thrombomodulin and a fragment of human thrombomodulin (TM456, residues 345–465 containing 4–6 EGF domains)[256] were evaluated for effect on the hydrolysis of Tos-gly-pro-arg-7-amido-4-methylcoumarin (a fluorogenic substrate).[257] Recombinant rabbit thrombomodulin decreases the K$_M$ for the fluorogenic substrate from 15.6 μM to 4.8 μM, with little effect on k$_{cat}$ providing a value for catalytic efficiency (k$_{cat}$/K$_m$) 11.4 μM^{-1}s^{-1} compared to 3.6 μM^{-1}s^{-1} in the

TABLE 3.10
The Effect of Fibrinogen Fragment E on the Hydrolysis of Two Peptide Nitroanilide Substrates by Thrombin[a]

Peptide	CBS-6525			S-2765		
	k_{cat} (s^{-1})	K_m (µM)	k_{cat}/K_M(s^{-1}µM^{-1})	k_{cat} (s^{-1})	K_m (µM)	k_{cat}/K_M(s^{-1}µM^{-1})
None	46.7	9.2	5.1	0.6	32	0.018
Fragment E	32	3.2	8.7	0.6	52	0.011
Hirudin (54–65)	35	2.6	13.4	0.9	67	0.013

[a] Data take from Bouton, M.-C., Jandrot-Perrus, M., Bezeaud, A., and Guillin, M.-C., Late fibrin(ogen) fragment E modulates human α-thrombin specificity, *Eur. J. Biochem.* 215, 143–149, 1993.

absence of thrombomodulin. Hirugen caused a decrease in both k_{cat} and K_M, resulting in a 50% decrease (1.8 µM^{-1}s^{-1}) in the specificity constant.

Thrombin binding to fibrin has been known since the early work on thrombin (fibrin ferment) in the 1800s. Thrombin is physically entrapped by fibrin and is also bound to fibrin-retaining activity.[253,258–260] The interaction of thrombin with fibrin is discussed in greater detail elsewhere in this work (Chapter 6). Digestion of a fibrin clot during therapeutic fibrinolysis does release active thrombin, using it bound to fibrin degradation products (FDPs).[261] Naki and Shafer[262] studied the interaction of thrombin with fibrin I. Fibrin I is the monomer product formed from fibrinogen by the release of fibrinopeptide A.[263] Naki and Shafer[262] determined the kinetic parameters for the hydrolysis of Tos-gly-pro-arg-7-amido-4-methylcoumarin. In the absence of fibrin I, the k_{cat} is 180 s^{-1} and K_m is 7.3 µM for the hydrolysis of the fluorogenic substrate; in the presence of fibrin I, the k_{cat} is 21 s^{-1} and the K_M is less than 0.23 µM. It can be seen that both k_{cat} and K_M are decreased. It is of interest that thrombin bound to fibrin I reacts with antithrombin more rapidly than free α-thrombin. I find it somewhat difficult to rationalize decreased amidase activity with increased reaction with antithrombin. The decreased K_M is consistent with other observations on the effect of ligand binding in exosite-1 on the hydrolysis of tripeptide nitroanilide substrates. I could not find any support for a difference between the nitroanilide substrate and fluorogenic substrates.[264] Most fluorogenic substrates use 7-amino-4-carbamoylmethyl coumarin (7-amino-4-methylcoumarinamide) as the signal.

Jabaiah and coworkers[265] used cellular libraries for peptide substrates (CLiPS)[266] to evaluate amino acid sequence on exosite modulation of thrombin activity, A fluorophore-labeled peptide is expressed on the surface of *Escherichia coli*. The cleavage of the peptide with concomitant release of the fluorophores is measured by fluorescence-activated cell sorting analysis (FACS); cleavage of the peptide yields a nonfluorescent cell. This technique was able to identify a consensus cleavage site for caspase-3 and enterokinase.[266] Jabaiah and coworkers[265] extended the CLiP technology to evaluate the effect of changes in exosite structure on the cleavage of a fluorophore-labeled peptide substrate (exosite cellular libraries of peptide substrates [eCLiPS]). The eCLiPs technique permits the screening of peptide libraries for exosite function. The peptide substrate is based on PAR-1, and the exosite sequence is based on the relevant domain in the PAR-1 receptor, a 25-mer sequence containing a hirudin-like sequence.[267] Two fluorescence resonance energy transfer (FRET)[268] peptides which were used to measure the influence of binding to exosite-1 on thrombin activity.[269] A 25-mer peptide contained the hirudin-like sequence from the PAR-1 receptor.[267] Cleavage of the 25-mer peptide was inhibited by bothroja-racin and glycyrrhizin; the cleavage was also inhibited by NaCl concentrations as low as 100 mM. Lars Hellmann's laboratory in Uppsala studied the effect of exosite occupation on the cleavage of peptides, representing the cleavage site in various sub-strates, including fibrinogen, factor V, factor VIII, and protein C.[269] A substrate was constructed that consisted of the thrombin cleavage site in a specific substrate (mini-mal cleavage site) and an upstream sequence containing a potential exosite cleavage was assessed by SDS-PAGE. The extent to which an exosite enhanced cleavage of a minimal sequence was variable but was as much as 50-fold.

APTAMER BINDING TO THROMBIN

The interaction of thrombin with aptamers has been of interest for some time. Aptamers are single-stranded RNA or DNA oligonucleotides that fold into a defined shape and have been shown to have high specificity in binding to proteins.[270–272] Aptamers were discovered by Jack Szostak's laboratory 1990. Professor Szostak is a distinguished investigator in genetics. He is a Nobel Laureate associated with Harvard Medical School and Massachusetts General Hospital. The discovery of aptamers as specific binding agents for proteins was truly a clever piece of work. Ellington and Szostak[273] used affinity selection of RNA molecules from a pool of RNA obtained by transcription of a pool of DNA molecules of random sequence. The transcribed RNA populations were selected by affinity chromatography on a dye column. It was possible to obtain RNA molecules specific for binding some organic dyes such as Cibacron Blue and reactive green 19. The appellation aptamer was derived from the Latin *aptos* which can mean fitted, suitable, connected, or fastened. Ellington and Szostak[274] subsequently prepared DNA aptamers from a ran-dom-sequence DNA pool.[275] DNA aptamers have seen considerable use in research and some therapeutic applications. The interaction of aptamers with thrombin was first described by Bock and coworkers at Gilead.[276] These investigators identified a 15-mer, GGTTGGTGGTTGTGG, which inhibited both fibrinogen clotting (169 seconds versus 25 seconds for no DNA) and the clotting of plasma at nanomolar concentrations. A scrambled 15-mer composition with same nucleotides but with a different sequence (GGTGGTGGTTGTGGT) had a clotting time similar to the control (26 seconds). The active 15-mer composition has become known as the thrombin binding aptamer (later known as 15-mer, HD-1, TBA). Subsequent work by Mocaya and coworkers[277] established that the DNA aptamer identified as binding thrombin—the thrombin binding aptamer—formed a quadraplex conformation in solution. These studies established that a specific DNA aptamer, thrombin binding aptamer, adopted a specific conformation in solution that inhibited thrombin. Wu and coworkers[278] then presented evidence suggesting that thrombin binding aptamer bound to exosite-1 in thrombin. These investigators showed the thrombin binding aptamer inhibited a K154A mutant but not an R70E mutant. Arg70 is located in exosite-1, while Lys154 is located outside exosite-1. The aptamer also inhibited platelet activation and thrombomodulin-dependent protein C activation (activation in the presence of calcium ions); the thrombin binding aptamer did not inhibit thrombomodulin-independent protein C activation (activation in the absence of calcium ions with EDTA). The scrambled 15-mer composition had no effect on platelet activation and thrombomodulin-dependent protein C activation. Subsequently Padmanabhan and coworkers[279] determined the crystal structure of the complex of thrombin binding aptamer and human α-thrombin. It was suggested that the aptamer was observed to bind to exosite-1 but also might interact with exosite-2. Later results from another laboratory[280] with a more refined analysis showed binding only to exosite-1; a finding supported by subsequent mass spectrometry data.[281]

The crystal structure of PPACK-thrombin with a variant of thrombin binding aptamer has been determined.[282] An abasic space[283,284] was inserted into the thrombin binding aptamer at T_3 and T_{12} (GGXTGGTGTGGXTGG) to provide a variant that bound to bovine thrombin with reduced affinity.[285] The association constant for thrombin binding aptamer is 7.5×10^7 M^{-1}, while the association constant for the T_3 variant is 1.8×10^7 M^{-1} and the T_{12} variant is 2.4×10 M^{-1}. The crystallography data support the interaction of the T_3T_4 loop with Arg70Glu72Arg73Asn74Ile79 (designated by the authors as the A region) and the interaction of the $T_{12}T_{13}$ with Arg70Tyr71 (designated by the authors as the B region). These authors advance the concept of adaptive binding to describe transition of the aptamer to a more ordered conformation.

Additional support for the binding of HD-1 to exosite-1 was obtained by observing the displacement of a fluorescent derivative of HD-1 (modified at the 5'-end with fluorescein 5'-isothiocynate).[285] There is an increase in fluorescence on the formation of a complex between the fluorophore-labeled HD-1 and thrombin, which decreases on dissociation of the labeled HD-1. The fluorophore-labeled HD-1 complexed with thrombin was displaced by antithrombin, heparin cofactor II, or a mutant (M358R) of α-antitrypsin. M358R α-antitrypsin is a mutant form of α-antitrypsin, where the mutation results in a potent inhibitor of thrombin.[286–288] A different aptamer, HD-22, was used to assess the integrity of exosite-2. These studies used the loss of fluorescence in a fluorophore-labeled thrombin. The fluorophore-labeled thrombin was prepared from prothrombin labeled with fluorescein-5-isothiocyanate and then activated to thrombin with prothrombinase. This process is necessary, as exosite-1 and exosite-2 are sensitive to chemical modification with pyridoxal phosphate[128] or fluorescein-5'-isothiocyanate.[289] Both of these reagents modify lysine residues in both exosite-1 and exosite-2. The fluorophore-labeled thrombin had catalytic activity comparable to unlabeled thrombin such that the coupling of the fluorophore did not have an adverse effect on the protein. HD-22 was added to complexes of the fluorophore-labeled thrombin with antithrombin, heparin cofactor II, or (M358)α$_1$-antitrypsin, resulting in a small change in fluorescence and suggesting that complex formation with the serpins had a minimal effect on exosite-2. Later work from a different laboratory[290] presented data supporting a significant change in the conformation of exosite-2 on complex formation with serpins. The Cambridge group also suggested that the formation of a serpin-thrombin complex I could be considered a reversal from active enzyme to zymogen, with disruption of the salt bridge between the amino-terminal isoleucine and the aspartic acid in the active site, a process they described as "rezymogenization."

Monovalent and divalent cations can influence aptamer performance by affecting aptamer conformation.[291–299] HD-1 aptamer requires a monovalent cation, K^{1+} or Na^{1+} for activity, with K^{1+} being optimal. There are a number of studies on the effect of various cations on the structure and activity of G-quadruplex aptamers. I have chosen one study[296] which examined the effect of monovalent and divalent cations on the structure and anticoagulant activity of HD-1 and several other G-quadruplex aptamers. A combination of CD, x-ray, and NMR were used to evaluate the effect of cations on aptamer structure, and a turbidimetric assay was used to measure the action of thrombin on fibrinogen. The K_i for HD-1 in the presence of potassium ions was 9.1 nM in the presence of 50 mM K^{+1} and 50 nM in the absence of K^{1+}. Other ions tested, NH_4^{1+}, Ba^{2+}, Sr^{2+} and Mn^{2+}, gave increased K_i values at all concentrations tested (5–50 mM). Similar results were obtained with two other aptamers (31-TBA, NU172[298]). The structural analyses are consistent with a disruption of the two TT loops, which are important for binding to exosite-1.[282.]

Aptamers have been advanced for therapeutic use as a thrombin inhibitor[272] or a tool for thrombin analysis.[300] An aptamer developed for analytical purpose can be described as an aptasensor.[301] As an aside, some years ago, when I was still at the University of North Carolina at Chapel Hill, I got a call from a fellow who asked me whether I knew of the inhibition of thrombin by aptamers. I knew a bit about aptamers but did not know of any work with thrombin. I suspect, but don't know, that hemorrhage was observed in a preclinical trial of an aptamer developed as a therapeutic. Tan and coworkers[302] showed that DNA aptamers which inhibited fibrinogen clotting activity enhanced the hydrolysis of Sar-Pro-Arg-pNA (*N*-methylGly-Pro-Arg-pNA) showing a 15-fold increase in initial rate. There was an decrease in K_M (0.040 mM vs. 0.112 mM) and an increase in k_{cat} (6.23 $nmol^{-1}min^{-1}$ vs 1.26 $nmol^{-1}min^{-1}$). Here are observations which were similar to those observed for fibrin I in that, in addition to the observed decrease in K_M, there was a change in k_{cat}, in this instance, an increase as opposed to the decrease observed with fibrin I.

There has been continued development of aptamers as inhibitors of thrombin.[303,304] The potential of aptamers as inhibitors of thrombin has been frustrated by instability (nuclease digestion) and rapid renal clearance, but progress has been made in addressing these problems.[305,306] There has been considerable interest in the use of aptamers for the assay of thrombin.[307–308] I am pessimistic about the value of aptamers as therapeutics for reasons discussed elsewhere, as well as the emergence of novel oral anticoagulants (NOACs). This is discussed in the section on therapeutic inhibition of thrombin. All of that said, aptamers can be of continuing value in the study of the relationship between structure and function and thrombin.[300,311,312]

EXOSITE-2

Exosite-2 in thrombin has not had the same degree of attention as exosite-1. As noted earlier, as far as I can tell, the term exosite-2 was introduced in 1995.[191] Also, to the best of my knowledge, the term anion-binding site has only been used to describe exosite-1. There was increasing use of the term exosite-2 in the decade following, in 1995, with use in review articles without citation.[313,314] One of these articles[313] concerned ximelagatran, an NOAC directed at thrombin. While this drug was withdrawn in 2006, this is an excellent article. The other cited article[314] is an excellent discussion of the development of NOACs.

THROMBIN AND HEPARIN

Heparin is known to bind to exosite-2, as discussed next; exosite-2 is frequently described as the heparin-binding site. The direct interaction of heparin with thrombin is frequently overlooked, reflecting the importance of heparin in the reaction of coagulation proteases with antithrombin. Heparin and other sulfated mucopolysaccharides do bind to thrombin, and it is likely that such interactions are responsible for the binding of thrombin to the extracellular matrix.[315,316] The following section is concerned with studies on the physical interaction of heparin with thrombin.

There were early studies on the interaction of heparin with bovine thrombin.[317-319] Pálos[317] showed that heparin did protect bovine thrombin from air (oxygen) oxidation. Previous work by this investigator had demonstrated that an equal weight of heparin protected bovine thrombin from inactivation at 100°C for 10 minutes. Subsequent work by Pálos, Machovich, and coworkers[320] showed that bovine thrombin was inactivated by 0.1 M 2-mercaptoethanol in 2.6 M urea at 22°C; the inactivated enzyme could be partially reactivated (0%–60%). Heparin accelerated reactivation, while 10 mM iodoacetamide prevented reactivation. Later work from a different laboratory[321] showed a time-dependent decrease in intrinsic fluorescence of bovine thrombin in the presence of heparin or dextran sulfate; the loss of thrombin activity (Boc-Val-Pro-Arg-MCA) was also observed under these conditions. Earlier work[322] had shown that dextran sulfate inhibited the fibrinogen clotting activity of purified bovine thrombin (1900 U/mg) in addition to fibrinopeptide A release in the absence of antithrombin; there was a very small effect of heparin. There was some dependence on the molecular weight of the dextran sulfate, with a greater effect with higher molecular weight (200 kDa) material than lower molecular weight material (3.5 kDa). Anomalous behavior was observed with 7.5 kDa material with little inactivation. Dextran sulfate was less effective than heparin in enhancing the inactivation of bovine thrombin by human antithrombin; here again the 7.5 kDa material was an outlier, being essentially ineffective. Dextran sulfate also inhibited the hydrolysis of H-D-Phe-Pip-Arg-pNA, with the greatest inhibition observed with the 200 kDa material. There was slight stimulation with heparin and significant enhancement with 7.5 kDa material. There was no comment by the authors on what appeared to anomalous behavior of the 7.5 kDa material. Later work showed that high-molecular-weight (500 kDa) dextran sulfate blocked the heparin-accelerated antithrombin reaction with thrombin.[323] A complex of dextran sulfate and thrombin was demonstrated by G-200 gel filtration and native gel electrophoresis. These studies support the direct interaction of heparin and thrombin. There are additional studies on the interaction of heparin with thrombin. The binding of heparin to thrombin was investigated with heparin labeled with a fluorophore[323] (fluorescein isothiocyanate[324]; excitation at 515 nm; emission at 490 nm). Fluorescence of the labeled heparin was quenched on binding, permitting the determination of an association constant (1.7×10^8 M^{-1}) with a stoichiometry of

0.52 thrombin to FTC-heparin, suggesting a heterotrimer. These measurements were consistent with previous results from Bob Rosenberg's laboratory,[325] who reported a dissociation constant of 8×10^{-7} M and 2:1 stoichiometry with low-molecular-weight heparin. Oshima and coworkers[326] observed that the binding of heparin to thrombin is sensitive to salt concentration with an association constant of $9.9 \times 1-0^5$ M^{-1} at 0.15 M NaCl, with an equimolar complex observed under these solvent conditions. Again, these observations were consistent with previous observations from the Rosenberg laboratory.[325] Later work from another laboratory[326] showed the dependence of the binding of heparin to thrombin in terms of ionic strength, as well as a second, weaker binding site for heparin on thrombin.

Machovich and coworkers[327] showed that heparin could protect thrombin from heat inactivation (60°). These investigators also showed that unfractionated heparin formed a complex with thrombin stable to gel filtration on G-200 Sephadex (50 mM Tris-HCl, pH 7.3). Griffith and coworkers[328] showed that heparin could be fractionated on a thrombin-agarose matrix, and it was observed that high-molecular-weight heparin was selectively removed by the thrombin-agarose column. The binding of thrombin to heparin-affinity matrices was studied by Björklund and Hearn.[329] A stoichiometry of 1.8 mole thrombin/mole heparin was obtained for a soft-gel (Sepharose CL-6B) heparin matrix, while a stoichiometry of 2.4 mole thrombin/mole heparin was obtained for a silica matrix (Fractosil) 1000. I don't think Fractosil, which is a porous silica matrix, is still available; Sepharose CL-6B is a crosslinked agarose matrix that is still available from Cytiva (formerly GE Healthcare, formerly Pharmacia). Binding studies suggested high affinity binding and low affinity binding of thrombin to heparin. The authors suggest that the high affinity binding ($K_D \approx 10^{-8}$–10^9 M) is biospecific, while the low affinity binding ($K_D \approx 10^{-6}$–10^{-7}M) is electrostatic.

These studies show that heparin does bind thrombin with high affinity and that the binding of heparin does affect thrombin conformation. While exosite-2 had not been identified as a specific domain at the time, Church and coworkers[128] used chemical modification of thrombin with PLP to identify lysine residues (K174, K252) important in binding heparin, a function later ascribed to exosite-2. These investigators used PLP to modify lysine residues in human α-thrombin. Four lysine residues were modified under these reaction conditions in the absence of heparin; two residues were protected in the presence of heparin. The lysine residues not protected by heparin (K21 and K65) were subsequently shown to be located in exosite-1. Sheehan and Sadler[330] used site-specific mutagenesis to further explore the role of certain basic residues (K89E, R245E, K248E, K252E) in binding heparin, and these residues are located in exosite-2. Heparin binding was assessed by retention on a heparin agarose column. The data cited earlier provides strong support for the binding of heparin to exosite-1 in thrombin. The heparin binding site is not available in prothrombin but is exposed on activation to thrombin.[331]

EFFECT OF HEPARIN AND RELATED COMPOUNDS ON THE ACTIVITY OF THROMBIN

Griffith and coworkers[328] observed that heparin increased the activity of thrombin in the hydrolysis of TosGlyProArgpNA. In subsequent work, Griffith and coworkers[332] presented a more detail study of the effect of heparin on the hydrolysis of two peptide nitroanilide substrates. The rate of hydrolysis of TosGlyProArgpNA and BzPheValArgpNA was enhanced. The results with BzPheValArgpNA were somewhat complicated, with what appeared to be substrate inhibition, which was enhanced by the presence of heparin. The effect of heparin on the hydrolysis of TosGlyProArgpNA was examined in more detail: heparin had a significant effect on K_m (11 μM in the absence of heparin; 3.8 μM in the presence of heparin), with a lesser effect on V_{max} (20.4 nmol/min in the absence of heparin; 16.4 nmol/min in the presence of heparin). The effect on TosGlyProArgpNA was used to assess the stoichiometry of binding (0.96 mole heparin/mole thrombin) and affinity ($K_D = 1.7 \times 10^{-9}$ M). As discussed later, this value was obtained in 0.1 M triethanolamine-0.1% PEG6000; no NaCl. Heparin was also observed to accelerate the rate of inactivation of human α-thrombin by TLCK but did not affect the rate of inactivation with PMSF. As discussed elsewhere in this work, the results with TLCK could be interpreted as resulting from increased reactivity of the active-site histidine residue. Griffith and coworkers[333] observed an interaction between BzPheValArgpNA and heparin and thrombin, which formed an aggregate that could be separated by centrifugation; this interaction was responsible for the observed increased substrate inhibition. Heparin did enhance the rate hydrolysis of BzPheValArgpNA at low (10^{-5} M) substrate concentrations. The results did depend on the order of addition of components of the assay system. At a substrate concentration of 1.1×10^{-4}M, if heparin was added to the substrate prior to thrombin addition, the initial rate of hydrolysis was decreased by approximately 10%; if the heparin was premixed with thrombin, the initial rate of hydrolysis was reduced by approximately 75%. Goodwin and coworkers[334] did not observe the enhancement of the catalytic activity of human α-thrombin with heparin (data not shown). An examination of the solvent conditions used by Griffith and coworkers[333] (0.1 M triethanolamine-0.1% PEG 6000) with those used by Goodwin and coworkers[334] (0.1 M triethanolamine-0.1 M NaCl-0.1% PEG 6000–0.02 % sodium azide) could possible explain the difference in the results. It is not unreasonable that the affinity of heparin for thrombin was reduced by the presence of 0.1 M NaCl, as Oshima and coworkers[323] showed the affinity of bovine thrombin was reduced from 1.7×10^8 M^{-1} to 1×10^6 M^{-1} in the presence of 0.15 M NaCl. A binary complex was observed in the presence of 0.15 M NaCl, while a ternary complex was observed in the absence of NaCl. Nordenman and Björk[331] also reported a lack of effect of heparin on the hydrolysis of S-2238 (H-D-PhePipArgpNA). While the specific composition of their solvent is difficult to determine, it is likely that it did contain substantial NaCl. Regardless of this specific issue, Goodwin and coworkers did report that heparin markedly influenced the activity of des-ETW thrombin, which is discussed in more detail later by Le Bonniec and coworkers.[334,335] The presence of heparin lowered the K_M for four peptide nitroanilide substrates. The rate of inactivation by PPACK was increased in the presence of heparin. The K_i for p-aminobenzamidine was modestly decreased from 9.7 mM to 7.5 mM in the presence of heparin, while a decrease to 3.5 mM was observed in the presence of pentosan polysulfate. The enhancement of activity is reduced by the presence of adenosine diphosphate (ADP) or ATP, leading the authors to suggest that heparin's effect on des-ETW thrombin activity is due to interaction at exosite-1 with des-ETW thrombin. Berliner and coworkers[179] showed that ATP decreased the NaCl concentration required for the elution of thrombin from a fibrin-agarose column. However, Fredenburgh and coworkers[336] showed that fibrin interacts with thrombin and both exosite-1 and exosite-2, so the ATP effect noted by Goodwin and coworkers[334] could represent an interaction at either or neither exosite.

These studies suggest that heparin and other sulfated polysaccharides/ glycosaminoglycans/proteoglycans can bind specifically to exosite-2 in thrombin. The data also suggest that the affinity of binding is high, sensitive to ionic strength, involves electrostatic forces, and the extent of sulfation (charge density) is an important attribute. There may be other forces in the binding of heparin and other biological materials such as chondroitin sulfate. It is not likely that the direct interaction of heparin with thrombin at exosite-2 is of importance in blood, other than the stimulation of a reaction with antithrombin. These interactions may well be of greater importance in the extracellular/interstitial space. There are other sulfated glycosaminoglycans, such as chondroitin sulfate, which interact with the exosite-2 domain in thrombin. Chondroitin sulfate is a component of the cell membrane and extracellular matrix in the form of a proteoglycan.[337] Thrombomodulin is a membrane-associated proteoglycan that contains chondroitin sulfate.[338] Thrombomodulin binds to thrombin, decreasing fibrinogen clotting activity while activating protein C and thrombin-activated fibrinolysis inhibitor (TAFI); the complex of thrombomodulin and thrombin forms "anticoagulant thrombin." The chondroitin sulfate moiety increases the affinity of thrombomodulin for thrombin.[339,340] Chondroitin sulfate can increase the rate of reaction with antithrombin.[341]

A sequence of three sulfated tyrosine residues in glycoprotein Ibα has been shown to bind to exosite-2 in thrombin.[342-344] The sulfated tyrosine sequence led one group[345] to synthesize a sulfate mimic, SbO4L, which could bind to exosite-2 and inhibit platelet aggregation.

Sucrose octasulfate, as an aluminum salt, is a drug (sucralfate, Carafate) that was developed as an anti–peptic ulcer drug[345,346] and is used today for diabetic ulcers.[347,348] An early study of sucralfate suggested a lack of anticoagulant activity. A later study[349] showed that sucrose octasulfate accelerated the inactivation of thrombin by heparin cofactor II, but lacked an effect on antithrombin. Sucrose octasulfate also accelerated the rate of inactivation of meizothrombin by heparin cofactor II; there was also a lack of effect of aptamer HD-22 on the effect of sucrose octasulfate. These results suggest that binding of sucrose octasulfate to exosite-2 was not important in the acceleration of the reaction with heparin cofactor II. These investigators also showed that there were two sites in thrombin which bound sucrose octasulfate with dissociation constants of 10 μM and 400 μM. In view of the template mechanism for the reaction of thrombin with antithrombin in the presence of heparin,[250] it is not surprising that sucrose octasulfate had no effect on the reaction of antithrombin with thrombin. A more recent study[351] obtained a dissociation constant of 1.4 μM for the binding of sucrose octasulfate to thrombin. Crystallography studies showed that sucrose octasulfate binds to exosite-2, with two molecules bound to a thrombin dimer; ultracentrifugation showed that human and bovine thrombin were monomers in the presence of sucrose octasulfate. The binding of sucrose octasulfate to thrombin is associated with a modest decrease in hydrolysis of CH_3SO_2 D-Leu-Gly-Arg-pNA

(CBS31.39). Sucrose octasulfate has been shown to interact with fibroblast growth factors. Early work showed that sucrose octasulfate binds acidic fibroblast growth factor 1 (FGF-1).[352] The binding results in stabilization of FGF-1. The binding of sucrose octasulfate to FGF-1 does not result in dimerization, but does form a complex which enhances cell proliferation.[353] Heparin and heparan sulfate are well known for their interaction with FGFs.[354] Other studies[355] showed that sucrose octasulfate interacts with FGF-2 (basic fibroblast growth factor), inducing dimerization of FGF receptors. As with FGF-1, sucrose octasulfate facilitates the action of FGF-2 on cell function.[356] Crystallographic analysis of the complex between sucrose octasulfate and FGF-2 showed binding by three lysine residues. Thus, as expected, there are similarities in the binding of sucrose octasulfate to thrombin and to other proteins.

Polyphosphate, a polymer of inorganic phosphate, can be a product of eukaryotic cells,[357] including blood platelets[358] and bacteria.[359] The polyphosphate produced by bacteria is heterogeneous, ranging from 3 phosphates to more than 1,000 phosphate units.[360] The high-molecular-weight material has been suggested to be involved in contact activation.[360,361] The polyphosphate from platelets is smaller, 60–100 phosphates, binds to exosite-2 in thrombin,[342–362] and is a cofactor for the activation of factor XI by thrombin.[363] The effect of polyphosphates in blood coagulation has been effectively reviewed by Morrissey and coworkers.[364,365]

GLYCOPROTEIN 1Bα

The binding of a sequence of three sulfated tyrosine residues found in glycoprotein 1bα (GP1bα) to exosite-2 in thrombin was discussed earlier.[342–354] The interaction of thrombin with blood platelets will be discussed elsewhere in this work (Chapter 7). Suffice it for the present to note that GP1bα was the first receptor for thrombin identified in blood platelets.[366,367] While the work on the sulfated tyrosine residues strongly suggested GP1bα bound to exosite-2 on blood platelets, additional studies provided additional support. De Cristofaro and collegues[368] mutated a number of basic amino acids in the exosite-2 domain to alanine and measured the binding of the mutant thrombin to purified GP1bα bound to a high-binding-capacity polystyrene Immulon microplate.[369] The bound thrombin was measured by the hydrolysis of a tripeptide nitroanilide substrate (H-D-Phe-Pip-Arg-pNA). These investigators noted that the several mutations of the exosite-2 domain and the one exosite-2 mutant (R63A) had little effect on the hydrolysis of this substrate. The K_D for wild type was 0.102 µM. The various mutants showed increased K_D values: R89A, 2.3 µM; R245A, 2.9 µM; and K248A, 2.11 µM. These investigators also compared the rate of hydrolysis of a soluble PAR-1 peptide (PAR-1P) to the hydrolysis of PAR-1 on intact blood platelets. This study evaluated the importance of GP1bα in the *in situ* action of thrombin on PAR-1.[370] As an example, the R89A mutant was similar to wild type in the hydrolysis of the PAR-1P peptide (k_{cat}/K_m 8.2×10^7 M^{-1}s^{-1} compared to 7.8×10^7 M^{-1}s^{-1} for wild type) but was less effective in the cleavage of PAR-1 on the intact platelet (k_{cat}/K_M 0.2×10^7 M^{-1}s^{-1} compared to 1.15×10^7 M^{-1}s^{-1} for wild type). The R89A mutant was also less effective in platelet aggregation. The R63A mutant showed essentially identical binding to immobilized GP1bα (K_D 0.12 µM compared 0.102 for wild type) but decreased activity in the hydrolysis of the PAR-1P peptide, PAR-1 on intact platelets, and platelet aggregation. Earlier work from this laboratory[369] had shown that ligands binding to exosite-2, heparin (discussed earlier), and prothrombin fragment 2 (see later) inhibited the binding of wild-type thrombin to immobilized GPIbα. Subsequent work from this group and others further established that GPIbα bound to exosite-2 in thrombin, and such binding was critical for the efficient cleavage of PAR-1 and platelet activation.[370–375] Li and coworkers[371] showed that either glycocalicin or GPIbα bound to thrombin and decreased the release of fibrinopeptide from fibrinogen by approximately 50% at the highest concentration of ligand tested and the hydrolysis of Bz-D-Phe-Pro-Arg-pNA by approximately 10%; low-molecular-weight heparin gave an approximate 20% decrease in amidase activity, while hirugen increased amidase activity. They also confirmed earlier observations that there was a decrease in the binding of exosite-2 mutant protein (R89E, K248E) to GPIbα. The interaction of thrombin with glycocalicin was observed to be sensitive to NaCl concentration; a K_D of 0.187 μM was observed at 100 mM NaCl, while a K_D of 1.04 μM was observed 150 mM NaCl. Two separate crystallographic studies[372,373] suggested that one GPIbα molecule bound two thrombin molecules—one at exosite-2 and the other at exosite-1. Later studies from two different laboratories using hydrogen-deuterium exchange,[374] NMR,[375,376] analytical ultracentrifugation,[375] and x-ray crytallography[375] strongly supported the formation of a heterodimer between GPIbα and thrombin, with binding of GPIbα exclusively to exosite-2 on thrombin. The binding of thrombin to GPIbα on the platelet surface puts thrombin in a position to cleave PAR-1, resulting in platelet activation.

FIBRIN

The physical interaction of thrombin with fibrinogen/fibrin has been known for some time (Chapter 1) and is discussed in further detail in Chapter 6. Plasma fibrinogen exists in two forms resulting from an elongated γ chain (γ′ chain or γB chain) with the addition of 20 amino acids (408–427).[376] This variant form is the result of differences in mRNA splicing.[377,378] Thrombin binds to this extended sequence in the γ′ chain.[379–381] Mike Mosesson's laboratory in Milwaukee identified the sequence (408–427; VRPEHPAETEYDSLYPEDDL) in the variant γ′ chain, which bound thrombin with high affinity.[378] It was suggested that sulfation of the tyrosine residues enhanced the binding of thrombin. A reverse sequence of the peptide or a change in the position of the sulfated tyrosine residues decreased the affinity of thrombin binding. A study from Farrell's laboratory[380] at the Oregon Health Science University (OSHU) in Portland showed that thrombin bound to the phosphorylated γ′ chain peptide (a change from the sulfated form found *in vivo*). The phosphorylated γ′-chain peptide was labeled with fluorescein (fluorescein succinimyl ester) and found to bind to thrombin with a $K_D = 6.3 \times 10^7$ M. The peptide lacking phosphorylated tyrosine residues did not bind thrombin. The OSHU group also found that the γ′ peptide (410–427) inhibited the intrinsic coagulation pathway,[381] most likely by inhibiting a specific cleavage in factor VIII activation, which requires thrombin exosite-2 function.[382] While not directly relevant to the current observations, Hemker and colleagues have reported that thrombin is the enzyme responsible for the *in situ* activation of factor VIII during the clotting of plasma.[383] A subsequent study from the Milwaukee group in collaboration with the OSHU group[384] extended the understanding of the role the binding of thrombin to the γ′ chain in the formation of and structure of fibrin, including the crosslinking by factor XIIIa. There is another study from the Maurer laboratory at the University of Louisville[385] that used NMR and hydrogen-deuterium exchange to study conformational changes in thrombin on binding to the γ′-chain. These investigators also showed that the phosphorylation of at least one tyrosine residue is required for binding to thrombin.

PROTHROMBIN FRAGMENT 2

There are biological ligands for exosite-2 other than heparin, GPIbα, and fibrin. Prothrombin fragment 2 (earlier known as intermediate 3) binds to thrombin at exosite-2.[386] Early work by Heldebrant and Mann[387] showed that bovine fragment 2 bound to bovine α-thrombin and was observed to enhance esterase (TosArgOMe) activity, with modest inhibition of fibrinogen clotting activity. Bovine fragment 2 had no effect on the activity of trypsin in the hydrolysis TosArgOMe. As noted earlier, Ken Mann and associates referred to fragment 2 as intermediate 3. This was another of the nomenclature battles fought in blood coagulation since the 19th century, which was settled, as others since the middle of the last century, by an august committee of the International Society for Thrombosis and Haemostasis (ISTH) (see the footnote in reference 326). Myrmel and coworkers[388] performed a more rigorous study of the interaction between bovine prothrombin fragment 2 and bovine thrombin. It was observed that bovine fragment 2 enhanced the TosArgOMe esterase activity (increase in k_{cat}) of both human and bovine thrombin, while human prothrombin fragment 2 was without effect. Inhibition (decrease in k_{cat}) was observed with BzArgOEt or BzArgpNA. Bovine prothrombin fragment 2 had no effect on the rate of inactivation of bovine α-thrombin with DFP or PMSF, but there was an increase in the rate of inactivation by TLCK. A dissociation constant of 10^{-10} was estimated from the effect of TosArgOMe hydrolysis. Walker and Esmon[389] observed that prothrombin fragment 2 inhibited the reaction of thrombin and antithrombin; at saturating concentrations of prothrombin fragment 2, the rate of reaction of thrombin with antithrombin was the same as that of an equivalent concentration of meizothrombin. Bock[390] showed that the binding of prothrombin fragment 2 to thrombin labeled at the active site with a fluorophore-labeled peptide chloromethyl ketone demonstrated changes at the enzyme-active site. Liaw and coworkers[391] showed that prothrombin fragment 2 bound to the exosite-2 domain of thrombin. The association of prothrombin fragment 2 with α-thrombin yields a complex with a structure similar to meizothrombin.[356–392]

OTHER SUBSTANCES INTERACTING WITH EXOSITE-2

ECOTIN

Ecotin is a periplasmic protease inhibitor obtained from *E. coli*.[393,394] Ecotin was shown to be a potent inhibitor of trypsin, chymotrypsin, and elastase but had little or no effect on subtilisin, plasmin, thrombin [*bis*-(N-carbobenzoxy-arginine) rhodamine], tryptase, or uronkinase.[393] The purified protein had a monomer molecular weight of 18,000 determined experimentally,[393] or 16,096 determined from sequence.[394] The Goldberg laboratory also determined an isoelectric point of 6.1.[393] Subsequent work from a group at UCSF high on Parnassus in Bagdad-by-Bay (Herb Cain) established the sequence of the monomer unit and identified that the "active site" was methionine 84.[394] The inhibitor forms a complex with the target protease and, unlike antithrombin, is not cleaved. The presence of a methionine residue at the "active site" accounts for the broad specificity of the inhibitor. α_1-Antitrypsin, another broad-spectrum protease inhibitor, also has a methionine at the "active site."[395] A group at UCSF subsequently reported the crystal structure of a mutant ecotin (M84R) and bovine thrombin.[396] The complex was stable to gel filtration and native gel electrophoresis; no complex was observed with native ecotin. These investigators suggested that ecotin mutant (M84R) had two sites of interaction with thrombin. Two sites on ecotin for interaction with target proteases has been previously suggested by another group at UCSF.[397] The region around the R84 bound to the active-site domain and another region of ecotin bound to the C-terminal region of thrombin, most likely the exosite-2 domain.[398] These investigators reported changes in the 60 loop (chymotrypsin numbering)[404] and the 148 loop (chymotrypsin numbering) on binding of the M84R mutant, reflecting increasing energy of binding. The M84R mutant of ecotin had been previously described by Pál and coworkers in 1994.[399] The change of methionine to arginine at position 84 did not cause a significant change in K_D of the inhibitor for trypsin, chymotrypsin, or elastase, suggesting that interactions other than at the S_1 site are important in the interaction of ecotin with target proteases. At the same time a group at Genentech in South San Francisco described the interaction of the native ecotin and the M84R ecotin mutant with factor Xa.[400] A K_i of 54 pM was determined for the inhibition of human factor Xa by native ecotin; a K_D of 11 pM was found for the M84R mutant. Factor Xa activity was determined with methoxycarboxy-D-cyclohexylglycyl-arginine-p-nitroanilide. As noted earlier, thrombin was not inhibited by native ecotin but is inhibited by the M84R mutant (96% inhibition); inhibition was also observed with the M84K mutant (84% inhibition). A similar effect was observed with plasmin and factor XIIa. While ecotin is a potent inhibitor of homologous proteases, as shown earlier, it had not been demonstrated to inhibit thrombin without the mutation of the methionine residue at position 84 to a basic amino acid residue. While data is not available, it may be that although the association reaction is fast, the complex of native ecotin with thrombin lacks stability in binding to the active-site domain. Castro and coworkers[401] were able to show that native ecotin did bind to human thrombin; however, there was no activity in the hydrolysis of peptide nitroanilide substrates. But instead of the 2:2 stoichiometry of native ecotin and other enzymes such as trypsin (one ecotin dimer and two enzyme molecules), complex formation between ecotin and human thrombin involved one ecotin dimer and one thrombin, a 2:1 stoichiometry, as determined by native gel electrophoresis and gel filtration. Complex formation was not observed with the ecotin monomer (monomer below K_D of dimer). The presumed binding of ecotin did influence the activity of human thrombin, enhanced fibrinogen clotting activity, protected thrombin from inactivation by heparin/antithrombin, and was inactivated by bovine pancreatic trypsin inhibitor (BPTI) to a greater extent than native thrombin. Ecotin modestly enhanced inactivation of thrombin by antithrombin. The binding of ecotin to human thrombin was blocked by either prothrombin fragment 2 or heparin. The fluorescence intensity of fluorescein-labeled PPACK-labeled thrombin was enhanced by the binding of ecotin. The results of this study supported the concept that ecotin binding the exosite-2 domain in the C-terminal portion of the thrombin B-chain and that such binding influences exosite-1 and the active-site region.

While most work has focused on either exosite-1 or exosite-2, there are some studies where a ligand binds to both sites or where binding to one site influences the behavior of the other site. Hogg and coworkers[402] showed that heparin binding to exosite-1 in human thrombin enhances binding of fibrin to exosite-1, forming a ternary complex that affects the environment of the enzyme active site. These studies used a derivative of thrombin, 2-anilinonaphthalene-6-sulfonyl (ANS), coupled to a thioacetyl derivative of PPACK, which is coupled to the active site histidine. The ANS moiety serves as a reporter group[403] for conformational changes in thrombin. Fluorescence intensity increases on the addition of either fibrin or heparin; the increase in fluorescence is associated with a small blue shift. The addition of both heparin and fibrin caused a further increase in fluorescence intensity. A model is presented where heparin binding at exosite-2 facilitates fibrin binding at exosite-1, forming a ternary complex. Inhibition of the hydrolysis of a tripeptide nitroanilide substrate (D-Ile-Pro-Arg-pNA) was not observed with heparin or fibrin alone, while the combination of heparin and fibrin showed an approximately 20-fold increase in the K_i for hirudin. There are other studies on the effect of ligand binding to exosites in thrombin that should be mentioned. Chuck Esmon and Pete Lollar showed that both exosite-1 and exosite-2 were involved into the processing of factor VIII to a functionally active form.[404] This study used a mutant of thrombin (RA) with a defective exosite-2, R93A, R97A, R101A, and synthetic N-acetylated dodecapeptide, Ac-Asn-Gly-Asp-Phe-Glu-Glu-Ile-Pro-Glu-Glu-Tyr(SO$_4$)-Leu (hirugen[405,406]), which binds to exosite-1. Early work suggested that β2-glycoprotein I inhibited thrombin by binding at both exosites.[407] Later work from this group[408] showed the previous work suggesting interaction at exosite-1, which was based on inhibition action on fibrinogen, reflected action of β2-glycoprotein I on fibrinogen, not exosite-1. This is not unique; inhibition of an enzyme-catalyzed reaction by an inhibitor-binding substrate has been known for some time.[409] The fact that there is communication between the two exosites complicates the conclusion that both are involved in binding a single ligand.

SODIUM ION BINDING SITE IN THROMBIN

Another functional site physically outside of the active-site region of thrombin is the sodium ion binding site. While there is clear evidence that the occupancy of this site does modulate thrombin function, it is unclear as to whether this would ever have physiological consequences. It is somewhat unlikely that thrombin would ever exist *in vivo* without bound sodium ion. Effects of sodium ion on thrombin were noted much earlier and have, with the exception of the excellent work of Orthner and Kosow,[112] been largely overlooked. I will acknowledge that these earlier studies lack the sophistication of later studies, but data is data, and it is imperative to understand the development of concepts. Credit for the first observation of the effect of sodium ions on the catalytic activity of thrombin belongs to Ehrenpreis and Scheraga for work published in 1957.[410] These investigators observed that 0.024 M NaBr inhibited the hydrolysis of TosArgOME in 0.15 KCl (titrimetric assay). Greater inhibition was observed at higher concentrations of NaBr. In a work published in 1959,[411] these investigators extended the study on the effect of sodium ions on thrombin (partially purified thrombin), observing approximately 50% inhibition of TosArgOMe hydrolysis at 0.22 NaCl (pH 6.8, 25°C). These investigators suggested the inhibition was due to ionic strength. It is of interest that inhibition was observed in the presence of 0.15 M KCl, which some later studies argue is antagonistic to the effect of sodium ions on thrombin. It should also be noted that the studies were performed with TosArgOMe which, at the time, was one of the few synthetic substrates available.

Curragh and Elmore[412] were likely the first (1964) to investigate the effect of sodium ions as a specific effect and not an ionic strength effect. These studies used a partially purified preparation of bovine thrombin—bovine topical thrombin—but it is likely that the observed effect represented action on thrombin and not a contaminant such as factor Xa. These investigators observed inhibition of thrombin catalyzed TosArgOMe hydrolysis by either sodium ions (NaCl) or potassium ions (KCl). The magnitude of inhibition by sodium ions was comparable to that observed by Scheraga and Ehrenpreis.[410,411] Neither sodium ions nor potassium ions had an effect on the hydrolysis of BzArgOEt. These investigators argued that the diacylation step was rate-limiting for BzArgOEt (generally accepted for hydrolysis of esters by serine proteases), while both acylation and diacylation contribute to observed k_3 with TosArgOMe. Substrate activation of the hydrolysis of TosArgOMe has been reported, as has activation of thrombin hydrolysis by indole.[413] Substrate activation by TosArgOMe has also been reported for trypsin, where activation with the D-isomer of TosAgrOMe was also observed.[414] Based on the log P values for benzene (2.13) and toluene (2.73), one could argue that TosArgOMe was more hydrophobic than BzArgOEt. It is possible that TosArgOMe binding to the S_3 binding site would enhance enzyme activity toward TosArgOMe. Nardini and coworkers[415] observed binding of N^α-(N,N-dimethylcarbamoyl)-α-azalysine-p-nitrophenyl ester (Dmc-azaLys-ONP) to the "aryl" binding site in human α-thrombin inhibited by Dmc-azaLys-ONP, which reacts with the active-site serine (Ser205) as an active-site titrant. Four years later, Roberts and Burkat,[416,417] at MCV in Richmond, described the inhibition of bovine or human thrombin by sodium ions, while potassium ions had no effect. There are some problems comparing these data to other work. The enzyme preparations contained substantial amounts of glycerol, which appears to have an independent effect on the assay system.

My laboratory became interested in this problem in 1978.[418] To be honest, we were skeptical of the observations, since the blood concentration of sodium ions is 135–145 milliequivalents (mEq)/L; below 135 mEq/L is a condition known as hyponatremia, which can have severe consequences.[425] As would be expected, the concentration of sodium ions in extracellular fluid is the same as that in blood. While we were aware of studies on substrate activation with TosArgOMe and thrombin, it was not clear to us why there should be a difference in the response of BzArgOEt and TosArgOMe to sodium ions. We observed that sodium ions were a noncompetitive inhibitor of the hydrolysis of TosArgOMe and BzArgNA by purified bovine thrombin but had no effect on the hydrolysis of BzArgOEt. Potassium chloride at a concentration of 0.2 M had no effect on the thrombin-catalyzed hydrolysis of these substrates; much weaker inhibition was observed with lithium chloride with either TosArgOMe (Ki = 0.7 M) or BzArgpNA (Ki = 0.6 M). We also observed that sodium ions reduced the rate of inactivation of bovine thrombin by TLCK. We speculated that the effect of sodium ions was on the histidine residue. A later study by Griffith and coworkers[419] using human α-thrombin observed the enhancement of amidase activity (TosGlyProArgpNA) with either lithium ions, sodium ions, or potassium ions. These monovalent cations also enhanced the rate of reaction of human thrombin with human antithrombin in the presence of heparin.

The work of Orthner and Kosow[112] has been mentioned earlier. These investigators studied the effect of monovalent cations on the catalytic activity of human α-thrombin. Orthner and Kosow observed that both sodium ions and potassium ions enhanced the rate of thrombin-catalyzed hydrolysis of D-Phe-Pip-Arg-pNA. As ionic strength controls, neither choline chloride or tetramethylammonium chloride enhanced activity. These investigators also observed that sodium ions enhanced the rate of inactivation of human thrombin by TLCK. Landis and coworkers[420] reported that the hydrolysis of TosArgOMe by human thrombin is complex but noted inhibition by sodium ions. The emphasis of this work was focused on other properties of human thrombin. Finally, Lundblad and Jenzano[421] reported on the effect of sodium ions and potassium ions on the rate of hydrolysis of TosGlyProArgpNA and H-D-Phe-Pip-ArgpNA by bovine α-thrombin (Table 3.11) and β-thrombin (Table 3.12). Both sodium ions and potassium ions enhanced the rate of hydrolysis of both tripeptide nitroanilide substrates but had only a marginal effect on the hydrolysis of BzArgpNA. The effect was more marked for Tos-Gly-Pro-ArgpNA than for H-D-Phe-Pip-ArgpNA.

Differing results have been obtained with different substrates and appear to be dependent on whether bovine or human thrombin was used in the experiments. This would suggest that there are species differences in the effect of sodium on thrombin. It has been reported that murine thrombin lacks sodium activation.[422] The use

---

TABLE 3.11
Effect of Monovalent Cations on the Hydrolysis of Two Peptide Nitroanilide Substrates by Bovine α-Thrombin[a]

	TosGlyProArgpNA			H-D-Phe-Pip-ArgpNA			BzArgpNA		
Addition[b]	K_M (μM)	k_{cat}(s⁻¹)	k_{cat}/K_m	K_M (μM)	k_{cat}(s⁻¹)	k_{cat}/K_m	K_M (μM)	k_{cat}(s⁻¹)	k_{cat}/K_m
None	6	25	4.1×10^6	8.1	12	1.5×10^6	370	0.132	3.6×10^2
0.1 M NaCl	13	42	6.4×10^6	5.1	39	7.6×10^6	230	0.046	1.4×10^2
0.1 M KCl	11	83	3.8×10^6	4.4	23	5.2×10^6	240	0.120	2.6×10^2

[a] Data taken from Lundblad, R.L. and Jenzano, J.W., Effect of monovalent cations on bovine α- and β-thrombin: importance of substrate structure, *Thromb. Res.* 37, 53–59, 1985.

[b] The solvent was 10 mM Tris-10 mM Hepes-0.1 % PEG 6,000, pH 7.8

TABLE 3.12
Effect of Monovalent Cations on the Hydrolysis of Two Peptide Nitroanilide Substrates by Bovine β-Thrombin[a]

	TosGlyProArgpNA			H-D-Phe-Pip-ArgpNA			BzArgpNA		
Addition[b]	K_M (µM)	k_{cat}(s^{-1})	k_{cat}/K_m	K_M (µM)	k_{cat}(s^{-1})	k_{cat}/K_m	K_M (µM)	k_{cat}(s^{-1})	k_{cat}/K_m
None	33	12	3.6×10^5	8.1	7	8.6×10^5	330	0.010	30
0.1 M NaCl	19	25	1.3×10^6	5.6	35	6.3×10^6	240	0.011	50
0.1 M KCl	26	16	6.2×10^5	5.5	22	4.0×10^6	230	0.012	48

[a] Data taken from Lundblad, R.L. and Jenzano, J.W., Effect of monovalent cations on bovine α- and β-thrombin: importance of substrate structure, *Thromb. Res.* 37, 53–59, 1985.

[b] The solvent was 10 mM Tris-10 mM Hepes-0.1 % PEG 6,000, pH 7.8

TABLE 3.13
Effect of Solvent on Thrombic Hydrolysis of Several Peptide Nitroanilide Substrates

	Tos-Gly-Pro-Arg-pNA (Chromozyme TH)		
Solvent/pH	K_M (µM)	k_{cat} (s^{-1})	k_{cat}/K_m
100 mM Tris/pH7.95	50	60	0.92
100 mM Tris/pH 8.58	44	50	1.1
50 mM Tris/pH 8.58	23	32	1.4
50 mM Tris-375 mM NaCl/pH 8.58	11	43	3.9
100 mM sodium phosphate/pH 8.0	8	50	6.3
100 mM sodium phosphate/pH 8.0	8.5	72	8.5
100 mM PIPES/pH 8.0	9	62	6.9
100 mM tricine/pH 8.0	8	94	12
100 mM MOPS/pH 8.0	12	108	9.0
100 mM HEPES/pH 8.0	11	93	8.5
100 mM TES/pH 8.0	9	102	11.3
D-Phe-Pip-Arg-pNA (S-2238)			
100 mM Tris/pH 7.95	18.5	23.3	1.3
50 mM Tris/pH 8.36	15	38.3	2.6
100 mM sodium phosphate/pH 8.0	4	63.3	16
100 mM PIPES/pH 8.0	3.7	37.9	10

of different solvents might also complicate the comparison of studies from different laboratories. Lottenberg and coworkers[423] compiled an extensive list of kinetic information for thrombin in various solvents (Table 3.13). It can easily be seen that the observed differences in the kinetic constants for hydrolysis of two commonly used tripeptide nitroanilide substrates in different solvents are not dissimilar to the sodium effects reported earlier.

There are observed differences in the conformation between sodium ion–free thrombin and thrombin containing sodium ions. Villanueva and Perret, in 1983,[424] observed an increase in ultraviolet (UV) absorbance in thrombin containing sodium ions, which they ascribed to the exposure of tyrosine residues. Change in near-UV absorbance in thrombin containing sodium ions was also observed by Orthner and Kosow.[112] These investigators observed a hyperchromic response to sodium ions with a red shift (bathochromic shift) in the absorption maximum, which would be consistent with tyrosine exposure, as suggested by the authors. A change in the environment of a tryptophan residue is also a possibility. These changes are not large, but there is a correlation between a change in UV absorbance and the effect on thrombin activity induced by sodium ions, with similar dissociation constants (approximately 20 mM) for both processes. Another laboratory[425] subsequently used changes in intrinsic fluorescence to measure sodium binding to thrombin.

This discussion described work performed in a number of different laboratories on the effect of sodium ions on the activity of thrombin. While several of these studies[112,420] were more extensive than the other studies, it is fair to say that there was a lack of understanding of the effect of sodium ions. A consideration of texts on bioorganic chemistry suggested that sodium ions functioned in ionic strength, membrane potential (including electrical transmission), and structural stabilization. In looking at the information obtained from 1958 to 1965, it is possible to conclude the following regarding the interaction of sodium ions and thrombin:

1. The effect of sodium ions on the hydrolysis of TosArgOme by thrombin is anomalous and may reflect substrate activation.
2. There are species differences in the response of thrombin to sodium ions and other monovalent cations. While the data obtained with bovine thrombin is of interest, it is unlikely to be of significant value in understanding the role of sodium in human thrombin.
3. The hydrolysis of most amide substrates is enhanced by sodium ions, implying an effect on the acylation step, and enhanced the nucleophilicity of the active-site serine.[426,427]
4. The effect of monovalent cations on the action of thrombin on synthetic substrates is modest. Similar effects on the activity of human thrombin and bovine thrombin toward synthetic substrates are also observed with different solvents.[424]
5. The data presented next suggests a structural, not regulatory, role for sodium ions in human thrombin. The binding data for sodium ions reported by Orthner and Kosow[112] suggest that under normal physiological conditions (c.f. 140 mM sodium ions), the relatively low levels of thrombin formed during normal hemostasis would be saturated with sodium ions. First, the binding of sodium ions to thrombin is very rapid.[428] Studies by Kroh and coworkers[429] suggest that the sodium binding site is formed during the activation of prothrombin, and binding of sodium ions occurs during activation, not at the time of final product formation. The normal concentration of prothrombin in plasma is 1.4 μM[430] and limits conversion of prothrombin to thrombin during hemostasis[431] in two phases: initiation, where initial clot formation results from platelet aggregation and fibrin

formation, and propagation/consolidation, where there is more fibrin formation and fibrin crosslinking (see Chapter 2). Even with maximal thrombin formation, plasma sodium ion concentration would suggest that thrombin would be fully saturated. That said, there has been some interesting work on the structural role of sodium in thrombin function, which is discussed next.

As far as I can tell, more interest in the interaction of sodium ions and thrombin appeared in the early 1990s with several quantitative papers[425,432] from Erico Di Cera's laboratory in St. Louis. Di Cera and colleagues[425] formulated the existence of a "slow" form of human thrombin in the absence of sodium ions, which was rapidly converted to a "fast" form in the presence of calcium ions, with the "slow" form having reduced activity toward peptide nitroanilide substrates and fibrinogen.[433,434]

It is argued that sodium ions act on thrombin in an allosteric manner. While not being critical of the quality of the observations, the fact that human thrombin does bind sodium ions, or the existence of "slow" thrombin and its relationship to thrombin bound to thrombomodulin,[435] I do raise issue with the use of the term allosteric. Granted, sodium ions are structurally distinct from peptide nitroanilide substrates, but sodium ions do not fit into my classical description of allosterism.[436] However, I also recognize that the term allosteric has been used to describe regulatory interactions of thrombin.[437,438] But the classical definition of allosteric suggests feedback regulation of a pivotal enzyme by a downstream product,[439] such as the inhibition by histidine of the first step in the biosynthetic pathway. In the case of a plant, *Medicago truncatula* (a model legume), histidine inhibits the first step in the biosynthetic pathway by binding to a regulatory subunit of ATP-phosphoribosyltransferase.[440] Thrombin does not participate in a metabolic pathway, nor is thrombin a heterosubunit protein with catalytic and regulatory subunits. All of that said, the author recognizes that the definition of allostery has changed. I refer the reader to an excellent recent paper[441] from Professor Emily Parker's laboratory at the University of Welllington, where the allosteric effect is described as "Allostery exploits the conformational dynamics of enzymes by triggering a shift in population ensembles toward functionally distinct conformational or dynamic states." It is my sense that the role of sodium ions in human thrombin is structural stabilization. An excellent review by Huntington[442] suggests that sodium ions stabilize the enzyme-active site in addition to substrate binding sites.

APOLAR/NUCLEOTIDE BINDING SITES

A number of studies have suggested that thrombin has an apolar binding site, or perhaps several apolar binding sites. Many proteins have hydrophobic patches, which can be useful in hydrophobic interaction chromatography (HIC) and related techniques.[443] HIC has been used to purify human thrombin.[444] Larry Berliner, while at Ohio State University, suggested an apolar substrate binding site in thrombin,[445] following on work by Art Thompson in Seattle on the use of *p*-chlorobenzylamine as an affinity matrix.[446] Earlier work[447] had identified 4-chlorobenzylamine (see also Chapter 8) as a potent inhibitor of thrombin and suggested a mechanism based on

S_1 site occupancy. Thompson correctly reasoned that the effectiveness of p-chloro-benzylamine was based on hydrophobic characteristics. The log P for p-chloroben-zylamine is 1.68 compared to 1.0 for benzylamine.[448] Thompson[446] observed that while 2.0 M NaCl at pH 8.0 did not elute thrombin, the protein was eluted with 25% dioxane. Thompson also observed that prothrombin did not bind to the p-chloroben-zylamido matrix; later work from another laboratory[449] showed that prethrombin-2 did not bind that matrix either. This work also showed that thrombin inactivated by TLCK did bind to the p-chlorobenzylamido matrix, while thrombin inactivated by PPACK did not bind to the matrix.

It is clear that Thompson's work with the p-chlorobenzylamido ligand was instructive in the understanding of an apolar binding site near the enzyme-active site—a number of other studies have also proved instructive in the development of the concept of an apolar binding site. Sherry and colleagues[450] can also be given credit for the importance of a hydrophobic binding site in thrombin from their use of tosyl-arginine methyl ester (TAMe) as a substrate for thrombin. Their data showed that both partially purified bovine and human thrombin were more active in the hydrolysis of TAMe than either arginine methyl ester or N^α-acetyl-arginine methyl ester, inferring the importance of the tosyl group. These investigators also noted high activity in the hydrolysis of tosyl-lysine methyl ester. As later studies will show, the increased activity is due to a higher V_{max}; the K_m is substantially higher, indicating that TAMe is, in fact, a better substrate. These studies also showed that trypsin had a preference for TAMe compared to arginine methyl ester. Later studies[122,125] with peptide nitroanilide substrates provided additional support for hydrophobic interactions in the P_2–P_4 positions.[451] Sherry's work[450] is cited as seminal in the development of synthetic inhibitors for thrombin.[452,453]

It has been suggested that the Trp loop contributes to the P_2 site, with Try47 and Trp50 as major determinants,[454] which would provide a base for the hydrophobic nature of the site. 2-Acetoxy-5-nitrobenzyl bromide, an analogue of a p-nitrophenyl ester,[455–457] was shown to modify a tryptophan residue in thrombin with a loss of catalytic activity.[458] It is suggested that the removal of the acetyl group by hydroly-sis provides 2-hydroxy-5-nitrobenzyl bromide, which would react with a tryptophan residue near the enzyme-active site.

In their argument for an apolar binding site near the thrombin enzyme site, Berliner and Shen[453] reported that indole displaced proflavine from thrombin. The binding of proflavine to trypsin and chymotrypsin is associated with a shift in the absorption spectrum of proflavine to a longer wavelength (red shift, 440 nm to ~470 nm).[459–461] This shift in spectrum is consistent with the transition from an aque-ous solution to a nonpolar environment. Koehler and Magnusson[462] had shown that thrombin also bound proflavin with a loss of catalytic activity ($K_i = 1 \times 10^{-5}$M) toward fibrinogen and TAMe, and the change in spectral properties observed with trypsin and chymotrypsin suggest that the thrombin-active site is close to a nonpolar environment. The proflavin was displaced by benzamidine, a competitive inhibi-tor of thrombin.[447] Other studies supporting a nonpolar environment for the throm-bin catalytic site are the reaction of thrombin with N-buryrylimidazole[463] and with hydrophobic p-nitrophenylacylates.[464]

Berliner and Shen[445] also observed TosArgOMe esterase activity was stimulated by indole at one substrate concentration (1 mM) but not a higher (5 mM) substrate concentration. This observation may be related to the reported substrate activation with thrombin, which is discussed later. Other work with spin-label derivatives suggested that there is a unique binding site for indole in the S_2/S_3 binding site of thrombin. Subsequent work from the Ohio State group[465] extended these observations to other indole derivatives, such as demonstrating that tryptophan was an activator of TosArgOMe esterase, while serotonin is an inhibitor. ATP was also shown to displace proflavin from the thrombin-active site and to be an inhibitor of TosArgOMe esterase. It was also shown that ATP, ADP, adenosine monophosphate (AMP), uridine diphosphate (UDP), and inorganic phosphate could inhibit fibrinogen clotting, with inhibition reduced by the presence of indole. Later work from this laboratory[466] demonstrated that nucleotide binding (and inorganic phosphate and pyrophosphate) occurs at the exosite-1 binding site. ATP binding did not influence the mobility of apolar spin-labels.

There are additional studies on TosArgOMe that relate to an apolar binding site. Substrate activation of TosArgOMe hydrolysis by thrombin and other enzymes was observed by Weinstein and Doolittle in 1972.[9] Substrate activation is a phenomenon where a higher rate of hydrolysis is observed at a higher substrate concentration than that which would be predicted from conventional kinetic plots. This was observed by Koehler and Magnusson[462] and more recently by Strukova and coworkers.[467] Another group[468] failed to observe substrate activation with TosArgOMe. To the best of my knowledge, there has not been a totally satisfactory explanation for the process of substrate activation, but it is likely that there is a second binding site(s) for TosArgOMe, and it is not unlikely that it is related to the putative apolar binding site.

BINDING TO THE EXTRACELLULAR MATRIX

Most investigators, including myself, who have worked on blood coagulation have not thought much outside of the vascular space. There is a subset of investigators who really don't think much about vascular function but consider proteins such as thrombin interesting. In this sense, I am reminded of a story from my graduate student days at the University of Washington where a graduate student at another institution, who had done elegant work on a liver enzyme, was unable to answer a question about liver function. The point here is that many investigators, including myself, have a problem thinking outside the box. I will say that my days at Baxter were sufficiently challenging and did enlarge my vision. I was a coauthor of a book on proteolysis in the extravascular space[469] published several years ago. While doing the library research for this book, I was surprised and embarrassed (see earlier) to find that there are substantial amounts of coagulation factors in lymph and, by extension, in interstitial fluid. Colleagues from Duke and I subsequently published a paper on thrombin in the interstitial space, which discussed various mechanisms for the formation of thrombin that are independent of the canonical prothrombinase.[470] These various mechanisms are discussed in Chapter 2. The point is that thrombin is present in the extravascular/interstitial space and may have physiological importance. It is acknowledged that there are likely inhibitors such as antithrombin, heparin cofactor II, and α_2-macroglobulin in the interstitial space that could modulate thrombin function.[469]

While I am not surprised, I am disappointed that there has not been more work on the interaction of thrombin with the extracellular matrix. It would also seem that there is similarity between the subendothelial matrix and the extracellular matrix.[471,472] I would also include the endothelial basement membrane in this discussion. Collagen IV within the endothelial basement membrane binds a substantial amount of coagulation factor IX.[473] For the purpose of simplicity, unless there is need for reference of a unique environment, such as the endothelial basement membrane, the term extracellular matrix will be used. It is not unlikely that thrombin is bound to components of the extracellular matrix in a manner similar to thrombin binding to fibrin or platelets with retention of activity. For the purpose of this discussion, the extracellular matrix is defined as the network of macromolecules, such as collagen, chondroitin sulfate, fibronectin, thrombospondin, laminin, and entactin,[474–476] and is subject to remodeling.[476]

The binding of human thrombin to the extracellular matrix has been known since 1989.[477] Bar-Shavit and coworkers studied the binding of thrombin to an extracellular matrix produced by cultured endothelial cells. The binding was saturable, reversible, and did not require the enzyme-active site. Binding was inhibited by a synthetic tetradecapeptide corresponding to a sequence in the 60s loop of thrombin (47–60). A shorter peptide (47–56) was without effect. Substitution of a tyrosine for a tryptophan eliminated the ability of the synthetic peptide to block thrombin binding. The text implies a substitution of a tryptophan for a tyrosine. The bound thrombin was active, as measured by the clotting of fibrinogen and the hydrolysis of a tripeptide nitroanilide substrate (Chromozym® TH). The bound thrombin was not inhibited by antithrombin but was inhibited by hirudin. Studies with enzymes that degraded glycosaminoglycans suggested that thrombin bound to dermatan sulfate. Degradation of heparan sulfate with heparitinase inhibited basic FGF binding but had no effect on thrombin binding; treatment of chondroitinase ABC inhibited thrombin binding but had no effect on the binding of basic FGF to the extracellular matrix. These results are consistent with thrombin binding to exosite-2. Somewhat later, Salatti and coworkers[478] reported that thrombin bound to extracellular matrix–induced fibrin formation in flowing blood in a perfusion virus. Hatton, in later work,[479] showed that purified bovine thrombin bound to dermatan sulfate in rabbit aorta subendothelium. This work also showed that the binding of thrombin to the extracellular matrix is independent of the enzyme-active site. Although not directly related to the current discussion, Okamura and coworkers[480] showed that only a small number of bound thrombin molecules are necessary to affect a cellular response. The resistance of thrombin bound to extracellular matrix to inactivation by antithrombin is similar to the resistance of thrombin bound to fibrin to inactivation by antithrombin.[481–483] Thrombin bound to fibrin can be inactivated by hirugen and PPACK.[482] There is little additional work on thrombin bound to the extracellular matrix. This is unfortunate, as there is ample reason to suggest the importance of thrombin function in the interstitial space.[470,484]

REFERENCES

1. Jenny, N.S., and Mann, K.G., Thrombin, in *Hemostasis and Thrombosis: Basic Principles and Clinical Practice*, 4th edn, ed. R.W. Colman, J. Hirsh, V.J. Marder, A.W. Clowes, and J.N. George, Chapter 10, pp. 171–189, Lippincott, Williams & Wilkins, Philadelphia, PA, 2001.

2. López, M.L., Bruges, C., Crespo, G., *et al.*, Thrombin selectively induces transcription of genes in human monocytes invoóved in inflammation and wound healing, *Thromb. Haemost.* 112, 992–1001, 2014.

3. Ziv-Polat, O., Topaz, M., Brosh, T., and Margel, S., Enhancement of incisional wound healing by thrombin conjugated iron oxide nanoparticles, *Biomaterials* 31, 741–747, 2010.

4. Stubbs, M.T., and Bode, W., A player of many parts: the spotlight falls on thrombin's structure. *Thromb. Res.* 69, 1–58, 1993.

5. Siller-Matula, J.M., Schwameis, M., Blann, A., Mannhalter, C., and Jilma, B., Thrombin as a multifunctional enzyme. Focus on *in vitro* and *in vivo* effects, *Thromb. Haemost.* 106, 1020–1033, 2011.

6. Szmula, R., Kukor, Z., and Sahin-Toth, M., Human mesotrypsin is a unique digestive protease specialized for the degradation of trypsin inhibitors, *J. Biol. Chem.* 278, 48580–48589, 2003.

7. Friedrich, P. and Bozoky, Z., Digestive versus regulatory proteases: on calpain action *in vivo*, *Biol. Chem.* 386, 609–617, 2005.

8. Lonsdale-Eccles, J.D., Hogg, D.H., and Elmore, D.T., The esterolytic specificities of bovine thrombin and factor Xa, *Biochim. Biophys. Acta* 258, 577–590, 1972.

9. Weinstein, M.S. and Doolittle, R.F., Differential specificities of thrombin, plasmin and trypsin with regard to synthetic and natural substrates, *Biochim. Biophys. Acta* 612, 395–400, 1980.

10. Mihalyi, E., Weinberg, R.W., Towne, D.W., and Friedman, M.E., Proteolytic fragmentation of fibrinogen. I. Comparison of the fragmentation of human and bovine fibrinogen by trypsin or plasmin, *Biochemistry* 15, 5372–5381, 1976.

11. Bode, W., The structure of thrombin: a Janus-headed proteinase, *Sem. Thromb. Hemost.* 32(Suppl 1), 16–31, 2006.

12. Le Bonniec, B.F., Thrombin, in *Handbook of Proteolytic Enzymes*, 3rd edn, ed. N.D. Rawlings and G. Salveson, Chapter 643, pp. 2915–2932, Elsevier, Amsterdam, Netherlands, 2013.

13. Henriksen, R.A., Owen, W.G., Nesheim, M.E., and Mann, K.G., Identification of a congenital dysthrombin, Thrombin Quick, *J. Clin. Invest.* 66, 934–940, 1980.

14. Henriksen, R.A. and Owen, W.G., Characterization of the catalytic defect in the dysthrombin, thrombin Quick, *J. Biol. Chem.* 262, 4664–4669, 1987.

15. Henriksen, R.A. and Mann, K.G., Identification of the primary structural defect in the dysthrombin thrombin Quick1: substitution for arginine 382, *Biochemistry* 27, 9160–9165, 1988.

16. Henriksen, R.A. and Mann, K.G., Substitution of valine for glycine-558 in the congenital dysthrombin thrombin Quick II alters primary substrate specificity, *Biochemistry* 28, 2078–2082, 1989.

17. Leong, L., Henriksen, R.A., Kermode, J.C., Rittenhouse, S.E., and Tracy, P.B., The thrombin high-affinity binding site on platelets is a negative regulator of thrombin-induced platelet activation. Structure-function studies using two mutant thrombins, Quick I and Quick II, *Biochemistry* 31, 2567–2576, 1992.

18. Phillips, J.E., Shirk, R.A., Whinna, H.C., Henriksen, R.A., and Church, F.C., Inhibition of dysthrombins Quick I and II by heparin cofactor II and antithrombin, *J. Biol. Chem.* 268, 3321–3327, 1993.

19. Henriksen, R.A., Identification and characterization of mutant thrombins, *Methods Enzymol.* 222, 312–327, 1993.

20. Astrup, T. and Alkjaersig, N., Classification of proteolytic enzymes by means of their inhibitors, *Nature* 169, 314–316, 1952.

21. Rawlings, N.D. and Barrett, A.J., Families of serine peptidases, *Methods Enzymol.* 244, 19–61, 1994.

22. Barrett, A.J., Proteolytic enzymes. Nomenclature and classification, in *Proteolytic Enzymes: A Practical Approach*, 2nd edn, ed. R. Benyon and J.S. Bond, Chapter 1, pp. 1–21, Oxford University Press, Oxford, UK, 2001.

23. McDonald, A.G., Boyce, S., and Tipton, K.F., ExplorEnz: the primary source of the IUBMB enzyme list, *Nucleic Acids Res.* D593–D597, 2008.

24. Di Cera, E., Serine proteases, *IUBMB Life* 61, 510–515, 2009.

25. Schomburg, D. and Schomburg, I., Enzyme databases, *Methods Mol. Biol.* 609, 113–128, 2010.

26. McDonald, A.G. and Tipton, K.F., Fifty-five years of enzyme classification: advances and difficulties, *FEBS J.* 281, 583–592, 2014.

27. Reeck, G.P., Hahn, C., Teller, D.C., *et al.*, "Homology" in proteins and nucleic acids: a terminology muddle and a way out of it, *Cell* 50, 667–668, 1987.

28. Rawings, N.D. and Barrett, A.J., MEROPS: the peptidase database, *Nucleic Acids Res.* 27, 325–331, 1999.

29. Rawlins, N.D., Protease families, evolution and mechanism of action, in *Proteases: Structure and Function*, ed. K. Prix and W. Stocker, Chapter 1, pp. 1–36, Springer-Verlag, Vienna (Wien), Austria, 2013.

30. Rawlings, N.D. and Barrett, A.J., Introduction: Serine peptidases and their clans, in *Handbook of Proteolytic Enzymes*, 3rd edn, ed. N.D. Rawlings and G. Salveson, Chapter 559, pp. 2491–2523, Elsevier, Amsterdam, Netherlands, 2013.

31. Polgar, L., The catalytic triad of serine peptidases, *Cell. Mol. Life. Sci.* 62, 2161–2172, 2005.

32. Dodson, G. and Wlodawer, A., Catalytic triads and their relatives, *Trends Biochem.* 23, 347–352, 1998.

33. Davie, E.W. and Kulman, J.D., An overview of the structural and function of thrombin, *Semin. Thromb. Hemost.* 32(Suppl 1), 3–15, 2006.

34. Hunkapiller, M.W., Smallcombe, S.H., Witaker, D.R., and Richards, J.H., Ionization behavior of the histidine residue in the catalytic triad of serine proteases. Mechanistic implications, *J. Biol. Chem.* 248, 8306–8308, 1973.

35. Carter, P. and Wells, J.A., Dissecting the catalytic triad of serine proteases, *Nature* 332, 564–568, 1998.

36. Bryan, P., Partoliano, M.W., Quill, S.G., Hisao, H.-Y., and Poulos, T., Site specific mutagenesis and the role of the oxyanion holei subtilisin, *Proc. Natl. Acad. Sci. USA* 83, 3743–3745, 1994.

37. Lonsdale-Eccles, J.D., Neurath, H., and Walsh, K.A., Probes of the mechanism of zymogen catalysis, *Biochemistry* 17, 2805–2809, 1978.

38. Whiting, A.K. and Peticolas, W.L., Details of the acyl-enzyme intermediate and oxyanion hole in serine protease catalysis, *Biochemistry* 33, 552–561, 1994.

39. Sprang, S., Standing, T., Fletterick, R.J., *et al.*, The three-dimensional structure of Asn[102] mutant of trypsin. Role of Asp[102] in serine protease catalysis, *Science* 237, 905–909, 1987.

40. HIbbard, L.S., Nesheim, M.E., and Mann, K.G., Progressive development of a thrombin inhibitor binding site, *Biochemistry* 21, 2285–2292, 1982.

41. Jing, H., Macon, K.J., Moore, D., *et al.*, Structural basis of profactor D activation: from a highly flexible zymogen to a novel self-inhibited serine protease, complement factor D, *EMBO J.* 18, 804–814, 1999.

42. Hendrix, H., Lindhout, T., Mertens, K., Engels, W., and Hemker, H.C., Activation of human prothrombin by Staphylocoagulase, *J. Biol. Chem.* 258, 3637–3644, 1983.

43. McAdow, M., Missiakas, D.M., and Schneewind, O., *Staphylococcus aureus* secretes coagulase and von Willebrand factor binding protein to modify the coagulation cascade and establish host infections, *J. Innate Immun.* 4, 141–148, 2012.

44. Panizzi, P., Friedrich, R., Fuentes-Prior, P., *et al.*, Novel fluorescent prothrombin analogs as probes of staphylocoagulase-prothrombin interactions, *J. Biol. Chem.* 281, 1169–1178, 2006.

45. Bustin, M. and Conway-Jacobs, A., Intramolecular activation of porcine pepsinogen, *J. Biol. Chem.* 246, 615–620, 1971.

46. Bradford, H.N and Krishnaswamy, S., Meizothombin as an unexpected zymogen-like variant of thrombin, *J. Biol. Chem.* 287, 30414–30425, 2012.

47. Ivanciu, L. and Camire, R.M., Hemostatic agents of broad applicability produced by selective tuning of factor Xa zymogenicity, *Blood* 126, 94–102, 2015.

48. *Oxford English Dictionary*, 2008.

49. Bellamy, J.F., On the agents concerned with the production of the trypsin ferment from its zymogen, *J. Physiol.* 27, 323–331, 1901.

50. Neurath, H., Mechanism of zymogen activation, *Fed. Proc.* 23, 1–7, 1964.

51. Morgan, P.H., Robinson, N.C., Walsh, K.A., and Neurath, H., Inactivation of bovine trypsinogen and chymotrypsinogen by diisopropylphosphorofluoridate, *Proc. Natl. Acad. Sci. USA* 69, 3312–3316, 1972.

52. Gertler, A., Walsh, K.A., and Neurath, H., Catalysis by chymotrypsinogen. Demonstration of an acyl enzyme intermediate, *Biochemistry* 13, 1302–1310, 1974.

53. Lonsdale-Eccles, J.D., Neurath, H., and Walsh, K.A., Probes of the mechanism of zymogen catalysis, *Biochemistry* 17, 2805–2809, 1978.

54. Lonsdale-Eccles, J.D., Kerr, M.A., Walsh, K.A., and Neurath, H., Catalysis by zymogens: increased reactivity at high ionic strength, *FEBS Lett.* 100, 157–160, 1979.

55. Gertler, A., Walsh, K.A., and Neurath, H., Catalysis by chymotrypsinogen. Demonstration of an acyl enzyme intermediate, *Biochemistry* 13, 1302–1310, 1974.

56. Neurath, H., Limited proteolysis and zymogen activation, in *Proteases and Biological Control*, ed. E. Reich, D.B. Rifkin, and E. Shaw, pp. 51–64, Cold Spring Harbor Laboratory, Cold Spring Harbor, New York, USA, 1975.

57. Neurath, H., Proteolytic enzymes, past and present, *Fed. Proc.* 44, 2907–2713, 1985.

58. Friedrich, P. and Bozóky, Z., Digestive versus regulatory proteases: on calpain action *in vivo*, *Biol. Chem.* 386, 609–612, 2005.

59. Elrod, J.P., Hogg, J.L., Quinn, D.M., Venkatshubban, J.S., and Schowen, R.L., Protonic reorganization and substrate in catalysis by serine proteases, *J. Am. Chem. Soc.* 102, 3917–3922, 1980.

60. Hunkapiller, M.W., Forgac, M.D., and Richards, J.H., Mechanism of action of serine proteases: tetrahedral intermediate and concerted proton transfer, *Biochemistry* 15, 5581–5588, 1976.

61. Anderson, V.E., Ruszczycky, M.W., and Harris, M.E., Activation of oxygen nucleophiles in enzyme catalysis, *Chem. Rev.* 106, 3236–3251, 2006.

62. Case, A. and Stein, R.L., Mechanistic origins of the substrate selectivity of serine proteases, *Biochemistry* 42, 3335–3348, 2003.

63. Gladner, J.A., and Laki, K., The inhibition of thrombin by diisopropylphosphorofluoridate, *Arch. Biochem. Biophys.* 62, 501–503, 1956.
64. Lundblad, R.L., A rapid method for the purification of bovine thrombin and the inhibition of the purified enzyme with phenylmethylsulfonyl fluoride. *Biochemistry* 10, 2501–2506, 1971.
65. Thompson, A.R., Inhibition of thrombin by sarin, *Biochim. Biophys. Acta* 198, 392–395, 1970.
66. Boyland, E., Biochemical reactions of chemical warfare agents, *Nature* 161, 225–227, 1948.
67. Wood, J.R., Medical problems in chemical warfare, *JAMA* 144, 606–609, 1950.
68. Ni, L.-M. and Powers, J.C., Synthesis and kinetic studies of an amidine-containing phosphonofluoridate: a new potent inhibitor of trypsin-like enzymes, *Bioorg. Med. Chem.* 6, 1167–1173, 1998.
69. Lundgren, H.P. and Ward, W.H., Chemistry of amino acids and proteins, *Annu. Rev. Biochem.* 18, 115–154, 1949.
70. Schaffer, N.K., Mass, S.C., Jr., and Summerson, W.H., Serine phosphoric acid from diisopropylphosphoryl derivative of ell cholinesterase, *J. Biol. Chem.* 206, 201–207, 1954.
71. Cohen, W. and Erlanger, B.F., Studies on the reactivation of diethylphosphorylchymotrypsin, *J. Am. Chem. Soc.* 82, 3928–3934, 1960.
72. Webb, J.L., Interaction of inhibitors with enzymes, in *Enzyme and Metabolic Inhibitors*, Chapter 6, pp. 193–317, Academic Press, New York, USA, 1965.
73. Strumeyer, D.H., White, W.N., and Koshland, D.E., Jr., Role of serine in chymotrypsin activity, *Proc. Natl. Acad. Sci USA* 50, 931–935, 1963.
74. Ashton, R.W., and Scheraga, H.A., Preparation and characterization of anhydrothrombin. *Biochemistry* 34, 6453–6463, 1965.
75. Wedemeyer, W., Ashton, R.W., and Scheraga, H.A., Kinetics of competitive binding with application to thrombin complexes, *Anal. Biochem.* 248, 130–140, 1997.
76. Nogami, K., Shima, M., Hosokawa, K., Nagata, M., Koide, T., Saenko, E.L., Tanaka, I., Shibata, M., and Yoshioka, A., Factor VIII C2 domain contains the thrombin-binding site responsible for the thrombin-catalyzed cleavage at Arg[1689.] *J. Biol. Chem.* 275, 25774–25780, 2000.
77. Pei, G., Baker, K., Emfinger, S.M., Fowlkes, D.M., and Lentz, B.R., Expression, isolation, and characterization of an active site (Serine 528 → Alanine) mutant of recombinant prothrombin. *J. Biol. Chem.* 266, 9598–9604, 1991.
78. Gan, Z.R., Li, Y., Connolly, T.M., *et al.*, Importance of the Arg-Gly-Asp triplet in human thrombin for maintenance of structure and function, *Arch. Biochem. Biophys.* 301, 228–236, 1993.
79. Carter, P., and Wells, J.A., Dissecting the catalytic triad of a serine protease. *Nature* 332, 564–568, 1988.
80. Braxton, S., and Wells, J.A., The importance of a distal hydrogen bonding group in stabilizing the transition state in subtilisin BPN'. *J. Biol. Chem.* 266, 11797–11800, 1991.
81. Bergstrom, R.C., Coombs, G.S., Ye, S., Madison, E.L., Goldsmith, E.J., and Corey, D.J., Binding of nonphysiological protein and peptide substrates to proteases: differences between urokinase-type plasminogen activator and trypsin and contributions to the evolution of regulated proteolysis. *Biochemistry* 42, 5395–5402, 2003.
82. Baglin, T.P., Carrell, R.W., Church, F.C., Esmon, C.T., and Huntington, J.A., Crystal structures of native and thrombin-complexed heparin cofactor II reveal a multistep allosteric mechanism. *Proc. Nat. Acad. Sci. USA* 99, 11079–11084, 2002.

83. Krem, M.M., and Di Cera, E., Dissecting substrate recognition by thrombin using the inactive mutant S195A. *Biophys. J.* 100, 469–479, 2003.

84. Dixon, G.H., Neurath, H., and Pechåre, J.-F., Proteolytic enzymes, *Annu. Rev. Biochem.* 27, 489–532, 1958.

85. Hartley, B.S., Proteolytic enzymes, *Annu. Rev. Biochem.* 29, 45–72, 1960.

86. Ong, E.B., Shaw, E., and Schoellmann, G., The identification of the histidine residue at the active center of chymotrypsin, *J. BIol. Chem.* 240, 694–698, 1985.

87. Shaw, E., Mares-Guia, M., and Cohen, W., Evidence for an active-center histidine in trypsin through use of a specific reagent, 1-chloro-3-tosylamido-7-amino-2-hepatnone, *Biochemistry* 4, 2219–2224, 1965.

88. Koshland, D.E., Jr., Properties of the active site of enzymes, *Ann. N. Y. Acad. Sci.* 103, 630–642, 1963.

89. Koshland, D.E., Jr., The active site and enzyme action, *Adv. Enzymol. Relat. Sub. Biochem.* 22, 45–97, 1960.

90. Baker, B.R., Lee, W.W., Tong, E., and Ross, L.O., Potential anticancer agents. LXVL. Non-classical antimebolites. III. 4-(iodoacetamido)-salicyclic acid, an *exo*-alklylating irreversible inhibitor of glutamic dehydrogenase, *J. Am. Chem. Soc.* 83, 3713–3714, 1961.

91. Sonder, S.A. and Fenton, J.W., II, Differential inactivation of human and bovine α-thrombin by exosite affinity-labeling reagents, *Thromb. Res.* 32, 623–629, 1983.

92. Bardon, P., Fenton, J.W., II, and Maraganore, J.M., Affinity labeling of lysine-149 in the anion-binding exosite of human α-thrombin with N^{α}-(dinitrofluorobenzoate)hirudin C-terminal peptide, *Biochemistry* 29, 6379–6384, 1990.

93. Gram, H.G., Mosher, C.W., and Baker, R.R., Potential anticancer agents. XVII. Alkylating agents to phenylalanine mustard, *J. Am. Chem. Soc.* 81, 3103–3108, 1959.

94. Ross, L.O., Goodman, L., and Baker, R.R., Potential anticancer agents XVIII Synthesis of substituted 4,5-trimethylenepyrimidines, *J. Am. Chem. Soc.* 81, 3108–3114, 1959.

95. Wofsy, L., Metzger, H., and Singer, S.J., Affinity labeling-a general method for labeling the active sites of antibody and enzyme molecules, *Biochemistry* 1m, 1031–1039, 1961.

96. Shaw, E., Chemical modification by active-site directed reagents, in *The Enzymes*, Volume 1, ed. P.D. Boyer, Chapter 2, pp. 91–146, Academic Press, New York, USA, 1971.

97. Schoellman, G. and Shaw, E., A new method for labeling the active center of chymotrypsin, *Biochem. Bioiphys. Res. Commun.* 7, 36–40, 1962.

98. Shaw, E., Site-specific reagents for chymotrypsin, trypsin, and other serine proteases, *Meth. Enzymol.* 25, 655–600, 1972.

99. Plapp, B.V., Application of affinity labeling for studying structure and function of enzymes, *Methods Enzymol.* 87, 469–499, 1982.

100. Mor, A., Maillard, J., Favreau, C., and Reboud-Ravaux, M., Reaction of thrombin and proteinases of the fibrinolytic system with a mechanism-based inhibitor, 3,4-dihydro-3-benzyl-6-chloromethylcoumarin, *Biochim. Biophys. Acta* 1038, 119–124, 1990.

101. Stangier, J., Clinical pharmacokinetics and pharmacodynamics of the oral direct thrombin inhibitor dabigatran etexilate, *Clin. Pharmacol.* 47, 286–295, 2008.

102. Sheffield, W.P., Lanbourne, M.D., Eltringham-Smith, L.J., *et al.*, $_{\gamma T}$-S195A thrombin reduces the anticoagulant effect of dabigatran *in vitro* and *in vivo*, *J. Thromb. Haemost.* 12, 1110–1115, 2014.

103. Glover, G. and Shaw, E., The purification of thrombin and isolation of a peptide containing the active site histidine, *J. Biol. Chem.* 246, 4594–4501, 1971.

104. Coggins, J.R., Kray, W., and Shaw, E., Affinity labeling of proteinases with tryptic specificity by peptides with C-terminal lysine chloromethyl ketones, *Biochem. J.* 138, 579–585, 1974.

105. Collen, D., Lijnen, H.R., DeCock, F., Durieux, J.F., and Loffet, A., Kinetic properties of tripeptide lysyl chloromethylketones and tripeptide *p*-nitroanilide derivatives toward trypsin-like serine proteinases, *Biochim. Biophys. Acta* 165, 158–166, 1980.

106. Kettner, C., and Shaw, E., D-PHE-PRO-ARGCH$_2$Cl. A selective affinity label for thrombin, *Thromb. Res.* 14, 969–973, 1974.

107. Kettner, C. and Shaw, E., Inactivation of trypsin-like enzymes with peptides or arginine chloromethyl ketone, *Methods Enzymol.* 80, 826–842, 1981.

108. Lundblad, R.L., Nesheim, M.E., Straight, D.L., *et al.*, Bovine α- and β-thrombin. Reduced fibrinogen-clotting activity of β-thrombin is not a consequence of reduced affinity for fibrinogen, *J. Biol. Chem.* 251, 6991–6995, 1984.

109. Hofsteenge, J., Braun, P.J., and Stone, S.F., Enzymatic properties of proteolytic derivatives of human α-thrombin, *Biochemistry* 27, 2144–2151, 1988.

110. Exner, T. and Koppel, J.L., Cholate enhancement of interaction between thrombin and tosyl lysine chloromethyl ketone, *Biochim. Biophys. Acta* 329, 233–240, 1973.

111. Myrmel, K.H., Lundblad, R.L., and Mann, K.G., Characteristics of the association between prothrombin fragment 2 and α-thrombin, *Biochemistry* 15, 1767–1773, 1976.

112. Orthner, C.L. and Kosow, D.P., Evidence that human α-thrombin is a monovalent cation-activated enzyme, *Arch. Biochem. Biophys.* 202, 63–75, 1980.

113. Brocklehurst, K. and Little, G., Reactivities of the various protonic states in the reactions of papain and of L-cysteine with 2,2'- and with 4,4-dipyridyl disulphide: evidence for nucleophilic reactivity in the un-ionized thiol group of the cysteine-25 residue of papain occasioned by its interactive with the histidine-159-asparagine-175 hydrogen-bonded system, *Biochem. J* 128, 471–474, 1972.

114. Micanovic, R., Procyk, R., LIn, W., and Matsueda, G.R., Role of histidine 373 in the catalytic activity of coagulation factor XIII, *J. Biol. Chem.* 269, 9190–9194, 1994.

115. Blow, D.M., The tortuous story of Asp . . . His . . . Ser: structural analysis of α-chymotrypsin, *Trends Biochem. Sci.* 22, 405–408, 1997.

116. Kossiakoff, A.A. and Spenser, S.A., Direct determination of the protonation states of aspartic acid-102 and histidine-57 in the tetrahedral intermediate of the serine proteases: neutron structure of trypsin, *Biochemistry* 20, 6462–6474, 1981.

117. Schecter, I. and Berger, A., On the size of the active site in proteases. I. Papain, *Biochem. Biophys. Res. Commun.* 27, 157–162, 1967.

118. Gallwitz, M., Enoksson, M., Thorpe, M., and Hellman, L . . ., The extended cleavage specificity of human thrombin, *PLos One* 7(2), e31756, 2012.

119. Kretz, C.A., Tomberg, K., Van Esboeck, A., Yee, A., and Ginsberg, D., High thoughput protease profiling comprehensively defines active site specificity of thrombin ADAMTS13, *Scientific Reports* 8, 2788, 2018.

120. Svendsen, L., Blombäck, B., Blombäck, M., and Olsson, P.I., Synthetic chromogenic substrates for determination of trypsin, thrombin, and thrombin-like enzymes, *Thromb. Res.* 1, 267–278, 1972.

121. Lottenberg, R., Hall, J.A., Blinder, M., Blinder, E.P., and Jackson, C.M., The action of thrombin on peptide *p*-nitroanilide substrates. Substrate selectivity and examination of hydrolysis under different reaction conditions, *Biochim. Biophys. Acta* 742, 539–557, 1983.

122. Hemker, H.C., *Handbook of Synthetic Substrates for the Coagulation and Fibrinolytic Systems*, Kluwer, Boston, MA, USA, 1983.

123. Le Bonniec, B.F., MacGillivray, R.T., and Esmon, C.T., Thrombin Glu-39 restricts the P'C specificitiy of nonacidic residues, *J. Biol. Chem.* 266, 13796–13803, 1991.

124. Pozgay, M., Szabó, G.S., Bajusz, S., *et al.*, Investigations of the substrate binding site of trypsin by use of tripeptide nitroanilide substrates, *Eur. J. Biochem.* 115, 497–502, 1981.

125. Inagami, T. and York, S.S., Effect of alkylguanidine and alkylamines on trypsin catalysis, *Biochemistry* 7, 4045–4052, 1968.

126. Inagami, T. and Hatano, H., Effect of alkylguanidine on inactivation of trypsin by alkylation and phosphorylation, *J. Biol. Chem.* 244, 1176–1182, 1969.

127. Ascenzi, P., Menegatti, E., Guareri, M., Bortolloti, F., and Antonini, E., Catalytic properties of serine proteases. 2. Comparison between human urinary kallikrein and human urokinase, bovine β-trypsin, bovine thrombin, and bovine α-chymotrypsin, *Biochemistry* 21, 2483–2490, 1982.

128. Harmison, C.R. and Seegers, W.H., Some physicochemical properties of bovine autoprothrombin II, *J. Biol. Chem.* 237, 3074–3076, 1962.

129. Olsen, P.H., Esmon, N.L., Esmon, C.T., and Laue, T.M. Ca^{2+} dependence of the interaction between protein C, Thrombin and the elastase fragment of thrombomodulin. Analysis by ultracentrifugation, *Biochemistry* 31, 746–754, 1992.

130. Singh, R.K., Kazansky, Y., Wathieu, D., and Fushman, D., Hydrophobic patch of ubiquitin is important for its optimal activation by ubiquitin activating enzyme E1, *Anal. Chem.* 89, 7852–7860, 2017.

131. O'Conner, B.F. and Cummins, P.M., Hydrophobic interaction chromatography, *Methods Mol. Biol.* 1485, 355–363, 2017.

132. Robinson, J., Roush, D., and Cramer, S.M., The effect of pH on antibody retention in multimodal cation exchange chromatographic systems, *J. Chromatog. A*, 1617, 460838, 2020, doi: 10.1016/j.chroma.2019.460838.

133. Del Vecchio, K., Frick, C.T., Gc, J.B., *et al.*, A cationic, C-terminal patch and structural rearrangements in Ebola virus matrix VP40 protein control its interactions with phosphatidylserine, *J. Biol. Chem.* 293, 3335–3349, 2018.

134. Scott, J.L., Frick, C.T., Johnson, K.A., *et al.*, Molecular analysis of membrane targeting by the C2 domain of the E3 ubiquitin ligase Smrf, *Biomolecules*, 10(2), 229, 2020, doi: 10.3390/biom1002029.

135. Mora-Obando, D., Fernández, J., Montecucco, C., Gutiérrez, J.M., and Lomonte, B., Synergism between basic Asp49 and Lys49 phospholipase A2 myotoxins of viperid snake venon *in vitro* and *in vivo*, *PLoS One* 9(10), e109846, 2014.

136. Baker, B.R., *Design of Active-Site-Directed Irreversible Enzyme Inhitors. The Organic Chemistry of the Enzymic Active Site*, Chapter 1, pp. 1–21, General considerations, John Wiley and Sons, New York, NY, USA, 1967.

137. Troeberg, L., Fushimi, K., Khokha, R., *et al.*, Calcium pentosan polysulfate is a multifaceted inhibitor of aggreganases, *FASEB J* 22, 35125–3524, 2008.

138. Santamaria, S., Yamamoto, K., Botkjaer, K., *et al.*, Antibody-based exosite inhibitors of ADAMTS5 (aggreganase-2), *Biochem. J.* 471, 391–401, 2015.

139. Apte, S.S., Anti-ADAMTS5 monoclonal antibodies: implications for aggreganase inhibition in osteoarthritis, *Biochem. J.* 473, e1–e4, 2016.

140. McQuibban, G.A., Butlet, G.S., Gong, J.-H., *et al.*, Matrix metalloproteinase activity inactivates the CXC chemokine stromal cell-derived factor-1, *J. Biol. Chem.* 276, 43503–43508, 2001.

141. Overall, C.M., McQuibban, G.A., and Clark-Lewis, I., Discovery of chemokine substrates for matrix metalloproteinases by exosite-scanning: a new tool for degradomics, *Biol. Chem.* 383, 1059–1066, 2002.

142. Robichaud, T.K., Steffersen, B., and Field, G.B., Exosite interactions impact matrix metalloproteinase collagen specificities, *J. Biol. Chem.* 286, 37335–37342, 2011.

143. Vandenbroucke, R.E. and Libert, C., Is there new hope for therapeutic matrix metalloproteinase inhibition?, *Nat. Rev. Drug. Discov.* 13, 904–927, 2014.

144. Gomis-Ruth, E.X., Third-time lucky? Getting a grip on matrix metalloproteinases, *J. Biol. Chem.* 292, 17975–17976, 2017.

145. Scannevin, R.H., Alexander, R., Haarlander, T.M., *et al.*, Discovery of a highly selective chemical inhibitor of matrix metalloproteinase-9 (MMP-9) that allosterically inhibits zymogen activation, *J. Biol. Chem.* 292, 17963–17974, 2017.
146. Lauer-Fields, J.L., Whitehead, J.K., Li, S., *et al.*, Selective modulation of matrix metalloproteinase 9 (MMP-9) functions via exocite inhibition, *J. Biol. Chem.* 283, 20087–20095, 2008.
147. Santamaria, S. and de Groot, R., Monoclonal antibodies against metzincin targets. *Br. J. Pharmacol.* 176, 52–66, 2019.
148. Santamaria, S., Yamamoto, K., Batkjaer, K., *et al.*, Antibody-based exosite inhibition of ADAMTS5(aggrecanase-2), *Biochem. J.* 471, 391–401, 2015.
149. Katschke, K.J., Jr., Wu, P., Ganesan, R., *et al.*, Inhibiting the alternative pathway complement activity by targeting the factor D exosite, *J. Biol. Chem.* 287, 12886–12892, 2012.
150. Breidenbach, M.A. and Brungenr, A., Substrate recognition strategy for botulinum neurotoxin A, *Nature* 432, 925–929, 2004.
151. Dong, J., Thompson, A.A., Fan, Y., *et al.*, A single domain Llama antibody potently inhibits the enzymatic activity of Botulinum neurotoxin by binding the non-catalytic α-exosite binding region, *J. Mol. Biol.* 397, 1106–1118, 2010.
152. Šilhár, P., Čapková, K., Salzmameda, N.T., *et al.*, Botulinum neurotoxin A protease: discovery of natural product exosite inhibitors, *JACS* 132, 2868–2869, 2010.
153. Panwar, P., Søe, K., Guido, R.V.C., *et al.*, A novel approach to inhibit bone resorption: exosite inhibitors against cathepsin K, *Brit. J. Pharmacol.* 173, 396–410, 2016.
154. MacPherson, D.J., Mills, C.L., Ondrecten, M.J., and Hardy, J.A., Tri-arginine exosite patch of cathepsin-6 recruits substrates for hydrolysis, *J. Biol. Chem.* 294, 71–88, 2019.
155. Kennan, R.M., Wong, W., Dhungyel, O.P., *et al.*, The subtilisin-likeprotease AprV2 is required for virulence and uses a novel disuphide-tethered exosite to bind substrate, *PLoS Pathogen* 6(11), e1001210, 2010.
156. Maianti, J.P., Tan, G.A., Vetere, A., *et al.*, Substrate-selective inhibitors that reprogram the activity of insulin-degrading enzyme, *Nat. Chem. Biol.* 15, 565–594, 2019.
157. Chu, Q., Chang, T., and Saghatelian, G., Substrate-selective enzyme inhibitors, *Trends Pharm. Sci.* 40, 716–718, 2019.
158. Bing, D.H., Cory, M., and Fenton, J.W., II, Exo-site affinity labeling of human thrombins. Similar labeling on the A chain and B chain/fragments of clotting α- and nonclotting β/γ-thrombins. *J. Biol. Chem.* 252, 8027–8034, 1977.
159. Bing, D.H., Andrews, J.M., and Cory, M., Affinity labeling of thrombin and other serine proteases with an extended reagent, in *Chemistry and Biology of Thrombin*, ed. R.L. Lundblad, J.W. Fenton, II, and K.G. Mann, pp. 159–177, Ann Arbor Science, Ann Arbor, MI, USA, 1977.
160. Cory, M., Andrews, J.M., and Bing, D.H., Design of exo affinity labeling reagents, *Methods Enzymol.* 46, 115–130, 1977.
161. Fenton, J.W., II and Singer, S.J., Affinity labeling of antibodies to the *p*-azophenyltrimethylammonium haptan and a comparison of affinity-labeled antibodies to two different specificities, *Biochemistry* 10, 1429–1437, 1971.
162. Baker, B.R., Lee, W.W., Tong, E., and Ross, L.O., Potential anticancer agents. LXVI. Non-classical antimetabolites III. 4-(iodoacetamido)-salicylic acids, an exo-alkylating irreversible inhibitor of glutamic dehydrogenase, *J. Am. Chem. Soc.* 83, 3713–3714, 1961.
163. Baker, B.R. and Erickson, E.H., Irreversible enzyme inhibitors. CXLIV. Proteolytic enzymes. VII. Additional active-site-directed irreversible inhibitors of trypsin derived from *m*- and *p*-(phenoxyalkoxy)benzamidines with a terminal sulfonyl fluoride, *J. Med. Chem.* 12, 112–117, 1969.

164. Baker, B.R., and Cory, M., Irreversible enzyme inhibitors. 186. Irreversible inhibitors of the C'1a component of complement derived from *m*-(phenoxypropoxy)benzamidine by bridging to a terminal sulfonyl fluoride, *J. Med. Chem.* 14, 805–808, 1971.

165. Markwardt, F., Landmann, H., and Walsmann, P., Comparative studies on the inhibition of trypsin, plasmin, and thrombin by derivatives of benzylamine and benzamidine, *Eur. J. Biochem.* 6, 502–506, 1968.

166. Gerazt, J.D. and Tidwell, R.R., The development of competitive reversible thrombin inhibitors, in *Chemistry and Biology of Thrombin*, ed. R.L. Lundblad, J.W. Feinman, II, and K.D. Mann, pp. 179–198, Ann Arbor Science, Ann Arbor, MI, USA, 1977.

167. Stürzebacher, J., Walmann, P., Voigt, B., and Wagner, G., Inhibition of bovine and human thrombin by derivatives of benzamidine, *Thromb. Res.* 36, 457–485, 1984.

168. Yang, B., Wu, H., Schnier, P.D., *et al.*, Proximity-enhanced SuFEx chemical cross-linker for specific and multitargeting cross-linking mass spectrometry, *Proc. Natl. Acad. Sci. USA* 115, 11162–11167, 2018.

169. De Cristofero, R., Akhavan, S., Altomare, C., *et al.*, A natural prothrombin mutant reveals an unexpected influence of A-chain structure on the activity of human α-thrombin, *J. Biol. Chem.* 279, 13035–13043, 2004.

170. Xiao, J., Melvin, R.L., and Salsbury, F.R., Jr., Probing light chain mutation effects on thrombin via molecular dynamics simulations and machine learning, *J. Biomol. Struct. Dyn.* 37, 982–999, 2019.

171. Bing, D.H., Laura, R., Andrews, J.M., and Cory, M., Exo-site affinity labeling of C1s, a component of the first component of complement by *m*-[*o*-(2-chloro-5-fluorosulfonyl-phenylureido)phenoxybutoxy]benzmidine, *Biochemistry* 17, 5713–5718, 1978.

172. Bar-Shavit, R., Kahn, A., Fenton, J.W., II, and Wilner, G.D., Chemotactic response of monocytes to thrombin, *J. Cell. Biol.* 96, 282–287, 1983.

173. Bar-Shavit, R., Kahn, A., Fenton, J.W., II, and Wilner, G.D. Receptor-mediated chemo-tactic response of a macrophage cell line (J774) to thrombin, *Lab. Invest.* 49, 702–707, 1983.

174. Sonder, S.A., and Fenton, J.W., II, Differential inactivation of human and bovine α-thrombins by exosite affinity-labeling reagents. *Thromb. Res.* 32, 623–629, 1983.

175. Kaminski, M. and McDonagh, J., Studies on the mechanism of thrombin. Interaction with fibrin, *J. Biol. Chem.* 258, 10530–10535, 1983.

176. Kaminski, M. and McDonagh, J., Inhibited thrombins. Interaction with fibrinogen and fibrin, *Biochem. J.* 242, 881–887, 1987.

177. Heene, D.L. and Matthias, F.R., Adsorption of fibrinogen derivatives on insolubilized fibrinogen and fibrin monomer, *Thromb. Res.* 2, 137–154, 1977.

178. Church, F.C., Pratt, C.W., Noyes, C.M., *et al.*, Structural and functional properties of human α-thrombin. Phosphopyridoxylated α-thrombin and γ-thrombin. Identification of lysyl residues in α-thrombin that are critical for heparin and fibrinogen interactions, *J. Biol. Chem.* 264, 18419–18425, 1989.

179. Berliner, L.J., Sugawara, Y., and Fenton, J.W., II, Human α-thrombin binding to non-polymerized fibrin-Sepharose: evidence for an anionic binding site, *Biochemistry* 24, 7005–7009, 1985.

180. Fenton, J.W., II., Olson, T.A., Zabinski, M.P., and Wilner, G.D., Anion-binding exosite of human α-thrombin and fibrin(ogen) recognition, *Biochemistry* 27, 7106–7112, 1988.193.

181 Rydel, T.J., Yin, M., Padmanabhan, K.P., *et al.*, Crystallographic structure of human γ-thrombin, *J. Biol. Chem.* 269, 22000–22006, 1994.

182. Gokulrangan, G., Zaidi, A., Michaelis, M.L., and Schöneich, C., Proteomic analysis of protein nitration in rat cerebellum: effect of biological aging, *J. Neurochem.* 100, 1494–1504, 2007.

183. Holligan, B.D., Ruotti, V., Jin, W., *et al.*, ProMoST (Protein Modification Screening Tool): a web-based tool for mapping protein modifications on two-dimensional gels, *Nucl. Acids Res.* 32, W638–W644, 2004.
184. Harmon, J.T. and Jamison, G.A., Activation of platelets by α-thrombin is receptor-mediated event. D-Phenylalanine-L-proly-L-arginine chloromethyl ketone-thrombin, but not N^{α}-tosyl-L-chloromethyl ketone-thrombin, binds to the high affinity thrombin receptor, *J. Biol. Chem.* 261, 15928–15933, 1986.
185. Wu, H.F., White, G.C., 2nd, Workman, E.F., Jr., Jenzano, J.W., and Lundblad, R.L. Affinity chromatography of platelets on immobilized thrombin: retention of catalytic activity by platelet-bound thrombin, *Thromb. Res.* 67, 429–427, 1992.
186. Puri, R.N., Zhou, F., Colman, R.F., and Coleman, R.W., Cleavage of a 100 kDa membrane protein (aggregin) during thrombin-induced platlet aggregation Is mediated by the high affinity thrombin recpetors, *Biochem. BIophys. Res. Commun.* 162, 1017–1024, 1989.
187. Griffith, M.J., Kinetics of the heparin-enhanced antithrombin III/thrombin reaction. Evidence for a template model for the mechanism of action of heparin, *J. Biol. Chem.* 257, 7360–7365, 1982.
188. Tsiang, M., Jain, A.K., Dunn, K.E., *et al.*, Functional mapping of the surface residues of human thrombin, *J. Biol. Chem.* 270, 16854–16863, 1995.
189. Bode, W. and Stubbs, M.T., Spatial structure of thrombin and guide to its multiple site of interaction, *Sem. Thromb.* 19, 321–333, 1993.
190. Cohen, G.H., Isola, V.J., Kuhns, J., Berman, P.W., and Eisenberg, R.J., Localization of discontinuous epitopes of herpes simplex virus glycoprotein D: use of a nondenaturing ("native" gel) system of polyacrylamide gel electrophoresis coupled with Western blotting, *J. Virol.* 60, 157–166, 1986.
191. Guillin, M.C., Bezeau, A., Bouton, M.C., and Jandrot-Perrus, M., Thrombin specificity, *Thromb. Haemost* 74, 129–133, 1995.
192. Bode, W., Mayr, I., Baumann, U., *et al.*, The refined 1.9 Å crystal structure of human α-thrombin: interaction with D-Phe-Pro-Arg chloromethylketone and significance of the tyr-pro-pro-trp insertion segment, *EMBO J.* 8, 3467–3475, 1989.
193. Chang, A. and Detwiler, T.C., The reaction of thrombin with platelet-derived nexin requires a secondary binding site in addition to the catalytic site, *Biochem. Biophys. Res. Commun.* 179, 11987–1204, 1991.
194. Evans, D.L., McGrogan, M., Scott, R.A., and Carrell, R.W. Protease specificity and heparin binding and activation of protease nexin 1, *J. Biol. Chem.* 266, 22307–22312, 1991.
195. Bouton, M.-C., Boulaftali, B., Richard, B., *et al.*, Emerging role of serpin E2/protease nexin-1 in hemostasis and vascular biology, *Blood* 119, 2452–2457, 2012.
196. Aymonnier, K., Kawecki, C., Venise, L., *et al.*, Targeting protease nexin-1, a natural anticoagulant serpin, to control bleeding and improve hemostasis in hemophilia, *Blood* 134, 1632–1644, 2019.
197. Tollefsen, D.M., Maimone, M.M., McGuire, E.A., and Peacock, M.E., Heparin cofactor II activation by dermatan sulfate, *Ann. N. Y. Acad. Sci.* 556, 116–122, 1989.
198. Rau, J.C., Mitchell, J.W., Fortenberry, Y.M., and Church, F.C., Heparin cofactor II: discovery, properties, and role In controlling vascular homeostasis, *Semin. Thromb. Hemost.* 37, 339–348, 2011.
199. Hortin, G.L., Tollefsen, D.M., and Benutto, B.M., Antithrombin activity of a peptide corresponding to residues 54–75 of heparin cofactor II, *J. Biol. Chem.* 254, 13979–13982, 1989.
200. Hortin, G.L. and Thorpe, B.L., Allosteric changes in thrombin's activity produced by peptides corresponding to segments of natural inhibitors and substrates, *J. Biol. Chem.* 266, 6866–6871, 1991.

201. Tollefsen, D.M., Majerus, D.W., and Blank, M.K., Heparin cofactor II. Purification and properties of a heparin-dependent inhibitor of thrombin in human plasma, *J. Biol. Chem.* 257, 2162–2169, 1992.

202. Maraganore, J.M., Bourdon, P., Jabelonski, J., Ramachadran, K.L., and Fenton, J.W., II., Design and characterization of hirulogs: a novel class of bivalent peptide inhibitors of thrombin, *Biochemistry* 29, 7095–7101, 1990.

203. Manithody, C., Yang, L., and Rezaie, A.R., Identification of exosite residues of factor Xa in response in recognition of PAR-2 on endothelial cells, *Biochemistry* 51, 2551–2557, 2012.

204. Bourdon, P., Fenton, J.W., II., and Maraganore, J.M., Affinity labeling of lysine-149 in the anion-binding site exosite of human α-thrombin with an N^{α}-(Dinitrofluorobenzyl) hirudin C-terminal peptide, *Biochemistry* 29, 6379–6384, 1990.

205. Dennis, S., Wallace, A., Hofsteenge, J., and Stone, S.R., Use of fragments of hirudin to investigate thrombin-hirudin interaction, *Eur. J. Biochem.* 188, 61–66, 1990.

206. Sheehan, J.P., Wu, Q., Tollefson, D.M., and Sadler, J.E., Mutagenesis of thrombin selectively modulates inhibition by serpins heparin cofactor II and antithrombin III. Interaction with the anion-binding exosite determines heparin cofactor II specificity, *J. Biol. Chem.* 268, 3639–3645, 1993.

207. Myles, T., Church, F.C., Whinna, H.C., Monard, D., and Stone, S.R., Role of thrombin anion-binding exosite-1 in the function of thrombin-serpin complexes, *J. Biol. Chem.* 273, 31203–31208, 1988.

208. Narendo, V., Zhu, L., Li., B., Wilken, J., and Weiss, M.A., Sex-specific gene regulation. The doublesex DM motifis a bipartite DNA-binding domain, *J. Biol. Chem.* 277, 43463–43473, 2002.

209. Schinzel, R., Active site lysine promotes catalytic function of pyridoxal-5'-phosphate In α-glucan phosphorylase, *J. Biol. Chem.* 266, 9429–9431, 1991.

210. Sheehan, J.P., Tollefsen, D.M., and Sadler, J.E., Heparin cofactor II is regulated allosterically and not primarily by template effects. Studies with mutant thrombins and glycosaminoglycans, *J. Biol. Chem.* 269, 32747–32751, 1994.

211. Cooper, S.T., Rezaie, A.R., Esmon, C.T., and Church, F.C., Inhibition of thrombin anion-binding exosite-2 mutant by the glycosaminoglycan-dependent serpins protein C inhibitor or heparin cofactor II, *Thromb. Res.* 107, 67–72, 2002.

212. Baglin, T.P., Carrell, R.W., Church, F.C., Esmon, C.T., and Huntington, J.A., Crystal structures of native and thrombin-complexed heparin cofactor II reveal a multistep allosteric mechanism, *Proc. Natl. Acad. Sci. USA* 99, 11079–11084, 2012.

213. Vogel, C.N., Kingdon, H.S., and Lundblad, R.L., Correlation of *in vivo* and *in vitro* inhibition of thrombin by plasma inhibitors, *J. Lab. Clin. Med.* 93, 661–673, 1979.

214. Arnaud, E., Lafay, M., Gaussen, P., *et al.*, An autoantibody directed against human thrombin anion-binding exosite in a patient with arterial thrombosis: effects on platelets, endothelial cells, and protein C activation, *Blood* 84, 1843–1850, 1994.

215. Baglin, T.P., Langdown, J., Frasson, R., and Huntington, J.A., Discovery and characterization of an antibody directed against exosite-1 of thrombin, *J. Thromb. Haemost.* 14, 137–142, 2015.

216. Noeske-Juneblut, C., Haendler, B., Donner, P., Triabin, a highly potent inhibitor of thrombin, *J. Biol. Chem.* 270, 28629–28034, 1995.

217. Zingali, R.B., Janrot-Perrus, M., Guillin, M.-C., and Bon, C., Bothrojaracin, a new thrombin inhibitor isolated from *Bothrops jararaca* venom. Characterization and mechanism of thrombin inhibition, *Biochemistry* 32, 10794–10802, 1993.

218. Assafim, R.B., Frattani, E.S., Ferreira, M.S., *et al.*, Exploiting the antithrombin effect of (pro)thrombin inhibitor bothrojaracin, *Toxicon* 119, 46–51, 2016.

219. Monteiro, R.G. and Zingali, R.B., Inhibition of prothrombin activation by bothrojaracin, a C-type lectin from *Bothrops jararaca* venom, *Arch. Biochem. Biophys.* 382, 123–128, 2000.

220. Zingali, R.B., Bianconi, M.L., and Monteiro, R.O., Interaction of bothrojaracin with prothrombin, *Haemostasis* 31, 273–278, 2001.

221. Arocas, V., Zingali, R.B., Guillin, M.-C., Bon, C., and Janrot-Perrus, M., Bothrojaracin: a potent two-site directed inhibition, *Biochemistry* 35, 9083–9089, 1996.

222. Arocas, V., Lemaire, C., Bouton, M.C., *et al.*, Inhibition of thrombin-catalyzed factor V activation by bothrojaracin, *Thromb. Haemost.* 79, 1157–1161, 1998.

223. Zingali, R.B., Ferreira, M.S., Assafim, M., Bothrojaracin, a *Bothrops jararaca* snake venom-derived (pro)thrombin inhibitor, as an anti-thrombotic molecule, *Pathophysiol. Haemost. Thromb.* 34, 160–163, 2005.

224. Iyer, J.K., Koh, C.Y., Kaziomirova, M., *et al.*, Avathrin, a novel thrombin inhibitor derived from a multicopy precursor in the salivary glands of the ixodid tick, *Amblyomma variegatumn*, *FASEB J.* 31, 2981–2995, 2017.

225. Seiler, S.M., Goldenberg, H.J., Michelk, I.M., Hunt, J.T., Zavoico, G.B., Multiple pathways of thrombin-induced platelet activation differentiated by desensitization and a thrombin exosite inhibitor, *Biochem. Biophys. Res. Commun.* 181, 636–643, 1991.

226. Glusa, E., Bretscheider, E., Daum, J., and Noeske-Jungblut, C., Inhibition of thrombin-mediated cellular effects by triabin, a highly potent anion-binding exosite thrombin inhibitor, *Thromb. Haemost.* 77, 1196–1200, 1997.

227. Arocas, V., Castro, H.C., Zingali, R.B., *et al.*, Molecular cloning and expression of bothrojaracin, a potent thrombin inhibitor from snake venom, *Eur. J. Biochem.* 248, 550–557, 1997.

228. Myles, T., Le Bonniec, B.F., and Stone, S.R., The dual role of thrombin's anion-binding exosite-1 in the recognition and cleavage of the protease-activated receptor 1, *Eur. J. Biochem.* 268, 70–77, 2001.

229. Pineda, A.O., Cantwell, A.M., Bush, L.A., Rose, T., and Di Cera, E., The thrombin epitope recognizing thrombomodulin is a highly cooperative hot spot in exosite I, *J. Biol. Chem.* 277, 32015–32109, 2002.

230. Bukys, M.A., Orban, T., Kim, P.Y., *et al.*, The structural integrity of anion binding exosite I of thrombin is required and sufficient for timely cleavage and activation of factor V and factor VIII, *J. Biol. Chem.* 281, 1869–1850, 2006.

231. Braun, P.J., Hofsteenge, J., Chang, J.-Y., and Stone, S.R., Preparation and characterization of proteolyzed forms of human α-thrombin, *Thromb. Res.* 50, 273–283, 1988.

232. Côté, H.C., Bajzar, L., Stevens, W.K., *et al.*, Functional characterization of recombinant human meizothrombin and Meizothrombin (desF1). Thrombomodulin-dependent activation of protein C and thrombin-activatable fibrinolysis inhibitor (TAFI), platelet aggregation, and antithrombin-III inhibition, *J. Biol. Chem.* 272, 6194–6200, 1997.

233. Corral-Rodriguez, M.A., Bock, P.E., Hernámdez-Caravajal, E., Gutiérez-Gallego, R., and Fuentes-Prior, P., Structural basis of thrombin-mediated factor V activation: the Glu^{666}–Glu^{672} sequence is critical for processing at the heavy chain-B domain junction, *Blood* 117, 7164–7173, 2011.

234. Billur, R., Ban, D., Sabo, T.M., and Maurer, M.C., Diciphering conformational changes associated with the maturation of the thrombin anion binding exosite, *Biochermistry* 56, 6343–6354, 2017.

235. Billur, R., Sabo, T.M., and Maurer, M.C., Thrombin exosite maturation and ligand binding at ABEI II help stabilize PAR-binding competent conformation at ABE I, *Biochemistry* 58, 1048–1060, 2019.

236. Myles, T., Le Bonniec, B.F., Betz, A., and Stone, S.R., Electrostatic steering and ionic tethering in the formation of thrombin-hirudin complexes: the role of the thrombin anion-binding exosite-1, *Biochemistry* 40, 4972–4979, 2001.

237. Cheng, Y., Slon-Usakiewicz, J.J., Wang, J., Pursima, E.O., and Konish, Y., Nonpolar interactions of thrombin and its inhibitors at the fibrinogen recognition exosite: thermodynamic analysis, *Biochemistry* 35, 13021–13209, 1996.

238. Devine, Z.H., Du, F., Li, Q., *et al.*, Pharmacological profile of JNJ-641-79375: a novel, long-acting exosite-1 thrombin inhibitor, *J. Parmacol. Exp. Ther.* 371, 375–384, 2019.

239. Wilson, S.J., Connolly, T.M., Peters, G., *et al.*, Exosite 1 thrombin inhibition with JNJ-64179375 inhibits thrombus formation in a human translational model of thrombosis, *Cardiovascular Res.* 115, 669–677, 2019.

240. Davie, E.W. and Ratnoff, O.D., Waterfall sequence for intrinsic blood clotting, *Science* 145, 1310–1312, 1964.

241. Akeson, A. and Theorell, H., Molecular weight and FMN content of crystalline old yellow enzyme, *Arch. Biochem. Biochem.* 65, 439–448, 1956.

242. Parry, M.A., Stone, S.R., Hofsteenge, J., and Jackman, M.P., Evidence for common structural changes in thrombin induced by active-site or exosite binding, *Biochem. J* 290, 665–670, 1993.

243. Han, J.-H. and Tollefsen, D.M., Ligand binding to thrombin exosite II induces dissociation of the thrombin-heparin cofactor II (L444R) complex, *Biochemistry* 37, 3203–3209, 1998.

244. Derechin, V.M., Blinder, M.A., and Tollefsen, D.M., Substitution of arginine for Leu444 in the reactive site of heparin cofactor II enhances the rate of thrombin inhibition, *J. Biol. Chem.* 265, 5623–5628, 1990.

245. Malovichko, M.V., Sabo, T.M., and Maurer, M.C., Ligand binding to anion-binding exosites regulates conformational properties of thrombin, *J. Biol. Chem.* 288, 8667–8678, 2013.

246. Cory, R.H., Koeppe, J.R., Bergquist, S., and Komives, E.A., Allosteric changes in solvent accessibility observed upon active site modification, *Biochemistry* 43, 5246–5255, 2004.

247. Yeh, C.H., Stafford, A.R., Leslie, B.A., Fredenburgh, J.C., and Weitz, J.I., Dabigatroban and argatroban diametrically modulate thrombin exosite function, *PLoS One* 11(6), e0157471, 2016.

248. Sabo, T.M. and Maurer, M.C., Biophysical investigation of GpIbα binding to thrombin anion binding exosite II, *Biochemistry* 48, 7110–7122, 2009.

249. Musci, G., Berliner, L.J., and Esmon, C.T., Evidence for multiple conformational changes in the active center of thrombin induced by complex formation with thrombomodulin: an analysis employing nitroxide spin-labels, *Biochemistry* 27, 769–773, 1988.

250. Hortin, G.L. and Benutto, B.M., Inhibition of thrombin clotting activity by synthetic peptide segments of its inhibitors and substrates, *Biochem. Biophys. Res. Commun.* 169, 437–443, 1990.

251. Lottenberg, R., Christensen, U., Jackson, C.M., and Coleman, P.L., Assay of coagulation proteases using peptide chromogenic and fluorogenic substrates, *Meth. Enzymol.* 80, 341–361, 1980.

252. Hortin, G.L. and Trimpe, B.L., Allosteric changes in thrombin's activity produced by peptides corresponding to segments of natural inhibitors and substrates, *J. Biol. Chem.* 266, 6866–6871, 1991.

253. Meh, D.A., Siebenlist, K.R., and Mosesson, M.W., Identification and characterization of the thrombin binding sites on fibrin, *J. Biol. Chem.* 271, 23121–23125, 1996.

254. Slade, C.L., Pizzo, S.V., Taylor, L.M., Jr., Steinman, H.M., and McKee, P.A., Characterization of fragment E from fibrinogen and cross-linked fibrin, *J. Biol. Chem.* 251, 1591–1596, 1976.

255. Bouton, M.-C., Jarrot-Perrus, M., Bezeaud, A., and Guillin, M.-C., Late fibrinogen fragment E modulates human α-thrombin specificity, *Eur. J. Biochem.* 215, 143–149, 1993.

256. Li, W., Adams, T.E., Nangalia, J., Esmon, C.T., and Huntington, J.A., Molecular basis of thrombin recognition by protein C inhibitor revealed by the 1.6 Å structure of the heparin-bridged complex, *Proc. Natl. Acad. Sci USA* 105, 4661–4666, 2008.

257. Ng, N.M., Quinsey, N.S., Matthews, A.Y., *et al.*, The effects of exosite occupancy on the substrate specificity of thrombin, *Arch. Biochem. Biophys.* 489, 48–54, 2009.

258. Kaminski, M. and McDonagh, J., Studies on the mechanism of thrombin. Interaction with fibrin, *J. Biol. Chem.* 258, 10530–10535, 1983.

259. Nilsen, D.W.T., Brosstad, F., Kieroff, P., and Godal, H.C., Binding of various thrombin fractions to fibrin and the influence of AT-III on their adsorption, *Thromb. Haemost.* 55, 352–356, 1986.

260. Bänninger, H., Lämmle, B., and Furlan, M., Binding of α-thrombin to fibrin depends on the quality of the fibrin network, *Biochem. J.* 298, 157–163, 1994.

261. Mirshahi, M., Soria, J., Soria, C., *et al.*, Evaluation of the inhibition by heparin and hirudin of coagulation activation during r-TPA thrombolysis, *Blood* 74, 1025–1031, 1989.

262. Naki, M.C. and Shafer, J.A., α-Thrombin-catalyzed hydrolysis of fibrin I. Alternative binding modes and the accessibility of the active site in fibrin I-bound α-thrombin, *J. Biol. Chem.* 265, 1401–1407, 1990.

263. Lewis, D.S., Shields, P.P., and Shafer, J.A., Characterization of the kinetic pathway for liberation of fibrinopeptides during assembly of fibrin, *J. Biol. Chem.* 260, 10192–10199, 1985.

264. Kanoaoka, Y., Takahashi, T., Nakayama, H., and Tanizawa, K., New fluorogenic substrates for subtilisin, *Chem. Pharm. Bull.* 33, 1721–1724, 1985.

265. Jabaiah, A.M., Getz, J.A., Witkowski, W.A., Hardy, J.A., and Daughterty, P.S., Identification of protein-interacting peptides that enhance substrate cleavage kinetics, *Biol. Chem.* 393, 933–941, 2012.

266. Boulware, K.T. and Daugherty, P.S., Protease specificity determination by using cellular libraries of peptide substrate (CLiPs), *Proc. Natl. Acad. Sci. USA* 103, 7583–7588, 2006.

267. Vu, T.-K.H., Wheaton, V.L., Hung, D.T., Charo, I., and Coughlin, S.R., Domains specifying thrombin-receptor interaction, *Nature* 353, 674–677, 1991.

268. Vieiva, S.M., dos Reis, F.G., Geraldo, R., *et al.*, Investigation of thrombin activity with PAR1-based fluorogenic peptides, *Prot. Pept. Lett.* 20, 1129–1135, 2013.

269. Chahal, G., Thorpe, M., and Hellman, L., The importance of exosite interactions for substrate cleavage by human thrombin, *PLoS One* 10(6), e0129511, 2015.

270. *The Aptamer Handbook: Functional Oligonucleotides and Their Applications*, Wiley-VCH Verlag GmbH, Weinheim, Germany, 2006.

271. *Nucleic Acid Aptamers, Selection, Characterization, and Application*, ed. G. Mayer, Springer, New York, USA, 2006.

272. Hughes, Q.W., Le, B.I., Gilmore, G., *et al.*, Construction of a bivalent thrombin binding aptamer and its antidote with improved properties, *Molecules* 22(10), E1770, 2017.

273. Ellington, A.D. and Szostak, JW., *In vitro* selection of RNA molecules that bind to specific ligands, *Nature* 346, 818–822, 1990.

274. Ellington, A.D. and Szostak, J.W., Selection *in vitro* of single-stranded DNA molecules that fold into specific ligand-binding structures, *Nature* 355, 850–852, 1992.

275. Kang, J., Lee, M.S., and Gorenstein, D.G., The enhancement of PCR amplification of a random sequence DNA library by DMSO and betaine: application to *in vitro* combinatorial selection of aptamers, *J. Biochem. Biophys. Methods* 64, 147–151, 2005.

276. Bock, L.C., Griffin, L.C., Lathan, J.A., Vermass, E.H., and Toole, J.J., Selection of single-stranded DNA molecules that bind and inhibit human thrombin, *Nature* 355, 564–566, 1992.

277. Mocaya, R.E., Schultze, P., Smith, F.N., Roe, J.A., and Feigon, J., Thrombin binding DNA aptamer forms a quadruplex structure in solution, *Proc. Natl. Acad. Sci. USA* 90, 3745–3748, 1993.

278. Wu, Q., Tsiang, M., and Sadler, J.E., Localization of the single-stranded DNA binding site in the thrombin anion-binding exosite, *J. Biol. Chem.* 267, 24408–24017, 1992.

279. Padmanabhan, K., Padmanabhan, K.P., Ferrarai, J.D., and Tulinsky, A., The structure of α-thrombin inhibited by a 15-mer single-stranded DNA aptamer, *J. Biol. Chem.* 268, 17651–17654, 1993.

280. Krauss, I.R., Merlino, A., Giancola, C., *et al.*, Thrombin-aptamer recognition: a revealed ambiguity, *Nucleic Acids Res.* 39, 7858–7867, 2017.

281. Zhang, J., Ogorzalek Loo, R.R., and Loo, J.A., Structural characterization of a thrombin-aptamer complex by high resolution native top-down mass spectrometry, *J. Am. Soc. Mass Spectrom.* 28, 1815–1822, 2017.

282. Pica, A., Krauss, I.R., Merlino, A., *et al.*, Dissecting the contribution of thrombin exosite I in the recognition of thrombin binding aptamer, *FEBS J.* 280, 6581–6588, 2013.

283. Xiang, Y., Tang, A., and Lu, Y., Abasic site-containing DNAzyme and aptamer for label-free fluorescent detection of Pb^{2+} and adenosine with high sensitivity, selectivity, and high dynamic range, *J. Am. Chem. Soc.* 131, 15352–15357, 2009.

284. Nagatoishi, S. and Sugimoto, N., Interaction of water with the G-quadruplex loop contributes to the binding energy of G-quadruplex to protein, *Mol. Biosystems* 8, 2766–2770, 2012.

285. Fredenbergh, J.C., Stafford, A.R., and Weitz, J.J., Conformational changes with thrombin when complexed by serpins, *J. Biol. Chem.* 276, 44828–44834, 2001.

286. Owen, M.C., Brennan, S.O., Lewis, J.H., and Carrell, R.W., Mutation of antitrypsin to antithrombin. α1 Antitrypsin Pittsburgh (358 Met leads to Arg), a fatal bleeding disorder, *N. Engl. J. Med.* 309, 694–698, 1983.

287. Emmerich, J., Alhenc-Gelas, M., Gandrille, S., *et al.*, Mechanism of protein C deficiency in a patient with arginine 358 α1-antitrypsin (Pittsburgh mutation): role in the maintenance of hemostatic balance, *J. Lab. Clin. Med.* 125, 531–539, 1995.

288. Sheffield, W.F. and Bhakta, V., The M358R variant of α_1-proteinase inhibitor inhibits coagulation factor VIIa, *Biochem. Biophys. Res. Commun.* 470, 710–713, 2016.

289. Paborsky, L.R., McCurdy, S.N., Griffin, L.C., Toole, J.J., and Leung, L.L.K., The single-stranded DNA aptamer-binding site of human thrombin, *J. Biol. Chem.* 268, 20808–20811, 1993.

290. Li, W., Johnston, D.J.D., Adams, T.E., *et al.*, Thrombin inhibition by serpins disrupts exosite II, *J. Biol. Chem.* 285, 38621–38629, 2010.

291. Kankia, B.L. and Marky, L.A., Folding of the thrombin aptamer into a G-quadruplex with Sr^{2+}: stability, heat, and hydration, *J. Am. Chem. Soc.* 123, 10799–10804, 2001.

292. Mondragon-Sanchez, J.A., Liquier, J., Shafer, R.H., and Taillandier, E., Tetraplex structure formation in the thrombin-binding DNA aptamer by metal cations measured by vibrational spectroscopy, *J. Biomol. Struct.* 22, 365–373, 2004.

293. Tarjkovski, M., Sket, P., and Plavec, J., Cation localization and movement within DNA thrombin binding aptamer in solution, *Org. Biomol. Chem.* 7, 4677–4684, 2009.

294. Reshetnikov, R.V., Sponer, J., Rasokhina, O.I., *et al.*, Cation binding to 15-TBA quadruplex DNA is a multiple-pathway cation-dependent process, *Nucleic Acids Res.* 39, 9789–9802, 2011.

295. Smestad, J. and Maher, L.J., 3rd, Ion-dependent conformational switch by a DNA aptamer that induces remyelination in a mouse model of multiple sclerosis, *Nucleic Acids Res.* 41, 1329–1342, 2013.

296. Zavyelov, E., Tagiltsev, G., Reshetnikov, R., Arutyunyan, A., and Kopylov, A., Cation coordination alters the conformation of a thrombin-binding G-quadruplex DNA aptamer, *Nucleic Acid Therapeutics* 26, 299–307, 2016.

297. Russo Krauss, I., Napolitano, V., Petraccone, L., *et al.*, Duplex/quadruplex oligonucle-otides: role of the duplex domain in the stabilization of a new generation of highly effective anti-thrombin aptamers, *Int. J. Biol. Macromol.* 107(Pt B), 1697–1705, 2018.

298. Zavylova, E., Golovin, A., Reshetnikov, R., *et al.*, Novel modular DNA aptamer for human thrombin with high anticoagulant activity, *Curr. Med. Chem.* 18, 3343–3350, 2011.

299. Zavyalova, E., Golovin, A., Pavlova, G., and Kopylov, A., Module-activity relationship of G-quadruplex based DNA aptamers for human thrombin, *Curr. Med. Chem.* 20, 4836–4843, 2013.

300. Troisi, R., Balasco, N., Vitagliano, L., and Sica, F., Molecular dynamics simulations of human α-thrombin in different structural contexts: evidence for an aptamer-guided cooperation between the two exosites, *J. Biomol. Struct. Dyn.* 23, 1–19, 2020.

301. Radi, A.E., Acero Sánchez, J.L., Baldrich, E., and O'Sullivan, C.K., Reagentless, reus-able, ultrasensitive electrochemical molecular beacon aptasensor, *J. Am. Chem. Soc.* 128, 117–124, 2006.

302. Tan, X., Dey, S.K., Telmer, C., *et al.*, Aptamers act as activators for the thrombin-mediated hydrolysis of a peptide substrate, *ChemBioChem* 15, 205–208, 2014.

303. Hughes, Q.W., Le, B.T., Gilmore, G., Baket, R.I., and Veedu, R.N., Construction of a bivalent thrombin binding aptamer and its antidote with improved properties, *Molecules* 22(10), E1770, 2017.

304. Riccardi, C., Meyer, A., Vasseur, J.J., *et al.*, Fine-tuning the properties of the thrombin binding aptamers through cyclization: effect of 5',3' connecting linker on the aptamer stability and anticoagulant activity, *Bioorg. Chem.* 94, 103379, 2020.

305. Nimjee, S.M., White, R.R., Becker, R.C., and Sullinger, B.A., Aptamers as therapeutics, *Annu. Rev. Pharmacol. Toxicol.* 57, 61–79, 2017.

306. Kovacevic, K.D., Gilbert, J.C., and Jilma, B., Pharmacokinetics, pharmacodynamics and safety of aptamers, *Adv. Drug Deliv. Rev.* 134, 36–50, 2018.

307. Jiang, N., Zhu, T., and Hu, Y., Competitive aptasensor with gold nanoparticle dimers and magnetic nanoparticlesfor SERS-based determination of thrombin, *Mikrochim. Acta* 186(12), 747, 2019.

308. Pol, L., Acosta, L.K., Ferre'-Borrull, J., and Marsal, L.F., Aptamer-based nanoporous anodic alumina interferometric biosensor for real-time thrombin detection, *Sensors* (Basal) 19(20), E4543, 2019.

309. Alamrani, N.A., Greenway, G.M., Pamme, N., Goodard, N.J., and Gupta, R., A feasibil-ity study of a leaky waveguide aptasensor for thrombin, *Analyst* 144, 6048–6054, 2019.

310. Cren, H.J., Chen, R.L.C., Hsieh, B.C., *et al.*, Label-free and reagentless capacitive apta-sensor for thrombin, *Biosens. Bioelectron.* 131, 53–59, 2019.

311. Lira, A.L., Ferreira, R.S., Torquato, R.J.S., *et al.*, Allosteric inhibition of α-thrombin enzymatic activity with ultrasmall gold nanoparticles, *Nanoscale Adv.* 1, 378–388, 2019.

312. Troisi, R., Balasco, N., Vitagliano, L., and Sica, F., Molecular dynamics simulation of human α-thrombin in different structural contexts: evidence for an aptamer-guided cooperation between two exosites, *J. Biomol. Struct. Dyn.*, 39(6): 2199–2209, 2021, doi: 10.1080/07391102.2020.1746693.

313. Kereiakas, D.I., Ximelagatran: pharmacokinetics and pharmacodynamics of a new strat-egy for oral direct thrombin inhibition, *Rev. Cardiovasc. Med.* 5(suppl 5), S4–S11, 2004.

314. Lepor, N.E., Anticoagulation for acute coronary syndromes: from heparin to direct thrombin inhibitors, *Rev. Cardiovasc. Med.* 8(suppl 3), S9–S17, 2007.

315. Bar-Shavit, R., Eldor, A., and Vlodarsky, I., Binding of thrombin to subencothelial extracellular matrix. Protection and expression of functional properties, *J. Clin. Invest.* 84, 1096–1104, 1989.

316. Bar-Shavit, R., Benezra, M., Eldor, A., *et al.*, Thrombin immobilized to extracellular matrix is a potent mitogen for vascular smooth muscle cells: nonenzymatic mode of action, *Cell Regul.* 1, 453–463, 1990.

317. Pálos, L.A., Protective effect of heparin on the inactivation of thrombin by heat, *Proc. Soc. Exptl. Biol. Med.* 71, 471–472, 1949.

318. Pálos, L.A., Protecting effect of heparin on the oxidation of thrombin, *Experentia* 5, 207–208, 1949.

319. Spitzer, J.J., The influence of heparin on the activation of energy of thrombin inactivation, *Can. J. Med. Sci.* 30, 122–124, 1952.

320. Pálos, L.A., Blasko, G., Kosztovics, A., and Machovich, R., Air oxidation and reactivation of reduced thrombin, *Acta Physiol. Acad. Sci. Hurg.* 51, 299–303, 1977.

321. Oshima, G., Time-dependent conformational change of thrombin molecules induced by sulfated polysaccharides, *Chem. Pharm. Bull.* 37, 1324–1328, 1989.

322. Suzuki, K. and Hashimoto, S., Effect of dextran sulphates on thrombin activity, *J. Clin. Pathol.* 32, 439–444, 1979.

323. Oshima, G., Nagai, T., and Nagasawa, K., Abolition by dextran sulfate of the heparin-accelerated antithrombin III/thrombin reaction, *Thromb. Res.* 35, 601–611, 1984.

324. Uchiyama, H. and Nagasawa, K., Preparation of biologically active fluorescent heparin composed of fluorescein-labeled species and its behavior to antithrombin III, *J. Biochem.* 89, 185–192, 1981.

325. Jordan, R.E., Oosta, G.M., Gardner, W.T., and Rosenberg, R.D., The binding of the low molecular weight heparin to hemostatic enzymes, *J. Biol. Chem.* 255, 10073–10080, 1980.

326. Oshima, G., Uchiyama, H., and Nagasawa, K., Effect of NaCl on the association of thrombin with heparin, *Biopolymers* 25, 527–537, 1986.

327. Olson, S.T., Halvorson, H.R., and Björk, I., Quantitative characterization of the thrombin-heparin interaction. Discrimination between specific an nonspecific binding models, *J. Biol. Chem.* 266, 6342–6352, 1991.

327. Machovich, R., Blaskó, G., and Pálos, L.A., Action of heparin on thrombin-antithrombin reaction, *Biochim. Biophys. Acta* 379, 193–200, 1975.

328. Griffith, M.J., Kingdon, H.S., and Lundblad, R.L., Fractionation of heparin on affinity chromatography on covalently-bound human α-thrombin, *Biochem. Biophys. Res. Commun.* 83, 1198–1205, 1978.

329. Björklund, M. and Hearn, M.T.W., Characterization of silica-based heparin affinity sorbents from equilibrium binding studies on plasma fractions containing thrombin, *J. Chromatog. A.* 762, 113–133, 1997.

330. Sheehan, J.P. and Sadler, J.E., Molecular mapping of the heparin-binding exosite of thrombin, *Proc. Natl. Acad. Sci. USA* 91, 5518–5522, 1994.

331. Nordenmann, B. and Björk, I., Studies on the binding of heparin to prothrombin and thrombin and the effect of heparin binding on thrombin activity, *Thromb. Res.* 12, 755–765, 1978.

332. Griffith, M.J., Kingdon, H.S., and Lundblad, R.L., The interaction of thrombin with human α-thrombin: effect on the hydrolysis of anilide peptide substrates, *Arch. Biochem. Biophys.* 195, 378–384, 1979.

333. Griffith, M.J., Kingdon, H.S., and Lundblad, R.L., Hydrolysis of N-α-benzoyl-L-phenylalanyl-L-valyl-arginine p-nitroanilide by human alpha thrombin in the presence of heparin, *Thromb. Res.* 17, 83–90, 1980.

334. Goodwin, C.A., Deadman, J.J., Le Bonniec, B.F., *et al.*, Heparin enhances the catalytic activity of des-ETW-thrombin, *Biochem. J.* 315, 77–83, 1995.

335. Le Bonniec, B.F., Guinto, E.R., and Esmon, C.T., Interaction of thrombin des-ETW with antithrombin, the Kunitz inhibitors, thrombomodulin and Protein C. Structural

link between the autolysis loop and the Tyr-Pro-Pro-Trp insertion of thrombin, *J. Biol. Chem.* 267, 19341–19348, 1992.

336. Fredenburgh, J.C., Stafford, A.R., Leslie, B.A., and Weitz, J.I., Bivalent binding to γ_A/γ'-fibrin engages both exosites of thrombin and protects it from inhibition by the antithrombin-heparin complex, *J. Biol. Chem.* 283, 2470–2477, 2008.

337. Mikami, T. and Kitagawa, H., Biosynthesis and functions of chondroitin sulfate, *Biochim. Biophys. Acta* 1830, 4719–4733, 2013.

338. Weiter, H. and Isemann, B.H., Thrombomodulin, *J. Thromb. Haemost.* 1, 1515–1524, 2003.

339. Preissner, K.T., Koyama, T., Müller, D., Tschopp, J., Müller-Berghaus, G., Domain structure of the endothelial cell receptor thrombomodulin. Evidence for a glycoaminoglycan-dependent secondary binding site for thrombin, *J. Biol. Chem.* 265, 4915–4927, 1990.

340. Ye, J., Rezaie, A.R., and Esmon, C.T., Glycosaminoglycan contributions to both protein C activation an thrombin inhibition involve a common arginine-rich site in thrombin that includes Arginine 93, 97, 101, *J. Biol. Chem.* 269, 17965–17970, 1994.

341. Ye, J., Esmon, C.T., and Johnson, A.E., The chondroitin sulfate moiety of thrombo-modulin binds a second molecule of thrombin, *J. Biol. Chem.* 268, 2373–2379, 1993.

342. Dong, J.-f.L., Li, C.Q., and López, J.A., Tyrosine sulfation of the glycoprotein Ib-IX complex: identification of sulfated residues and effect on ligand binding, *Biochemistry* 33, 13496–13953, 1994.

343. Marchese, P., Murata, M., Mazzucato, M., *et al.*, Identification of three tyrosine residues of glycoprotein Ibα with distinct roles in von Willebrand factor and α-thrombin binding, *J. Biol. Chem.* 270, 9571–9576, 1995.

344. Mehta, A.Y., Thakkar, J.N., Mohammed, B.M., *et al.*, Targeting the GPIbα binding site in thrombin to simultaneously induce dual anticoagulant and antiplatelet effects, *J. Med. Chem.* 57, 3030–3039, 2014.

345. Nagashima, R., Development and characteristics of sucralfate, *J. Clin. Gastroenterol.* 3(suppl 2), 103–110, 1981.

346. Samloff, I.M., Inhibition of peptic aggression by sucralfate. The view from the ulcer crater, *Scand. J. Gastroenterol Suppl.* 83, 7–11, 1983.

347. Edmonds, M., Lázaro-Martinez, J., Alfayata-Garcis, J.M., *et al.*, Sucrose octasulfate dressing versus control dressings in patients with neuroischemic diabetic foot ulcers (Explorer): an interational multicenter double-blinded randomized controlled trials, *The Lancet Diabetes & Endocrinology* 6, 186–196, 2018.

348. Lohmann, R., Augustin, M., Lowell, H., *et al.*, Cost-effectiveness of TLC-sucrose octa-sulfate versus control dressings in the treatment of diabetic foot ulcers, *J. Wound Care* 28, 808–816, 2019.

349. Sarilla, S., Habib, S.Y., Kravtsov, D.V., *et al.*, Sucrose octasulfate selectively accelerate thrombin inactivation by heparin cofactor II, *J. Biol. Chem.* 285, 8278–8289, 2010.

350. Li, W., Johnson, D.J.D., Esmon, C.T., and Huntington, J.A., Structure of the antithrom-bin-thrombin-heparin ternary complex reveals the antithrombotic mechanism of hepa-rin, *Nature Struct. Mol. Biol.* 11, 857–869, 2004.

351. Desai, B.J., Boothello, R.S., Mehta, A.Y., *et al.*, Interaction of thrombin with sucrose octasulfate, *Biochemistry* 50, 6973–6982, 2011.

352. Volkin, D.B., Verticelli, A.M., Marfia, K.E., *et al.*, Sucralfate and soluble sucrose octa-sulfate bind and stabilize acidic fibroblast growth factor, *Biochim. Biophys. Acta* 1203, 18–26, 1993.

353. Arunkumar, A.I., Kumar, T.K., Kathir, K.M., *et al.*, Oligomerization of acidic fibro-blast growth fator is not a prerequisite for its cell proliferation activity, *Protein Sci.* 11, 1050–1061, 2002.

354. Pelligrini, L., Role of heparan sulfate in fibroblast growth factor signalling: A structural view, *Curr. Opin. Struct. Biol.* 11, 629–633, 2003.

355. Yeh, B.K., Eliseenkova, A.V., Plotnikov, A.N., *et al.*, Structural basis for activation of fibroblast growth factor signaling by sucrose octasulfate, *Molec. Cell Biol.* 22, 7184–7192, 2002.

356. Fannon, M., Forsten-Williams, K., Nugent, M.A., *et al.*, Sucrose octasulfate regulates fibroblast growth factor-2 binding, transport, and activity: potential for regulation of tumor growth, *J. Cell Physiol.* 215, 434–441, 2008.

357. Desfougéres, Y., Saiardi, A., and Azevedo, C., Inorganic polyphosphate in mammals: where's Wally?, *Biochem. Soc. Trans.* 48, 95–101, 2020.

358. Silcox, D.C., Jacobelli, S., and McCarty, D.J., Identification of inorganic pyrophosphate in human platelets and its release on stimulation with thrombin, *J. Clin. Invest.* 52, 1595=1600, 1973.

359. Xie, L. and Jakob, U., Inorganic polyphosphate, a multifunctional polyanionic protein scaffold, *J. Biol. Chem.* 294, 2180–2190, 2019.

360. Puy, C., Tucker, E.I., Wong, Z.C., *et al.*, Factor XII promotes blood coagulation independent of factor XI in the presence of long-chain polyiphosphates, *J. Thromb. Haemost.* 11, 1341–1352, 2013.

361. Gaisiewicz, J.M., Smith, S.A., and Morrissey, J.H., Polyphosphate and RNA differentially modulate the contact pathway of blood clotting, *J. Biol. Chem.* 292, 1808–1814, 2017.

362. Mutch, N.J., Myles, T., Leung, L.L.K., and Morrissey, J.H., Polyphosphate binds with high affinity to exosite II of thrombin, *J. Thromb. Haemost.* 8, 548–555, 2009.

363. Choi, S.H., Smith, S.A., and Morrissey, J.H., Polyphosphate is a cofactor for the activation of factor XI by thrombin, *Blood* 118, 6983–6970, 2011.

364. Morrissey, J.H., Polyphosphate multi-tasks, *J. Thromb. Haemost.* 10, 2313–2314, 2012.

365. Baker, C.J., Smith, S.A., and Morrissey, J.H., Polyphosphate in thrombosis, hemostasis, and inflammation, *Res. Pract. Thromb. Haemost.* 3, 18–25, 2019.

366. Okumura, T. and Jamieson, G.A., Platelet glycocalicin: a single receptor for platelet aggregation induced by thrombin and ristocetin, *Thromb. Res.* 8, 701–706, 1976.

367. Okumura, T., Hasitz, M., and Jamieson, G.A., Platelet glycocalicin—interaction with thrombin and role as thrombin receptor on platelet surface, *J. Biol. Chem.* 253, 3435–3443, 1978.

368. De Cristofaro, R., De Candia, F., Landolfi, R., Rutello, S., and Hall, S.W., Structural and functional mapping of the thrombin domain involved in the binding to platelet glycoprotein 1b, *Biochemistry* 40, 13268–13273, 2001.

369. de Cristofaro, R. de Canada, E., Croce, G., Morosetti, R., and Landolfi, R., Binding of human α-thrombin to platelet Gp1b: energetics and functional effects, *Biochem. J.* 332, 643–650, 1998.

370. De Candia, E., Hall, S.W., Ruttela, S., *et al.*, Binding of thrombin to glycoprotein 1b accelerates the hydrolysis of Par 1 on intact platelets. *J. Biol. Chem.* 276, 4692–4698, 2001.

371. Li, C.Q., Vindigni, A., Sadler, J.E., and Wardell, M.R., Platelet glycoprotein Ibα binds to thrombin anion binding exosite II induced allosteric changes in the activity of thrombin, *J. Biol. Chem.* 276, 6161–6168, 2001.

372. Celikel, R., McClintock, R.A., Roberts, J.R., *et al.*, Modulation of α-thrombin function by distinct interactions with platelet glycoprotein Ibα, *Nature* 301, 218–221, 2003.

373. Dumas, J.J., Kumar, R., Seehra, J., Somers, W.S., and Mosyak, L., Crystal structure of the GpIbα-thrombin complex essential for platelet aggregation, *Nature* 301, 222–226, 2003.

374. Sabo, T.M. and Mauer, M.C., Biophysical investigation of GPIbα binding to thrombin anion binding exosite II, *Biochemistry* 48, 7110–7122, 2009.

375. Lechtenberg, B.C., Freund, S.M.V., and Huntington, J.A., GpIbα interacts with exosite II of thrombin, *J. Mol. Biol.* 426, 881–893, 2014.

376. Uitte de Willige, S., Standeven, K.F., Philippou, H., and Ariëns, R.A., The pleiotropic role of the fibrinpgen γ' chain in hemostasis, *Blood* 114, 3994–4001, 2009.

377. Chung, D.W. and Davie, E.W., γ and γ' chain of human fibrinogen are produced by an alternative mRNA processing, *Biochemistry* 23, 4232–3236, 1984.

378. Fornace, A.J. Jr., Cummings, D.E., Comeau, C.M., Kant, J.A., and Crabtree, G.R., Structure of the human γ-fibrinogen gene. Alternative mRNA splicing near the 3' end of the gene produces γA and γB forms of γ-fibrinogen, *J. Biol. Chem.* 259, 12826–12830, 1984.

379. Meh, D.A., Siebenlist, K.R., Brennan, S.L., Holyst, T., and Mosesson, M.W., The amino acid sequence in fibrin responsible for high affinity thrombin binding, *J. Thromb. Haemost.* 85, 470–474, 2001.

380. Lovely, R.S., Moaddel, M., and Farrell, D.H., Fibrinogen γ' chain binds to thrombin exosite II, *J. Thromb. Haemost.* 1, 124–131, 2003.

381. Lovely, R.S., Boshkov, L.K., Marzec, U.M., Hanson, S.R., and Farrell, D.H., Fibrinogen γ' chain carboxy terminal peptide selectively inhibits the intrinsic coagulation pathway, *Brit. J. Haematol.* 139, 494–503, 2007.

382. De Cristofaro, R. and De Filippis, V., Interaction of the 268–282 region of glycoprotein Ibα with the heparin-binding site of thrombin inhibits the enzyme activation of factor VIII, *Biochem. J.* 373, 593–601, 2003.

383. Pieters, J., Lindhout, T., and Hemker, H.C., In situ-generated thrombin is the only enzyme that effectively activates factor VIII and factor V in thromboplastin-activated plasma, *Blood* 74, 1021–1024, 1989.

384. Siebenlist, K.R., Mosesson, M.W., Hernandez, I., *et al.*, Studies on the basis for the properties of fibrin produced from fibrinogen-containing γ' chains, *Blood* 106, 2730–2736, 2005.

385. Sabo, T.M., Farrell, D.H., and Maurer, M.C., Conformational analysis of γ' peptide (410–427) interactions with thrombin anion binding exosite II, *Biochemistry* 45, 7434–7445, 2006.

386. Arni, R.K., Padmanabhan, K., Padmanabhan, K.P., Wu, T.P., and Tulinsky, A., Structures of the noncovalent complexes of human and bovine prothrombin fragment 2 with human PPACK-thrombin, *Biochemistry* 32, 4727–4737, 1993.

387. Heldebrant, C.M. and Mann, K.G., The activation of prothrombin. I Isolation and preliminary characterization of intermediates, *J. Biol. Chem.* 248, 3642–3652, 1973.

388. Myrmel, K.H., Lundblad, R.L., and Mann, K.G., Characteristics of the association between prothrombin fragment 2 and α-thrombin, *Biochemistry* 15, 1767–1773, 1976.

389. Walker, F.J. and Esmon, C.T., The effect of prothrombin fragment 2 on the inhibition of thrombin by antithrombin III, *J. Biol. Chem.* 254, 5618–5622, 1979.

390. Bock, P.E., Active-site-selective labeling of blood coagulation proteinases with fluorescence probes by the use of thioester peptide chloromethyl ketones. II Properties of thrombin derivatives as reporters of prothrombin fragment 2 binding and specificity of the labeling approach for other proteinases, *J. Biol. Chem.* 267, 14974–14981, 1992.

391. Liaw, P.C.Y., Fredenburgh, J.C., Stafford, A.R., *et al.*, Localization of the thrombin-binding domain on prothrombin fragment 2, *J. Biol. Chem.* 273, 8932–8239, 1988.

392. Martin, P.D., Makowski, M.G., Box, J., *et al.*, New insights into the regulation of the blood clotting cascade derived from x-ray crystal structure of meizothrombin-des1 in complex with PPACK, *Structure* 5, 1681–1693, 1997.

393. Chung, C.H., Ives, H.E., Almeda, S., and Goldberg, A.L., Purification from *Escherichia coli* of a periplasmic protein that is a potent inhibitor of pancreatic proteases, *J. Biol. Chem.* 258, 11032–11038, 1983.

394. McGrath, M.E., Hines, W.M., Sakanan, J.A., Fletterick, R.J., and Craik, C.S., The sequence and reactive site of ecotin: a general inhibitor of pancreatic serine proteases from *Escherichia coli*, *J. Biol. Chem.* 6620–6625, 1991.

395. Hall, P.K. and Roberts, R.C., Methionine oxidation and inactivation of alpha 1-proteinase inhibitor by Cu^{2+} and glucose, *Biochim. Biophys. Acta* 1121, 325–330, 1992.

396. Wang, S.X., Esmon, C.T., and Fletterick, R.J., Crystal structure of thrombin-ecotin reveals conformational changes and external interactions, *Biochemistry* 40, 10038–10046, 2001.

397. Yang, S.Q., Wang, C.-i., Gilmor, S.A., *et al.*, Ecotin: a serine protease inhibitor with two distinct and interactive binding sites, *J. Mol. Biol.* 279, 945–957, 1998.

398. De Cristofaro, R. and De Candia, E., Thrombin domains: structure, function and interaction with platelet receptors, *J. Thrombosis Thrombolysis* 15, 151–163, 2003.

399. Pál, G., Sprengel, G., Pathy, A., and Gráf, L., Alteration of the specificity of ecotin, a *E. coli* serine proteinase inhibitor, by site directed mutagenesis, *FEBS Lett.* 342, 57–60, 1994.

400. Seymour, J.L., Lindquist, R.N., Dennis, M.S., *et al.*, Ecotin is a potent anticoagulant and reversible tight-binding inhibitor of factor Xa, *Biochemistry* 33, 3949–3958, 1994.

401. Castro, H.C., Montiero, R.G., Assafim, M., *et al.*, Ecotin modulates thrombin activity though exosite-2 interactions, *Int. J. Biochem. Cell Biol.* 38, 1893–1900, 2006.

402. Hogg, P.J., Jackson, C.M., Labanowski, J.K., and Bock, P.E., Binding of fibrin monomer and heparin to thrombin in a ternary complex alters the environment of the thrombin catalytic site, reduces affinity for hirudin, and inhibits cleavage of fibrinogen, *J. Biol. Chem.* 271, 26088–26095, 1996.

403. Kirtley, M.E., and Koshland, D.E., Jr., The introduction of a "reporter" group at the active site of glyceraldehyde-3-phosphate dehydrogenase, *Biochem. Biophys. Res. Commun.* 23, 810–815, 1966.

404. Esmon, C.T. and Lollar, P., Involvement of thrombin anion-binding exosites 1 and 2 in the activation of factor V and factor VIII, *J. Biol. Chem.* 271, 13882–13887, 1996.

405. Jakubowski, J.A., and Maragonore, J.M., Inhibition of coagulation and thrombin-induced platelet activities by a synthetic dodecapeptide modeled on the carboxy-terminus of hirudin, *Blood* 75, 399–406, 1990.

406. Skrzypczak-Jankun, E., Carperos, V.E., Ravichandran, K.G., *et al.*, Structure of the hirugen and hirulog 1 complexes of α-thrombin, *J. Mol. Biol.* 221, 1379–1393, 1991.

407. Pozzi, N., Acquasaliente, L., Frasson, R., *et al.*, $β_2$-gllycoprotein I binds to thrombin and selectively inhibits the enzyme procoagulant functions, *J. Thromb. Haemost.* 11, 1093–1102, 2013.

408. Acquasaliente, L., Peterle, D., Tescari, S., *et al.*, Molecular mapping of α-thrombin (αT)/β2-glycoprotein I (β2GPI) interaction reveals how β2GPI affects αT functions, *Biochem. J.* 473, 4629–4650, 2016.

409. Webb, J.L., *Enzyme and Metabolic Inhibitors, General Principles of Inhibition*, Vol. 1., pp. 85–88, Academic Press, New York, USA, 1963.

410. Ehrenpreis, S. and Scheraga, H.A., Observations on the analysis for thrombin and the inactivation of fibrin monomer, *J. Biol. Chem.* 227, 1043–1061, 2957.

411. Scheraga, H.A., and Ehrenpreis, S., Kinetics of the fibrinogen-fibrin conversion, in *Proc. Forth Intern. Cong. Biochem. Vienna, 1958*, Vol. 1, ed., E. Deutsch, pp. 212–227, Pergammon Press, New York, USA, 1959.

412. Curragh, E.F., and Elmore, D.T., Kinetics and mechanism of catalysis by proteolytic enzymes. 2.   Kinetic studies of thrombin-catalyzed reactions and their modification by bile salts and other detergents. *Biochem. J.* 93, 163–171, 1964.

413. Berliner, L.J. and Shen, Y.Y., Physical evidence for an apolar binding site near the catalytic center of human α-thrombin, *Biochemistry* 16, 4622–4626, 1977.

414. Trowbridge, C.G., Krehbiel, A., and Laskowski, M., Jr., Substrate activation of trypsin, *Biochemistry* 2, 843–850, 1963.

415. Nardini, M., Pesce, A., Rizzi, M., *et al.*, Human α-thronbin inhibition by the active site titrant N^{α}-(N,N-dimethylcarbamoyl)-α-azalysine p-nitrophenyl ester: a comparative kinetic and X-ray crystallographic study, *J. Mol. Biol.* 258, 851–859, 1996.

416. Roberts, P.S., and Burkat, R.K., Inhibition of the esterase activity of thrombin by Na⁺, *Proc. Soc. Exptl. Biol. Med.* 127, 447–450, 1968.

417. Roberts, P.S. and Fleming, P.B., Effects of salts and ionic strength on hydrolysis of TAME (p-toluenesulfonyl-L-arginine methyl ester) by thrombin and thrombokinase, *Thromb. Diaht. Haemmorh.* 27, 573–583, 1973.

418. Workman, E.F., Jr., and Lundblad, R.L., The effect of monovalent cations on the catalytic activity of thrombin, *Arch. Biochem. Biophys.* 185, 544–548, 1978.

419. Griffith, M.G., Beavers, G., Kingdon, H.S., and Lundblad, R.L., Effect of monovalent cations on the heparin-enhanced antithrombin III/thrombin reaction, *Thromb. Res.* 17, 29–39, 1980.

420. Landis, B.H., Koehler, K.A., and Fenton, J.W., II, Human thrombin. Group IA and IIA salt dependent properties of α-thrombin, *J. Biol. Chem.* 256, 4604–4614, 1981.

421. Lundblad, R.L., and Jenzano, J.W., Effect of monovalent cations on bovine α- and β-thrombin: importance of substrate structure. *Thromb. Res.* 37, 53–59, 1985.

422. Bush, L.A., Nelson, R.W., and Di Cera, E., Murine thrombin lacks Na⁺ activation but retains high catalytic activity, *J. Biol. Chem.* 281, 7183–7187, 2006.

423. Lottenberg, R., Christensen, U., Jackson, C.M., and Coleman, P.L., Assay of coagulation proteases using peptide chromogenic and fluorogenic substrates, *Methods Enzymol.* 80, 341–360, 1981.

424. Villanueva, G.B., and Perret, V., Effects of sodium ions and lithium salts on the conformation of human α-thrombin. *Thrombosis Res.* 29, 489–498, 1983.

425. Wells, C.M., and Di Cera, E., Thrombin is a Na⁺-activated enzyme. *Biochemistry* 31, 11721–11730, 1992.

426. Fastrez, J. and Fersht, A.R., Demonstration of the acyl-enzyme mechanism for the hydrolysis of peptides and anilides by chyrmotrypsin, *Biochemistry* 12, 2025–2034, 1973.

427. Fersht, A.R., *Enzyme Structure and Mechanism*, 2nd edn, pp. 405–409, W.H. Freeman, New York, USA, 1985.

428. Bah, A., Garvey, L.C., Ge, J., and Di Cera, E., Rapid kinetics of Na⁺ binding to thrombin, *J. Biol. Chem.* 281, 40049–40056, 2016.

429. Kroh, H., Tans, G., Nicolas, G.A., Rosing, J., and Bock, P.E., Expresion of allosteric linkage between the sodium binding site and exosites of thrombin during prothrombin activation, *J. Biol. Chem.* 282, 16095–16104, 2007.

430. Butanes, S. and Mann, K.G., Blood coagulation, *Biochemistry* (Mosc.) 67, 3–12, 2002.

431. Mann, K.G., Brummel, K., and Butenas, S., What is all that thrombin for?, *J. Thromb. Haemost.* 1, 1504–1514, 2003.

432. Di Cera, E., De Cristofero, R., Albright, D.J., and Fenton, J.W., II, Linkage between proton binding and amidase activity in human α-thrombin. Effect of ions and temperature, *Biochemistry* 30, 7913–7924, 1991.

433. Dang, O.D., Vindigni, A., and Di Cera, E., An allosteric switch controls the procoagulant and anticoagulant activities of thrombin, *Proc. Natl. Acad. Sci. USA* 92, 5977–5981, 1995.

434. Vindigni, A. and Di Cera, E., Release of fibinopeptides by the slow and fast forms of thrombin, *Biochemistry* 35, 4417–4426, 1996.

435. Adams, T.E., Li, W., and Huntington, J.A., Molecular basis of thrombomodulin activation of slow thrombin, *J. Thromb. Haemost.* 7, 1688–1695, 2009.

436. Monod, J., Changeux, J.P., and Jacob, F., Allosteric proteins and cellular control systems, *J. Mol. Biol.* 6, 306–329, 1963.

437. Gasper, P.M., Fuglestad, B., Komives, E.A., *et al.*, Allosteric networks in thrombin dintinguish procoagulant vs. anticoagulant activities, *Proc. Natl. Acad. Sci. USA* 109, 21216–21222, 2012.

438. Duffy, E.J., Anglisker, H., Le Bonniec, B.F., and Stone, S.R., Allosteric modulation of the activity of thrombin. *Biochem. J.* 321, 361–365, 1997.

439. Cornish-Bowden, A., Control of enzyme activity, in *Fundamentals of Enzyme Kinetics*, Chapter 9, pp. 203–237, Portland Press, London, UK, 1995.

440. Ruzkdowski, M., Guarding the gateway to histidine biosynthesis in plants: *Medicgo truncatula* ATP-phosphoribosyltransferase in relaxed and tense states, *Biochem. J.* 475, 2681–2687, 2018.

441. Jiao, W., Fan, Y., Blackmore, N.J., and Parker, E.J., A single amino acid substitution uncouples catalysis and allostery in an essential biosynthetic enzyme in *Mycobacterium tuberculosis*, *J. Biol. Chem.* 295, 6252–6262, 2020.

442. Huntington, J.A., How Na$^+$ activates thrombin-a review of the functional and structural data, *Biol. Chem.* 289, 1025–1035, 2008.

443. Fekete, S., Veuthey, J.L., Beck, A., and Guillarme, D., Hydrophobic interaction chromatography for the characterization of monoclonal antibodies and related products, *J. Pharm. Biomed. Anal.* 130, 3–18, 2016.

444. Karlsson, G., Analysis of human α-thrombin by hydrophobic interaction high-performance liquid chromatography, *Prot. Exp. Purif.* 27, 171–174, 2003.

445. Berliner, L.J. and Shen, Y.Y.L., Physical evidence for an apolar binding site near the catalytic center of human α-thrombin, *Biochemistry* 16, 4622–4626, 1977.

446. Thompson, A.R., High affinity binding of human and bovine thrombin to *p*-chlorobenzyamido-ε-aminocaproyl agarose, *Biochim. Biophys. Acta* 422, 200–209, 1976.

447. Markwardt, E., Landmann, H., and Walsmann, P., Comparative studies on the interaction of trypsin, plasmin, and thrombin by derivatives of benzylamine and benzylamidine, *Eur. J. Biochem.* 6, 502–506, 1968.

448. Lei, B., Li, J., Liu, H., and Yau, X., Accurate prediction of aquatic toxicity of aromatic compounds based on genetic algorithm and least squares vector machine, *QSAR Comb. Sci.* 27, 850–855, 2008.

449. Lundblad, R.L., Tsai, J., Wu, and H.-f., Hydrophobic affinity chromatography of human thrombin, *Arch. Biochem. Biophys.* 302, 109–112, 1983.

450. Sherry, S., Alkjaersig, U., and Fletcher, A.P., Comparative activity of thrombin on substituted arginine and lysine esters, *Am. J. Physiol.* 209, 577–583, 1965.

451. Schechter, I. and Berger, A., On the size of the active site in proteases, I Papain, *Biochem. Biophys. Res. Commun.* 27, 157–162, 1967.

452. Hijikata-Okunemiya, A. and Okamoto, S., A strategy for a rational approach to designing synthetic selective inhibitors, *Sem. Thromb. Hemost.* 18, 135–149, 1922.

453. Hilpert, K., Ackermann, J., Banner, D., *et al.*, Design and synthesis of potent and highly selective thrombin inhibitors, *J. Med. Chem.* 37, 3889–3911, 1994.

454. Srivastava, S., Goswami, L.N., and Dikshit, D.K., Progress in the design of low molecular weight thrombin inhibitors, *Med. Res. Rev.* 25, 66–92, 2005.

455. Horton, H.R. and Young, G., 2-Acetoxy-5-nitrobenzyl chloride. A reagent intended to introduce a reporter group near the active site of chymotrypsin, *Biochim. Biophys. Acta* 194, 777–778, 1969.

456. Horton, H.R. and Koshland, D.E., Jr., Modificaiton of proteins with active benzyl halides, *Methods Enzmmol.* 25, 468–484, 1972.

457. Josse, D., Xie, W., Mason, P., Schopfer, L.M., and Lockridge, O., Tryptophan residue(s) as major components of the human paraoxonase active site, *Chem. Biol. Interact.* 79–84, 119–120, 1999.
458. Uhteg, L.C. and Lundblad, R.L., The modification of tryptophan in bovine thrombin, *Biochim. Biophys. Acta* 491, 551–557, 1977.
459. Glazer, A.N., Spectral studies on the interaction of α-chymotrypsin and trypsin with proflavin, *Proc. Natl. Acad. Sci. USA* 54, 171–176, 1965.
460. Bernhard, S.A., Lee, B.E., and Tashijan, Z.H., On the interaction of the active site α-chymotrypsin with chromophores: proflavin binding and enzyme conformation during catalysis, *J. Mol. Biol.* 18, 405–420, 1966.
461. Feinstein, G. and Feeney, R.E., Binding of proflavin to α-chymotrypsin and trypsin and its displacement by avian ovomucoids, *Biochemistry* 6, 749–753, 1967.
462. Koehler, K.A. and Magnusson, S., The binding of proflavin to thrombin, *Arch. Biochem. Biophys.* 160, 175–184, 1974.
463. Lundblad, R.L., The reaction of bovine thrombin with *N*-butyrylimidazole. Two different reactions resulting in the inhibition of catalytic activity, *Biochemistry* 14, 1033–1037, 1975.
464. Lundblad, R.L., Observations on the hydrolysis of *p*-nitrophenyl acylates by purified bovine thrombin, *Thromb. Diath. Haemorrh.* 30, 248–254, 1973.
465. Conery, B.G., and Berliner, L.J., Binding subsites in human thrombin, *Biochemistry* 22, 369–375, 1983.
466. Berliner, L.J., Sugawara, Y., and Fenton, J.W., II, Human α-thrombin binding to non-polymerized fibrin Sepharose: evidence for an anionic binding site, *Biochemistry* 24, 7005–7009, 1985.
467. Stukova, S.M., Umarova, B.A., and Keveeva, E.G., Substrate activation in thrombin-catalyzed hydrolysis of synthetic esters of arginine, *Thromb. Res.* 12, 1123–1133, 1978.
468. Takaseki, S., Kasai, K. and Ishii, S., Comparison of the catalytic properties of thrombin and trypsin by kinetic analysis on the basis of active enzyme concentration, *J. Biochem.* 78, 1275–1285, 1975.
469. Pizzo, S.V., Lundblad, R.L., and Willis, M.S., *Proteolysis in the Interstitial Space*, CRC Press/Taylor & Francis, Boca Raton, FL, USA, 2017.
470. DeRidder, G.G., Lundblad, R.L., and Pizzo, S.V., Actions of thrombin in the interstitium, *J. Thromb. Haemost.* 14, 40–47, 2015.
471. Orr, A.W., Sonder, J.M., Bevrd, M., *et al.*, The subendothelial extracellular matrix modulates NF-κB activation by flow: a potential role in atherosclerosis, *J. Cell Biol.* 109, 191–202, 2004.
472. Béguin, E.P., Janssen, E.F., Hoogenboezem, M., *et al.*, Flow-induced reorganization of laminin-integrin networks with the endothelial basement membrane uncovered by proteomics, *Mol. Cell. Proteomics* 19, 1179–1192, 2020.
473. Wolberg, A.S., Stafford, D.W., and Erie, D.A., Human factor IX binds to specific sites on the collagenous domains of collagen IV, *J. Biol. Chem.* 272, 16717–16720, 1977.
474. Vissers, M.C.M. and Thomas, C., Hypochlorous acid disrupts the adhesive properties of subendothelial matrix, *Free Rad. Biol. Med.* 23, 401–411, 1997.
475. Healy, A.M. and Herman, M., Density-dependent accumulation of basic fibroblast growth factor in the subendothelial matrix, *Eur. J. Cell Biol.* 59, 46–67, 1992.
476. Deleon-Pennell, K.Y., Barker, T.H., and Lindsey, M.L., Fibroblasts: the arbiters of extracellular matrix remodeling, *Matrix Biol.* 92, 1–7, 2020.
477. Bar-Shavit, R., Eldor, A., and Vlodavsky, I., Binding of thrombin to subendothelial extracellular matrix. Protection and expression of functional properties, *J. Clin. Invest.* 84, 1096–1104, 1989.

478. Salatti, J.A., Fenton, J.H., II, Anton, P., and Sakarissen, K.S., α-Thrombin bound to extracellular matrix induces pronounced fibrin deposition and platelet thrombus growth in flowing non-anticoagulated human blood, *Blood Coag. Fibrnolysis* 5. 561–566, 1996.

479. Hatton, M.W.C., Evidence for thrombin binding to dermatan sulfate in the rabbit aorta subendothelium *in vitro*, *Blood Coag. Fibrinolysis* 4, 927–933, 1993.

480. Okamura, Y., Schmidt, R., Raschke, M., *et al.*, A few immobilized thrombins are sufficient for platelet spreading, *Biophys. J.* 100, 1855–1863, 2011.

481. Hogg, P.J. and Jackson, C.M., Fibrin monomer protects thrombin from inactivation by heparin-antithrombin III: implications for heparin efficacy, *Proc. Natl. Acad. Sci. USA* 86, 3619–3623, 1989.

482. Weitz, J.L., Hudoba, M., Massel, D., Maraganore, J., and Hirsh, J., Clot-bound thrombin is protected from inhibition by heparin-antithrombin but is inactivated by antithrombin-III-independent inhibitors, *J. Clin. Invest.* 86, 385–391, 1990.

483. Meddahi, S. and Samana, M.M., Is the inhibition of both clot-associated thrombin and factor X more clinically relevant than either one alone?, *Blood Coag. Fibrinolysis* 20, 207–214, 2009.

484. Yang, C.-C., Hsioao, L.-D., Yang, C.-M., and Lin, C.-C., Thrombin enhanced matrix metalloproteinase-9 expression and migration of SK-N-SH cells via PAR-1,s-src, PYK2, EGFR, Erk1/2, and AP-1, *Mol. Neurobiol.* 54, 3476–3491, 2017.

4 The Use of Protein Engineering (Mutagenesis) to Study Functional Regions in Thrombin Expression and Characterization of Recombinant Thrombins

Recombinant DNA technology has provided the ability to "engineer" protein structure by the process of target mutagenesis[1-4] and has been useful in the study of structure-function relationships in thrombin. Thrombin is a two-chain protein that results from a series of activation steps from prothrombin. As was the situation with insulin many years ago, it was difficult to assemble the separate A-chains and B-chains of thrombin to form functional α-thrombin. In addition, the B-chain of thrombin proved both difficult to express and to subsequently fold into an active conformation. Thus, the protein engineering work on thrombin has mostly used either prethrombin-2,[5] a precursor of thrombin (see Chapter 2), or prothrombin. In one early study[5] expression of human prethrombin-2 was accomplished in dihydrofolate reductase-deficient carbohydrate (CHO) cells using methotrexate gene amplification. Purification was accomplished using a hirudin affinity column. Activation of prethrombin-2 was accomplished with ecarin, the coagulant protein from *Echis carinatus* venom. The recombinant thrombin was essentially identical to human α-thrombin purified from human plasma by several different assay systems. It was noted that 25% of the recombinant prethrombin was two amino acids shorter at the amino terminal than wild-type prethrombin-2; however, both forms of recombinant prethrombin-2 yielded the same species of α-thrombin. Recombinant bovine thrombin can also be prepared from prethrombin-2 using ecarin activation.[6] This study showed that CHO was not required for the function of thrombin, as expression of the recombinant protein was accomplished in *Escherichia coli*. DiBella and coworkers[6] showed that their bovine α-thrombin was similar to wild-type bovine thrombin in the hydrolysis of H-D-Phe-Pip-Arg-pNA (S-2238) (k_{cat}/K_m = 42 μM^{-1} s^{-1} for recombinant versus 35.8 μM^{-1} s^{-1} for native) and fibrinopeptide A (FPA) release 117.2 μM^{-1} s^{-1} versus 16.7 μM^{-1} s^{-1} for native). Other studies have used prothrombin[7] or

DOI: 10.1201/b22204-4

prethrombin-1[8] for mutagenesis studies. As suggested in these studies and at least one other study,[8] the recombinant human α-thrombin was shown to be essentially identical to the plasma-derived protein. Bishop and coworkers[8] at Zymogenetics on the shore of Lake Union in Seattle presented a detailed comparison of recombinant human α-thrombin and the plasma-derived protein. Prethrombin 1 was expressed in CHO cells and converted to α-thrombin using a solid phase activator. Other than glycosylation, there were no significant differences between the recombinant protein and the plasma-derived protein. There are likely differences in glycosylation, but as cited earlier and in the work of Dibella and coworkers,[6] this should not pose a problem. The work of Soejima and coworkers[9] described the expression of human prethrombin-2 in E. coli and the activation of the product to human α-thrombin using ecarin. This study presents an excellent discussion of refolding conditions necessary to obtain functional proteins from inclusion bodies, as well as a detailed functional comparison of recombinant human α-thrombin and plasma-derived α-thrombin. It is safe to assume that the various recombinant proteins described next are also essentially identical, except for the changes in sequence and glycosylation, to the plasma-derived protein.

ACTIVE-SITE MUTANTS

Arcone and coworkers[10] prepared several thrombin mutants where active-site residues and an exosite residue were modified. The expression system used for prethrombin-2 used a dihydrofolate reductase–deficient CHO cell line.[11] The recombinant prethrombin-2 proteins were purified by chromatography on a hirudin-agarose affinity column and activated with ecarin to form human α-thrombin. Arcone and coworkers[10] showed that mutations at the active site (H43N, D99N) were essentially inactive. The effect on catalytic activity is shown in Table 4.1. This study also examined the effect of these mutations on mouse fibroblasts. The various mutations, with the exception of the S205A mutation, all enhanced fibroblast proliferation, although some at a higher concentration than the native thrombin. Receptor occupancy by the S205A mutant did block the effect of native thrombin. The S205A mutant had a very low level of activity in the hydrolysis of S-2238 but was totally inactive in the cleavage of fibrinogen: mutation at the active-site histidine and aspartic acid had much lower activity in the hydrolysis of the tripeptide nitroanilide substrate and, as with the active-site serine, no detectable activity in the clotting of fibrinogen. No activity in the hydrolysis of either substrate was observed with the oxyanion mutant. It is useful to consider the results of another study that explored the effect of mutagenesis of active-site residues in subtilisin. Carter and Wells[12] showed that mutagenesis of active-site residues in subtilisin retained a low level of activity in the hydrolysis of a peptide nitroanilide substrate (approximately 10^3 greater than the nonenzymatic rate of hydrolysis). Carter and Wells[13] also showed that mutation of the oxyanion hole in subtilisin resulted in a 150-fold decrease in k_{cat} with no change in K_m with a tripeptide nitroanilide substrate. Arcone and coworkers[9] showed a mutation in exosite-1, R73E, resulted in an 80% loss of fibrinogen clotting activity with little effect on the hydrolysis of a tripeptide nitroanilide substrate.

TABLE 4.1
Effect of Selected Mutations on Thrombin Activity

Mutation	S-2238 k_{cat}/K_m (μM^{-1} sec^{-1})[a]	Fibrinogen clotting[b]
Native α-thrombin	13.2	100
Recombinant α-thrombin	13.8	98
H43N (active-site histidine)	0.0009	not detected
D99N (active-site aspartic acid)	0.0014	not detected
S205A (active-site serine)	not detected	not detected
S205T (active-site serine)	0.0161	0.15
R73E (anion-binding exosite-1)	17.5	15.4
G203A (oxyanion hole)	not detected	not detected

Source: Arrone, R., Pagliuca, M.G., Chinali, A., *et al., Biochim. Biophys. Acta* 1451, 173–186, 1999
[a] 50 mM Tris, 0.1 M NaCl, 0.1% PEG 6000, pH 8.0
[b] Percentage of native α-thrombin fibrinogen clotting activity

Krem and Di Cera[14] used changes in intrinsic fluorescence to study the interaction of the S205A thrombin mutation with various substrates. Binding of a tripeptide nitroanilide substrate used for thrombin, H-D-Phe-Pro-Arg-pNA, caused a decrease in intrinsic fluorescence, while binding of a tetrapeptide nitroanilide substrate, *N*-succinyl-Ala-Ala-Pro-Phe-PNA, a substrate used for cathepsin G and subtilisin, had no effect. The binding of FPRCH$_2$Cl also caused a decrease in fluorescence. Binding of a PAR-1 peptide caused an increase in intrinsic fluorescence, as did a PAR-3 peptide, while a PAR-4 peptide caused a decrease in intrinsic fluorescence. The increase in intrinsic fluorescence was suggested to reflect interaction with exosite-1.

THE 60S LOOP (TRYPTOPHAN LOOP)

There are several sequences in thrombin which are not homologous to the sequence of chymotrypsin and are labeled insertion loops or exosites.[15–18]

The 60s loop is a sequence in the amino terminal segment of thrombin consisting of residues 48–55 (these residues are designated 60_a–60_i in the chymotrypsin sequence). Le Bonniec and coworkers[19] had shown that the deletion of Pro48Pro49Trp50 (desPPW mutant) resulted in changes in the hydrolysis of tripeptide nitroanilide substrates (Table 4.2). In general, there was an increase in K_m, indicating a decrease in affinity for the substrate, but the extent of the increase depended on the substrate, with a 1.6-fold increase in K_m for H-Val-Leu-Arg-PNA to a 20.5-fold increase in K_m for H-D-Ile-Pro-Arg-pNA. There were also changes in k_{cat} ranging from a small (0.2-fold) decrease with Tos-Gly-Pro-Arg-pNA to a 27-fold increase in k_{cat} for H-D-Val-Leu-Arg-pNA. In the case of H-D-Val-Leu-Arg-PNA, the value for k_{cat}/K_m is slightly larger for the mutant (1.91×10^5 M^{-1}s^{-1}) than for native thrombin

TABLE 4.2

Activity of DesPPW Thrombin in the Hydrolysis of a Series of Tripepeptide
***p*-Nitroanilide Substrates**

	Wild-Type		DesPPW	
Substrate	k_{cat} (sec^{-1})	K_m (µM)	k_{cat} (sec^{-1})	K_m (µM)
H-D-Phe-Pip-Arg-*p*NA[a]	106	3	82	49
Tosyl-GlyProArg-*p*NA[b]	119	10	22	91
H-D-Val-Leu-Arg-*p*NA	18	141	43	224
H-D-Ile-Pro-Arg-*p*NA	62	2	142	41
H-D-Hht-Ala-Arg-*p*NA[c]	54	3	95	8

Source: Le Bonniec, B.F., Guinto, E.M., MacGillvray, R.T.A., Stone, S.R., and Esmon, C.T., *J. Biol. Chem.* 268, 19055–19063, 1993

[a]   S-2238
[b]   Chromozym TH
[c]   Hht, hexahydroxytyrosyl

$(1.13 \times 10^5$ M^{-1}s$^{-1})$. The desPPW mutant released one peptide from denatured casein, while two peptides were released from denatured casein by native thrombin. The fibrinogen clotting activity of desPPW thrombin was 3% that of native thrombin. The K_m of desPPW for FPA release was similar to native thrombin, while there was a 50-fold decrease in k_{cat} for the mutant enzyme. Bovine pancreatic trypsin inhibitor (BPTI) was a much more potent inhibitor of the desPPW thrombin than native thrombin but less susceptible to antithrombin. The rate of inactivation of the desPPW by DFP was similar (12.6 M^{-1}s^{-1}) to wild-type (11.7 M^{-1}s^{-1}). The K_i for benzamidine was slightly lower for the desPPW (119 µM) than wild-type (250 µM). DesPPW has 3% of the fibrinogen clotting activity of wild-type. There was no significant change in K_m for fibrinopeptide A release (9.2 vs. 8.8 µM) but a major decrease in k_{cat} (0.93 vs. 47.5). There was also a decrease in the rate of protein C activation by the desPPW mutant. As shown by other studies, there was one remarkable observation. DesPPW was rapidly inactivated by bovine pancreatic trypsin inhibitor, while native thrombin was resistant to inactivation; there was a much smaller effect on reaction with soybean trypsin inhibitor. Le Bonniec, Guinto, and Stone [20] reported on the reaction of desPPW thrombin and other mutant thrombins with several serpins. Some of the results are shown in Table 4.3. The reduced rate of reaction of desPPW-thrombin with antithrombin observed in the earlier study[19] was confirmed. It was noted that the rate of reaction with antitrypsin Pittsburgh was affected to the same extent, while the deletion had a major effect on the reaction with protease nexin 1.

In work cited earlier, Le Bonniec and coworkers[19,20] showed that antithrombin reacted more slowly than the desPPW mutant than native thrombin but did show a similar enhancement of rate in the presence of heparin. Rezaie[21] extended this work on the role of the 60s loop by demonstrating that the W50A mutant also reacted with antithrombin much more slowly than the wild-type enzyme. Rezaie showed that while

TABLE 4.3
Reaction of DesPPW Thrombin and DesETW Thrombin with Some Serpins

Serpin	−Heparin			+Heparin		
	desPPW IIa	desETWIIa	wild-type IIa	desPPW IIa	desETWIIa	wild-type IIa
Antitrypsin	$<10^a$	<10	<10	<10	<10	<10
Antitrypsin-Pittsburgh[b]	2.2×10^4	1.0×10^4	1.2×10^6	1.3×10^4	9.9×10^3	5.9×10^5
α_1-Antichymotrypin	<10	<10	1.1×10^1	<10	<10	3.4×10^1
α_1-Antichymotrypin-Arg[c]	1.2×10^2	1.1×10^1	4.2×10^3	1.1×10^3	4.1×10^2	2×10^4
Antithrombin	6.1×1^1	3.7×10^1	1.3×10^4	4.9×10^6	2.1×10^6	1.7×10^8
Protease Nexin 1	8.7×10^5	1.3×10^3	1.5×10^6	1.1×10^8	1.1×10^6	7×10^8

Source: Le Bonniec, B.F., Guinto, E.F., and Stone, S.R., *Biochemistry* 34, 12241–12248, 1995

[a] k_{on} (second-order rate constant, $M^{-1}s^{-1}$)
[b] Replacement of P_1 methionine with arginine
[c] Replacement of P_1 leucine with arginine

there was a linear relationship between antitrypsin concentration and the rate of reaction with wild-type thrombin (first-order rate constant), the linear relationship between antithrombin concentration and rate of reaction was absent with either W50A thrombin or desPPW-thrombin. A similar effect of heparin was observed with all three forms of thrombin. The results suggest a unique role for the 60s loop in the association of thrombin with antithrombin. Rezaie and Olson[22] suggested that the results obtained with the K52A mutant demonstrated a role for K52 in the S'$_1$ binding site. They showed that antithrombin Denver has a P'$_1$ leucine reaction with K52A thrombin faster than wild-type thrombin. They proposed that K52 partially blocks the S'1 binding site. Rezaie and Yang[23] prepared a thrombin deletion mutant lacking the 60s loop (Tyr47 through Phe54; desYPPWDLNF-thrombin). The deletion had little effect on the K_M for several tripeptide nitroanilide substrates, but there was a decrease in k_{cat} for S-2235 but no change for S-2238. The change in reaction with antithrombin previously observed with the desPPW thrombin or the W50A thrombin was not seen with the des60s loop thrombin. There was a decrease in the rate of activation of factor V or factor VIII. It was noted that there was a decrease in the affinity of heparin for exosite-2.

Crystallographic analyses[24] show that the 60s loop is on one side of the active-site cleft and, as suggested by several of the studies noted previously, restricts access to the active site. There is data that suggests that the 60s loop does move to accommodate inhibitors such as PPACK.[25] Movement of the 60s loop is also observed on the binding of ecotin,[26] which is thought to interact with exosite-2. The work with various 60s loop mutants suggests that the loss of structure increases accessibility to "nonspecific inhibitors" such as BPTI, but the loss of specific structure could also result in decreased binding energy for specific inhibitors such as antithrombin. While both the desPPW and W50A mutants fit into this model, the des60s loop thrombin does not. The study on the des60s loop mutant[23] showed that desYPPW thrombin also had a greatly reduced reaction with antithrombin.

GLUTAMIC ACID 202 (GLUTAMIC ACID 192 IN CHYMOTRYPSIN NUMBERING)

Le Bonniec and coworkers[27] have characterized the D202Q mutant. This mutation had little effect on the release of FPA from fibrinogen and improved the catalytic efficiency of the release of fibrinopeptide B (FPB). The E202Q mutant was more effective than native thrombin in the activation of protein C in the absence of thrombomodulin but less effective in the presence of thrombomodulin. The mutant thrombin also cleaved a peptide corresponding to the sequence around the protein C activation cleavage site (Glu-Asp-Gln-Val-Asp-pPo-Arg-Leu-Ile-Asp-Gly-Lys) more rapidly (640 nM m^{-1}) than native thrombin (33 nM m^{-1}). These investigators also compared the activity of E202Q and native thrombin with a number of peptide nitroanilide substrates (Table 4.4). Rezaie and Esmon[28] examined the effect of mutation of a glutamic acid residue E202, which is thought to contribute to the specificity of thrombin. Two mutations were considered: Glu202Gln and Glu202M. This homologous position is occupied by methionine in chymotrypsin. Mutagenesis of human prethrombin 1 was performed to obtain the two mutants of human thrombin. Activation was accomplished by prothrombinase. This was part of a study with other coagulation proteases, so the information on thrombin was limited. They did measure the activity of the E202M mutant with several peptide nitroanilide substrates (Table 4.5). It would seem like the E202M mutants have substantially more activity, with at least one factor Xa substrate suggesting a change in specificity. Both E192M ($k_2 = 3.0 \times 10^3$ M^{-1}s^{-1}) and E192Q ($k_2 = 5.2 \times 10^3$ M^{-1}s^{-1}) react with antithrombin more slowly than wild-type thrombin ($k_2 = 9.5 \times 10^3$ M^{-1}s^{-1}). Both mutants were inactivated by α_1-antitrypsin more rapidly than native thrombin (1.3×10^2 M^{-1}s^{-1}),

---

TABLE 4.4
Hydrolysis of Some Peptide Nitroanilide Substrates by E192Q Thrombin

Substrate	Wild-Type		E192Q	
	K_m (µM)	k_{cat} (s^{-1})	K_m (µM)	k_{cat} (s^{-1})
Fibrinogen (fibrinopeptide A release)	8.8	28	9.6	33
H-D-Phe-Pip-Arg-pNA	8	277	9	254
Tos-Gly-Pro-Arg-pNA	10	368	12	472
D-Val-Leu-Arg-pNA	189	38	186	30
D-Hht^a-Ala-Arg-pNA	6	174	6	163
Ba-Phe-Val-Arg-pNA	69	87	83	229
Bz-Ile-Glu-Gly-Arg-pNA	67	2	55	16
MeO-CO^b-D-Chg^c-Gly-Arg-pNA	40	44	42	162

Source: Le Boniec, B.F. and Esmon, C.T., *Proc. Natl. Acad. Sci. USA* 88, 7371–7375, 1991

[a] Hexahyroxytyrosyl

[b] Methoxycarbonyl

[c] Cyclohexylglycyl

TABLE 4.5
The Effect of E202M on the Hydrolysis of Some Peptide
p-Nitroanilide Substrates

Substrate	% Native
Cbz-LysProArg-pNA (activated protein C substrate)	100
Hexahydrotyrosyl-AlaArg-pNA (thrombin substrate)	70
Cbz-D-ArgGlyArg-pNA (factor Xa substrate)	470
Methylsulfonyl-D-PheGlyArg-pNA (tPA substrate)	120

Source: Rezaie, A.R., and Esmon, C.T., *Eur. J. Biochem.* 242, 177–184, 1996

with the E202Q mutant inactivated more rapidly (9.2×10^4 M^{-1}s^{-1}) than the E202M mutant (2.2×10^4 M^{-1}s^{-1}). In later work, Marque and coworkers[29] evaluated the effect of E202Q mutation on specificity of thrombin using a series of fluorescence resonance energy transfer (FRET) substrates with variation in the P$'_2$ and P$'_3$ sites Abz-VGPRSXXLK(Dnp)D with abz being *o*-aminobenzyl and Dnp being dintrophenyl (substitution of lysine). These investigators studied a larger number of substrates showing that the D202Q mutant was less selective than native thrombin with respect to variation in the P$'_2$ and P$'_3$ positions. The data suggested that the E202Q mutant was less sensitive to changes in the P$'_2$ and P$'_3$ position in the sequence. Heparin, which is considered to bind to exosite-2, enhanced the cleavage of all peptides by native thrombin, while hir,[52–65] a hirudin peptide, is considered to bind to exosite-2. Both heparin and hir[52–65] enhanced peptide cleavage by E202Q.

AUTOLYSIS LOOP (γ-LOOP)

The other major insert loop in thrombin is the autolysis loop (γ-loop), and it contains the sequence of Asn143 through Pro157. The term autolysis loop refers to the fact that this sequence contains the peptide bond in human thrombin associated with the cleavage to form γ-thrombin.[30] The autolysis loop is considered more flexible than the 60s loop.

Di Bella and Scherage[31] replaced the autolysis loop in bovine thrombin with the homologous sequence from trypsin (Table 4.6). The mutant was designated as thrombin-148. Fibrinogen clotting activity was reduced from 2500 NIH U/mg native thrombin to 1760 NIH U/mg. The k_{cat}/K_m for the release of FPA is reduced from 25 μM^{-1}s^{-1} for native thrombin to 4.5 μM^{-1}s^{-1} for thrombin-148. Thus, it is reasonable to suggest that approximately 70% of the fibrinogen clotting activity was retained with a 20% decrease in k_{cat}/K_m. Thus, it is possible that a decrease of 20% in catalytic efficiency results in a 25% loss of fibrinogen clotting activity. Hirudin bound thrombin 148 with reduced affinity (Ki 500) compared to recombinant native bovine thrombin (Ki 12 pM). There were several studies with peptide p-nitroanilide substrates (Table 4.7). Insertion of a trypsin sequence into thrombin did not convert thrombin to a trypsin.

TABLE 4.6
Thrombin Autolysis Loop Sequences

Bovine thrombin	Asn143Arg144Arg145Glu146Thr147Trp148Thr149Thr150
Bovine thrombin 148	Asn143Arg144Lys145Ser46Ser147Gly148Thr149Thr150
Bovine trypsin	Asn143Thr144Lys145Ser146Ser147Gly148Thr149
Human thrombin	

Source: Dibella, E. and Scheraga. H.A., *Biochemistry* 35, 4427–4432, 1996

TABLE 4.7
A Comparison of the Hydrolysis of Several Peptide Nitroanilide Substrates by Bovine Thrombin, a Mutant Bovine Thrombin, and Bovine Trypsin

Enzyme	K_m (µM)	k_{cat} (s⁻¹)	k_{cat}/K_m
	S-2238		
Thrombin	2.9	123	42
Thrombin-148	30	116	3.9
Trypsin	41	70	1.7
S-2222 (BzIleGluGlyArgpNA)			
Thrombin	133	2.1	0.019
Thrombin-148	930	1.1	0.0012
Trypsin	45	611	1.3
CbzGlyProArgpNA			
Thrombin	77	119	1.5
Thrombin-148	422	84	0.2
Trypsin	36	116	3.2

Source: Dibella, E.E. and Scheraga, H.A., *Biochemstry* 35, 4427–4433, 1996

Le Bonniec and coworkers[32] reported on the effect of the deletion of Glu146Thr147Trp148 (desETW thrombin). This study was based on crystallographic data that suggested that this region of the thrombin sequence in the autolysis loop is directed toward the enzyme-active site. DesETW thrombin had 5% of the clotting activity of wild-type and reacted at a reduced rate of inactivation with several canonical inhibitors of serine proteases (Table 4.8). Antithrombin did not form a stable complex with desETW thrombin, indicating that the substrate (antithrombin) bound to thrombin but did not form the stable inhibitor-thrombin complex. Soybean trypsin inhibitor (SBTI) and BPTI have been shown to be inhibitors of desETW thrombin but do not form a stable complex. Rather, SBTI and BPTI are slow, reversible inhibitors of thrombin. Several studies suggest that SBTI can act as a competitive inhibitor of human α-thrombin. These investigators showed the Ki of SBTI from 2.6 µM for native recombinant thrombin to 34 nm for desETW thrombin and for BPTI

TABLE 4.8
The Interaction of desETW Thrombin with Inhibitors and Substrates

Inhibitor	Wild-Type	des-ETW
Diisopropylphosphorofluoridate (DFP)	$K_{on} = 694$ M^{-1}min^{-1}	$K_{on} = 73$ M^{-1}min^{-1a}
Tosyl-lysyl chloromethyl ketone (TLCK)	$K_{on} = 18$ M^{-1}min^{-1gui}	$K_{on} = 0.35$ M^{-1}min^{-1}
Benzamidine	$K_i = 0.36$ mM	$K_i = 14$ mM
Antithrombin	$K_{on} = 6500$ μM^{-1}s^{-1}	$K_{on} = 18$ μM^{-1}s^{-1}

Source: Le Bonniec, B.F., Guinto, E.R., and Esmon, C.T., *J. Biol. Chem.* 267, 19341–19348, 1992

[a] As noted by the authors, the serine residue in des-ETW thrombin appears to be a more potent nucleophile in reaction with DFP than factor Xa ($K_{on} = 1.3$ M^{-1}-min^{-1})

TABLE 4.9
The Hydrolysis of Peptide Nitroanilide Substrates by DesETW Thrombin

	Wild-Type		des-ETW	
Substrate	K_m (μM)	k_{cat} (s^{-1})	K_m (μM)	k_{cat} (s^{-1})
H-D-HisAlaArgpNA	3	54	254	91
TosGlyProArgpNA	10	119	583	104
H-D-PhePipArgpNA	6	82	239	76
H-D-ValLeuArgpNA	141	18	2,640	17
H-DIleProArgpNA	4	62	239	80
MeOCO-D-ChgGlyArgpNA	42	10	734	1

Source: Le Bonniec, B.F., Guinto, E.R., and Esmon, C.T., *J. Biol. Chem.* 267, 19341–19348, 1992

from 5.2 μM for native recombinant thrombin to 1.8 μM for desETW thrombin. As an aside, I was never able to show that SBTI was an inhibitor of bovine thrombin.[33] However, these assays were dependent on the rapid formation of a stable complex. Lanchantin and coworkers[34] did report inhibition of thrombin using a different assay system. I can recall some lively discussions with Gerry at the old Federation of American Societies for Experimental Biology (FASEB) meetings in Atlantic City. On balance, it would seem that SBTI can form a complex with thrombin, but this is quite different from the canonical complex formed with trypsin and SBTI. The formation of a complex between trypsin and SBTI is a rapid process, resulting in the formation of a stable complex.[35–37] SBTI does inhibit factor Xa.[38] However, the complex, again, is different from that formed between trypsin and SBTI in that SBTI can be used as a ligand in the affinity chromatography for factor Xa.[39] These investigators reported that thrombin did not bind to an SBTI column. LeBonniec and coworkers[32] also showed that there were significant changes in the hydrolysis of *p*-nitroanilide substrates (Table 4.9). Only several of the substrates tested are shown, but the overall trend was for the greatest effect on K_m with a lesser effect on k_{cat}.

The mutation in desETW thrombin is a deletion and may result in more of an effect on conformation than would be observed in a substitution mutant such as E202M. One common complaint of chemical modification studies was that there was conformational change secondary to the modification responsible for the observed changes in function. The same complaint can be raised for some mutagenesis studies. It is to this group's credit that they did address the effect of conformation changes using the reaction with hirudin mutants as a signal.[40] The use of hirudin mutants provided evidence that exosite-1 (anion binding site) was not affected by the deletion mutation, but that the mutation affected the enzyme-active site, the S_1 binding site, and the S_2 binding site. It was suggested that the effects of the desETW mutation may be due in part to the loss of a salt bridge between Glu146 and Arg223. A subsequent study[41] by Goodwin and coworkers studied the effect of heparin on des-ETW thrombin as compared to wild-type thrombin on the hydrolysis of several peptide nitroanilide substrates. No effect of heparin was observed on wild-type human thrombin under the experimental conditions of these studies (100 mM triethanolamine/HCl, pH 7.6. containing 0.1 M NaCl, 0.5%(w/v) polyethylene glycol 6000, 0.02% sodium azide) at 25°C), but there was an effect of the activity of the desETW mutant (Table 4.10). Previous studies on the desETW mutant[40] had shown that the deletion resulted in an increased K_m toward peptide nitroanilide substrates In this study, the effect of heparin and other sulfated polysaccharides on several p-nitroanilide substrates was predominantly on K_m with only a minor effect on k_{cat}. The effect of heparin was blocked by the presence of ADP or adenosine triphosphate (ATP), which have previously been shown to interact with thrombin.[42]

Other work showed that 6-kDa heparin and 8-kDa heparin had a lesser effect on K_m: a heparin pentasaccharide had no effect on K_m. Heparin also enhanced the rate of inactivation of desETW thrombin by antithrombin and FPRCH$_2$Cl. Studies on the interaction of heparin with thrombin have yielded conflicting results.[33] It is generally accepted that heparin binds to exosite-2 and accelerates the reaction of thrombin with antithrombin. Griffith and coworkers[43] showed that heparin decreased the K_M with a smaller effect on V_{max} for TosGlyProArgpNA by human α-thrombin.

TABLE 4.10
The Effect of Heparin on the Rate of Hydrolysis of Several Peptide Nitroanilide Substrates by DesETW-Thrombin

	des-ETW			des-ETW with 10 nM heparin[a]		
Substrate	K_m (μM)	k_{cat} (s^{-1})	k_{cat}/K_m (μM^{-1}s^{-1})	K_m (μM)	k_{cat} (s^{-1})	k_{cat}/K_m (μM^{-1}s^{-1})
H-D-PhePipArgpNA	194	45	0.23	53	54	1.02
H-D-ChgButArgpNA	273	32	0.12	129	38	0.29
H-D-ChgAlaArgpNA	314	38	0.12	188	59	0.31

Source: Goodwin, C.A., Deadman, J.J., Le Bonniec, B.F., *et al.*, *Biochem. J.* 315, 77–83, 1996
[a] Unfractionated heparin

These studies were performed in 0.1 M triethanolamine-0.1 % PEG, pH 8.0 at 23°C. It was possible to measure heparin binding to thrombin by the effect of the rate of hydrolysis of TosGlyProArgpNA. While there was an effect of heparin on the rate of hydrolysis of BzPheValArgpNA, the interpretation was complicated by substrate inhibition. Goodwin and coworkers[31] performed their studies in 0.1 M triethanol-amine-0.1 M NaCl-0.5% PEG 6000–0.2% sodium azide, pH 7.6. The increased ionic strength in the buffer used by Goodwin and coworkers[22] might explain the differences observed with the effect of heparin on native thrombin. Griffith and coworkers also observed an increase in the rate of inactivation of human α-thrombin by TLCK but no effect on the rate of inactivation with PMSF.

In addition to their work on the desPPW mutant,[20] Bonniec and coworkers studied the desETW mutant. There was a decrease in the rate of reaction of the desETW mutant with antitrypsin Pittsburgh, antithrombin, protease nexin 1, and α_1-antichymotrypsin-arg (Table 4.3).

The effect of the deletion of segments of the autolysis loop in human thrombin has been studied by Dang and coworkers.[44] Deletion of part of the loop (T^{149}-K^{154}) had little effect, while the inclusion of E^{146}-W^{148} in the deletion resulted in marked loss of fibrinogen clotting activity (Table 4.11). Deletion of E^{146}-K^{154} also reduced protein C activation and the rate of inactivation by antithrombin. The increase in the rate of protein C activation compared to the decrease in fibrinogen clotting activity suggests that the autolysis loop is a target for therapeutics. The Glu146-Lys154 deletion showed a large decrease in the ability to bind sodium ions, with a lesser effect shown by the shorter (T149-K154) deletion.

Additional work from this group provided more support for the importance of binding sodium ions in the activity of thrombin.[45] They studied the effect of the mutation of three amino acid residues: arginine 233 (R233A), lysine 236 (K236A), and tyrosine 237 (Y237P). The results are shown in Table 4.12. All three of the mutants showed decreased binding of sodium ions, with the greatest decrease shown by the W237P mutant. This mutant also showed the greatest increase in the ratio of the rate of protein C activation to fibrinogen clotting activity. Subsequent work from this group[46] replaced the autolysis loop with the same sequence from murine thrombin (murine thrombin does not bind Na+ ions), obtaining a more rigid loop.

TABLE 4.11
Effect of Autolysis Loop Deletion on Thrombin Activity

Enzyme	FPA[a]	FPB[a]	Protein C Activation[a]	Antithrombin[b] k_{on}
Wild type	17	9.1	0.22	13
ΔT^{149}-K^{154}	19	7	0.58	21
ΔE^{146}-K^{154}	0.071	0.018	0.11	0.55

Source: Dang, E., Sobetta, M., and Di Cera, E., *J. Biol. Chem.* 272, 19649–19651, 1997

[a] k_{cat}/K_m, $\mu M^{-1}s^{-1}$

[b] k_{on}, $\mu M\ s^{-1}$

TABLE 4.12
The Effect of Three Site-Directed Mutations on the Activity of Human α-Thrombin

Enzyme	FPA[a]	FPB[a]	Protein C Activation[a]	Thrombomodulin Binding[b]	Antithrombin[c]
Wild-type	17	9.4	0.22	0.99	13
R233A	1.1	0.49	0.13	3.00	1.1
K236A	0.53	0.36	0.08	8.2	1.3
Y237P	2.2	1.1	1.6	3.8	1.4

Source: Dang, Q.D., Quinto, E.R., and Di Cera, E., *Nat. Biotechnol.* 15, 146–149, 1997
[a] k_{cat}/K_m, $\mu M^{-1}s^{-1}$
[b] k_d, nM,
[c] k_{on} $\mu M^{-1}s^{-1}$

TABLE 4.13
The Effect of Mutagenesis of Glycine 189 on Some Activities of Human α-Thrombin

Enzyme	Fibrinopeptide A Release[a]	PAR-1 Peptide Cleavage[a]	Protein C Activation[a]	Rate of Reaction with Antithrombin[a]
Wild-type	17	30	0.22	13
Gly189Lys	0.082	0.13'	0.023	0.45
Gly189Arg	0.012	0.14	0.018	0.99

Source: Roy, D.B., Rose, I., and Di Cera, E., *Protein Struct. Funct. Genet.* 43, 315–318, 2001
[a] k_{cat}/K_m, $\mu M^{-1}s^{-1}$

The chimeric enzyme had increased activity towards a number of substrates and an increased affinity for sodium ions. This study used the binding of DAPA[38] to study the binding of sodium ions to thrombin. The chimeric thrombin bound sodium ions somewhat more tightly than wild type. Further work from the Di Cera laboratory used directed site-mutagenesis to substitute arginine or lysine for glycine 189.[47] This substitution essentially eliminated the binding of sodium ions with inhibition of fibrinogen clotting activity, cleavage of the PAR-1 extracellular domain peptide, and protein C activation, as well as a reduction in the rate of reaction with antithrombin (Table 4.13).

Niu and coworkers,[48] in a further work from the Di Cera laboratory, report on the effect of the mutagenesis of asparagine 151. This residue forms a hydrogen bond with glutamic acid 202, stabilizing the oxyanion hole. A Asn151Pro mutant showed an enhancement in hydrolysis of three tripeptide nitroanilide substrates with sodium ions.

TRYPTOPHAN

Several studies were discussed earlier in this chapter that describe the importance of a specific tryptophan residue in thrombin function. Rezaie[21] showed the importance of W50 in the interaction of thrombin with antithrombin. Di Bella and Scheraga[31] used the W148G mutant to show the importance of this residue in fibrinogen clotting activity. Other studies used the desPPW mutant[21] or the desETW mutant[32,40,41] to show the importance of these peptide segments in the interaction of thrombin with inhibitors and substrates. There were also studies on the chemical modification of tryptophan in thrombin described in Chapter 5. More definitive work on the importance of tryptophan residues in thrombin was provided by Bell and coworkers.[49] This study represented a collaborative effort between Ross MacGillivray at the University of British Columbia in Vancouver and the late Mike Nesheim at Queens University Kingston, Ontario. Ross and I share a common background, having both trained with the late Earl Davie at the University of Washington, while Mike trained with my long-time friend and occasional collaborator, Ken Mann, first at the Mayo Clinic and subsequently at the University of Vermont. Bell and coworkers used site-directed mutagenesis to replace five tryptophan residues in thrombin, Trp50, Trp92, Trp148, Trp219, and Trp227, with phenylalanine. The substitution of phenylalanine for tryptophan can be considered a conservative mutation but still represents a change in the nonpolar environment, as phenylalanine is twice as soluble in water as tryptophan. It was possible to show that Trp219 was the largest contributor to intrinsic fluorescence (approximately 35%), while Trp92 contributed approximately 11%. The intrinsic fluorescence of thrombin decreased 40% in the presence of DAPA. The largest individual decrease in intrinsic fluorescence was provided by W50, W148, and W227. Only W50F and W227F mutants showed a decrease in the activation (aggregation) of washing human blood platelets. The Trp227Phe mutant showed a decrease in protein C activation (in the presence of calcium ions and thrombomodulin), while the Trp50Phe mutant showed a significant decrease in thrombin-activated fibrinolysis inhibitor (TAFI) activation (in the presence of calcium ions and thrombomodulin). It was concluded that Trp50 was very important for TAFI activation and Trp227 was very important for protein C activation. There was no significant effect on any of the mutations of the hydrolysis of TosArgOMe or S-2238 (H-D-Phe-Pip-Arg-PNA) or on the binding of DAPA, suggesting that none of the mutants caused a gross change in the extended active site. It should again be recognized that these were conservative mutations. The Trp50Phe and Trp227Phe mutants had reduced fibrinogen clotting activity, with the greatest decrease shown by the Trp227Phe mutant. While fibrin(ogen) was used as a competitive inhibitor of the hydrolysis of S-2236 (PyroGlu-Pro-Arg-PNA), the K_i values for both the Trp50Phe mutant and the Trp227Phe mutant were increased approximate 3-fold over wild-type, while there was a slight decrease in the K_i value for the Trp219Phe mutant. Amidolytic activity of the Trp5oPhe mutant was greatly decreased in the hydrolysis of the S-2236 substrate with a lesser effect with the Trp227Phe mutant, thus differing from the S-2238 substrate. It is further noted that the S-2238 substrate is a much better substrate for bovine thrombin than the S-2366 substrate.[50] Since Mike Nesheim had been involved the study of the role of factor V in coagulation, it was only reasonable that they look at the effect of the various mutations on the activation of factor V, which involved three peptide bonds. It was possible to obtain first-order rate constants for the cleavage of each of the individual three peptide bonds. The Trp50Phe mutant and Trp227Phe mutant showed a decreased rate of cleavage of the Arg1006 peptide bond. The Trp227Phe mutant showed a decreased rate of cleavage of the Arg1536 peptide bond. The first-order rate constants for the other mutants were close to wild-type or elevated; the Trp148Phe mutant showed a 3-fold greater rate of cleavage of the Arg713 peptide bond, while the Trp50Phe mutant and the Try227Phe mutant showed greater than a 2-fold increase in the first-order rate constant.

Arosio and coworkers[51] studied the mutation of W227. The effect of a mutation of W227 on W227F, W227Y, and W227A is shown in Table 4.14 and Table 4.15. As would be expected, a conservative mutation (W227F) had less effect on activity, while a less conservative but still bulky mutation (W227Y) had a greater effect, and a much smaller mutation (W227A) had an even greater effect. The log P value for phenylalanine is −1.38 compared to −1.06 for tryptophan, while the value for tyrosine is −2.26 and that for alanine is −2.86. The various mutations had a greater effect on FPA release than on protein C activation, as well as reaction with antithrombin or a PAR-1 peptide (33–62). Later work from Komives and coworkers[52] showed that, using hydrogen-deuterium exchange, mutation of

TABLE 4.14
Effect of Mutagenesis of Tryptophan 227 in Human α-Thrombin on Catalytic Activity

Substrate[a]	k_{cat}/K_m ($\mu M^{-1}s^{-1}$)			
	Wild-Type	W227F	W227Y	W227A
FPRpNA	88	44	5.4	0.091
LDPRpNA	4.7	1.2	0.040	0.025
FPA	17	2.3	0.19	0.034
Protein C	0.22	0.13	0.032	0.075
PAR-1[b]	26	8.0	1.2	1.0

Source: Arosio, D., Ayala, Y.M., and Di Cera, E., *Biochemistry* 39. 8095–8101, 2010
[a] DFPRpNA, DPhrProArg*p*-nitroanilide; LDPR*p*NA, LeuAspProArg*p*-nitroanilide; FPA, fibrinopeptide A
[b] Hydrolysis of a peptide corresponding to residues 32–62 in the extracellular domain of the PAR-1 receptor

TABLE 4.15
Effect of Mutation of Tryptophan 227 on Reaction with Antithrombin

Rate of reaction with antithrombin ($\mu M^{-1}s^{-1}$)	13	12	1.2	0.56

Source: Arosio, D., Ayala, Y.M., and Di Cera, E., *Biochemistry* 39, 8095–8101, 2010

W227 in human α-thrombin resulted in a conformational change with changes at the enzyme-active site. DiBella and Scheraga[53] evaluated several mutants of bovine thrombin; W92S, a mutant where a portion of the autolysis loop (Y47-K42, which contains W48) is replaced by a tripeptide sequence (GGG) referred to as thrombinGGG, and a double mutant, W92S/thrombinGGG. Fibrinogen clotting activity of the various thrombins was evaluated by an increase in turbidity. There was a modest decrease in fibrinogen clotting activity with the W92S, a greater loss with thrombinGGG, and almost complete loss with the double mutant. There was a large loss of activity in the hydrolysis of the tripeptide nitroanilide substrate with the thrombinGGG and the double mutant, with a lesser loss with the W92S mutant. Both the thrombinGGG and the double mutant were rapidly inactivated by BPTI, with a smaller rate of inactivation with W92S. Little, if any, loss is observed with wild-type bovine thrombin.

Substitution of G189 with either lysine or arginine was performed in an attempt to duplicate the Na^+ effect at the thrombin-active site and produce the "fast" form of thrombin.[54] The two mutants had greatly reduced activity in hydrolysis of a chromogenic substrate (D-PheProArgpNA). The resulting derivatives (either the arginine mutant or lysine mutant) also had markedly reduced activity in the release of FPA from fibrinogen and the cleavage of PAR-1. The decrease in ability to activate protein C was reduced to a lesser extent. Le Bonniec and coworkers[55] showed the Glu25 restricted the P'_3 position to neutral amino acid residues. The G25K thrombin showed small changes in the hydrolysis of a number of peptide nitroanilide substrates. There was a small decrease in the rate of release of FPA and FPB. The hydrolysis of some peptides derived from the protein C activation sequence showed changes with alterations in the P' sequence.

Alanine scanning is a well-accepted technique in protein mutagenesis. Cummings and Wells,[56] working at the time at Genentech, developed the technique of alanine scanning to study the individual contributions of potentially each amino acid in the protein or peptide to function. In their study, they made 62 unique mutations in three segments of human growth hormone in a study to identify individual amino acid contributions to receptor binding. Another group, also working at Genentech, extended this study using alanine scanning to measure the individual contributions of each of 19 amino acids which had been shown to be at the interface between human growth hormone and its receptor.[57] Determination of changes in free energy binding for each mutant allowed the identification of seven residues critical for the interaction. Tang and Fenton,[58] working at the University of Kansas Medical Center, used whole-protein alanine scanning to study allosteric interactions in human liver pyruvate kinase. Whole-protein alanine scanning is an approach where every amino acid in a protein, other than alanine or glycine, is replaced by alanine. Human liver pyruvate kinase is inhibited by alanine and activated by 1,6-fructose bisphosphate. Whole-protein alanine scanning showed that only a small number of residues were involved in inhibition by alanine, but a large number of residues were in activation by 1,6-fructose bisphosphate. What this study shows is that a large portion of a protein could participate in allostery.

The use of large-scale alanine scanning has been used to study the function of surface residues in human α-thrombin by Leung and coworkers, initially at Gilead on the bay in Foster City and later down Highway 101 in Palo Alto at Stanford.[59] Some results of alanine scanning of surface residues in human thrombin are shown in Table 4.16. Subsequent work from this group used these mutants to identify residues important in binding fibrinogen (fibrinogen clotting) and fibrin. Hall and coworkers[60] used the mutants developed by Tsiang and coworkers[59] to study the importance of thrombin's surface residues in fibrinogen clotting and binding to fibrin, as shown in Table 4.17. It has been known for some time that thrombin would bind to both fibrinogen and fibrin. As described in Chapter 1, binding to fibrin was used for the purification of thrombin, and in Chapter 5, affinity chromatography on fibrin was used to study the chemical modification of thrombin. The interaction of thrombin with fibrin is discussed in detail in Chapter 6. Hall and coworkers were able to identify four residues in thrombin that are important in binding to fibrin: Lys65, His66, Tyr71, and Arg73. There is not absolute specificity, but there is a clear difference between fibrinogen and fibrin.

TABLE 4.16
The Effect of Some Selected Amino Acid Mutations on the Activity of Human α-Thrombin in Fibrinogen Clotting, Hydrolysis of a Peptide Nitroanilide Substrate, and Protein C Activation

		% Wild Type		
Mutation	Chtg#	Fibrinogen Clotting[a,b]	S-2238[c]	Protein C[b,d]
S205A	195	0	0	0
K21A	28	28	136	28
W50D	60D	5[d]	49	30[d]
D51A	60E	98	122	70
K65A	71	2	108	6
H66A	73	2	121	10
R68A	74	22	108	33
Y71A	76	1	128	12
R73A	77A	18	95	30
N74A	78	55	95	80
K77A	81	25	80	12
K106A	109	50	87	70
K107A	110	48	107	45
E229A	217	1	43	12
R233A	221	2	76	26

Source: Tsiang, M., Jain, A.E., Dunn, K.E., *et al., J. Biol. Chem.* 270, 16854–16863, 1995
[a] Fibrinogen clotting in 20 mM Tris-acetate, pH 7.5 containing 140 mM NaCl, 1 mM CaCl$_2$, 1 mM MgCl$_2$
[b] Approximated from bar graph
[c] 100 μM S-2238 (H-D-Phe-Pip-Arg-pNA)
[d] In the presence of thrombomodulin

TABLE 4.17
Effect of Selected Mutations in Human α-Thrombin on
Fibrinogen Clotting and Binding to Fibrin

Mutant	Fibrinogen Clotting[a]	Fibrin Binding[b]
R20A	95	68
K21A	48	85
W50A	12	80
R62A	43	73
K65A	5	25
H66A	8	20
R68A	28	65
Y71A	4	18
R73A	12	32
E229	5	70
R233A	10	53

Source: Hall, S.W., Gibbs, C.S., and Leung, L.L.K., *Thromb. Haemost.* 86,
1466–1474, 2001

[a] Data expressed a percentage of wild-type activity and is estimated from
a bar graph

[b] Fibrin binding was assessed by retention of thrombin in a washed
fibrin clot and by adsorption of added thrombin to a preformed fibrin
clot

Subsequent work from this group focused on the identification or a mutant with reduced or absent coagulant activity that retained the ability to activate protein C and, as such, could be used as an anticoagulant. Leung and Hall[61] summarized the work showing that a E339K mutant that had low clotting activity had retained substantial ability to activate protein C.[55,56] Gibbs and coworkers[62] extended the early work from Gilead, which showed that the E229A mutant had an increased specificity for protein C activation compared to fibrinogen clotting, showing a 22-fold difference between protein C activation and fibrinogen clotting.[59] Gibbs and coworkers[62] showed that the E229A mutant had potential as an anticoagulant. This mutant had 2.5% of the activity of native thrombin in fibrinogen clotting and reduced (14.4%) inhibition by antithrombin. The circulatory half-life of the mutant was modestly extended (180 seconds) compared to native thrombin (30 seconds). Later work from this group[63] showed the E299K mutant was an even more effective activator of protein C. This study used saturation mutagenesis[64] of four residues, Trp50, Lys32, Glu229, and Arg233, which had previously been shown to be important in protein C activation, to prepare a group of mutants for study. The Glu229Lys mutant had a 130-fold difference between fibrinogen clotting (FPA release) and protein C activation (Table 4.18).

TABLE 4.18
Activity of Selected Thrombin Mutants on Fibrinopeptide A Release and Protein C Activation

| Thrombin Species | k_{cat}/K_M ($\mu M^{-1}s^{-1}$) | |
	FPA Release	Protein C Activation
Native	18.7	0.20
E229A	0.33	0.095
E229K	0.07	0.098
E229W	0.059	0.047

Source: Tsiang, M., Paborsky, L.R., Li, W.-X., *et al.*, *Biochemistry* 35, 16449–16457, 1996

This study also showed that the Glu229Lys mutant did effectively activate protein C *in vivo*. Hall and coworkers[65] presented additional information showing that some functional activities of thrombin can be separated from each other by site-directed mutagenesis.

EXOSITE I

There has been considerable work on the characterization of exosites in thrombin discussed in Chapter 3. While some of those studies did use site-directed mutagenesis, I thought it useful to include several studies that used protein engineering to characterize exosite function in thrombin. The effect of exosite-1 mutations on reaction with serpins has been examined by Myles and coworkers,[66] with some data shown in Table 4.19. There was no effect of the various mutants on the reaction of thrombin with antithrombin in the presence or absence of heparin. There was a small effect of the mutations on the reaction with protease nexin-1, with a greater effect in the presence of heparin; the greatest effect was seen with R68Q, with a lesser effect seen with R70Q. The greatest effect of the mutations was observed with heparin cofactor II. Here, again, the greatest effect was seen with R68Q. It is not clear why the effect is of greater magnitude in the presence of heparin. A subsequent study extended these observations to include the effect of mutations in the exosite-1 of human thrombin on the cleavage of a peptide comprising residues 38–60 of the extracellular domain of the PAR-1 receptor, platelet aggregation, and fibrinogen clotting.[67] Some of the results of this study are shown in Table 4.20. Mutation of Arg62 has the greatest effect with the various ligands, with the Arg68Gln mutation also showing a significant effect. Both PAR peptide cleavage and platelet aggregation share the same determinants. It would have been interesting to see if the same pattern was followed with thrombin bound to GplB.

TABLE 4.19
Effect of Site-Directed Mutation of Amino Acid Residues in Exosite-1 on the Inactivation of Thrombin by Antithrombin, Protease Nexin-1, and Heparin Cofactor II in the Presence and Absence of Heparin

Mutant	AT[a] $\mu M^{-1}s^{-1} \times 10^{-4}$	AT[a] + Heparin $\mu M^{-1}s^{-1} \times 10^{-8}$	PN[b] $\mu M^{-1}s^{-1} \times 10^{-6}$	PN[b] + Heparin $\mu M^{-1}s^{-1} \times 10^{-9}$	HCII[c] $\mu M^{-1}s^{-1} \times 10^{-3}$	HCII[c] + Heparin $\mu M^{-1}s^{-1} \times 10^{-7}$
pIIa	0.92	0.75	1.42	2.56	1.10	4.48
rIIa	1.23	1.19	1.23	2.52	1.11	7.50
R20Q	1.10	1.29	1.23	0.68	0.56	3.00
K21Q	1.19	1.35	1.36	1.74	0.46	3.03
R68Q	0.82	0.92	0.89	1.23	0.34	0.54
R70Q	1.23	1.18	1.26	1.42	0.52	3.52
K77Q	1.13	2.04	0.93	2.04	0.56	2.56
K106Q	1.21	1.47	1.60	2.71	0.42	2.55
K107Q	1.30	1.56	1.52	2.15	0.47	2.65
K154Q	1.82	1.57	1.84	2.11	0.93	4.85

Source: Myles, T., Church, F.C., Whinna, H.C., Monard, D., and Stone, S.R., J. Biol. Chem. 273, 31203–31208, 1998
[a] Antithrombin
[b] Protease nexin-1
[c] Heparin cofactor II

TABLE 4.20
The Effect of Mutations of Basic Amino Acid Residues in Thrombin Exosite-1 on the Cleavage of a PAR-1 Peptide, Platelet Aggregation, Fibrinogen Clotting, and Hirudin Binding

Mutation	PAR 38–60 Cleavage[a] K_m (μM)	k_{cat} (s^{-1})	k_{cat}/K_m ($M^{-1}s^{-1}$)	Platelet Aggregation	Fibrinogen Clotting	Hirudin Binding
pIIa	1.3	118	9.1×10^7	0.5/3.5[b]	81[c]	88[d]
rIIa	1.3	119	9.2×10^7	0.5/3.5	100	100
R20Q	2.1	128	4.1×10^7	1.0/3.0	47	19
K21Q	3.6	130	3.6×10^7	1.0/3.0	80	23
R62Q	35.3	67	1.8×10^6	3.0/4.2	<1	<1
R68Q	25.8	119	4.6×10^6	3.0/3.3	22	2
R70Q	2.7	131	4.8×10^7	1.0/3.5	51	31
R73Q	3.9	130	3.3×10^7	1.0/3.3	19	29
K106Q	2.3	128	5.8×10^7	1.0/2.9	46	57
K107Q	2.3	129	5.6×10^7	1.0/3.0	48	78
K154Q	0.8	124	15.5×10^7	0.5/3.2	61	64

Source: Myles, T., Le Bonniec, B.F., and Stone, S.R., Eur. J. Biochem. 268, 70–77, 2001
[a] Cleavage of the PAR peptide was measured by RP-HPLC on a C_4 column with trifluoroacetic acid gradient in acetonitrile
[b] [Concentration of thrombin derivative, nM/time required for 50% aggregation]
[c] Percentage of recombinant thrombin (rIIa) as determined from a standard curve
[d] Percentage of rate constant determined with recombinant thrombin (rIIa)

CHAPTER 4 SUMMARY OF DATA

The use of site-directed mutagenesis is an elegant approach to exploring structure-function relationships in proteins. The approach, however, suffers from a problem similar to that faced by those of us who used site-specific chemical modification for the same purpose; that is, unless the results are complete, such as the modification of the active serine residue in serine proteases, there can be some ambiguity in interpretation. Regardless, the combination of such data with crystallographic analysis and chemical modification data can be quite useful. I have tried to summarize some of the data in the sections that follow.

ACTIVE-SITE MUTANTS

Active-site mutants, such as the substitution of alanine for serine 205, together with the chemical conversion to dehydroalanine, confirmed the importance of this residue in catalysis. The mutants also proved useful in binding studies. It is of interest that the S205A mutant was not totally inactive, as will be discussed in Chapter 7. It was necessary to react the S205A mutant with PPACK to obtain total inactivation. Mutation of the active-site histidine and aspartic acid greatly decreased but did not totally inactivate thrombin. Mutation at the oxyanion hole results in total loss of activity.

TRYPTOPHAN

Tryptophan has been an attractive target for both chemical modification and site-directed mutagenesis in thrombin.

TRYPTOPHAN 50 (60D IN CHYMOTRYPSIN NUMBERING)

Work on the desPPW mutant[19] supported a role for Trp50 in thrombin function, although deletion mutants can be problematic. Rezaie[21] showed that Trp50 was important in interaction with antithrombin. More clarity was provided by Bell and coworkers,[49] who showed the W50F mutant had greatly decreased both fibrinogen clotting activity and the hydrolysis of S-2366 (pyroGluProArgpNA). An increase in K_M for S-2366 was largely responsible for the loss of activity in the hydrolysis of S-2366, suggesting a defect in substrate binding. These investigators noted a red shift in the emission spectra for Trp50, suggesting surface exposure. Trp50 was responsible for 16% of energy transfer (fluorescence quenching) to DAPA; Trp96 was also responsible for 16% of the energy transfer, while Trp227 was responsible for 18% of the energy transfer. Crystallographic data[68,69] support the location of Trp50 in the extended binding site. Guinto and Di Cera[70] showed that the W50S mutant did not bind sodium ions and thus is similar to the "slow" form of thrombin. The change from tryptophan to serine is a major change in polarity; the log P value for serine is −3.07, while the log P value for tryptophan is −1.06. DiBella and Scheraga[71] replaced Tyr47-Lys52 with Gly-Gly-Gly [thrombin (GGG)]. This mutant was 3,000-fold less effective in cleaving fibrinogen to fibrin as measured with a turbidimetric assay.

Thrombin (GGG) was also less effective in the hydrolysis of a thrombin substrate (S-2238) or with the major difference in K_M (1000 μM vs. 3 μM) with a smaller change in k_{cat} (90 s^{-1} vs. 120 s^{-1}); a similar change was observed with a tripeptide nitroanilide substrate for trypsin (S-2222).

TRYPTOPHAN 92 (96 IN CHYMOTRYPSIN NUMBERING)

A mutant of Trp92 in bovine thrombin, W92S, was less effective in clotting fibrinogen than native bovine thrombin (it is difficult to convert the turbidimetric assay used by these investigators into quantitative terms).[71] They did show that there was a 4-fold decrease in the k_{cat}/K_m for FPA release and a 7-fold decrease for FPB release. These investigators suggest that Trp92 stabilizes the 60s loop by interaction with Pro48Pro49. Bell and coworkers[49] also studied a site-directed mutant at Try92: W92F. This is a more conservative mutation than the W92S mutation. These investigators reported that this residue was the second-largest contributor to the intrinsic fluorescence of thrombin and was equal to Trp50 in the contribution (16%) to the quenching of DAPA fluorescence. The contribution to the energy transfer to DAPA suggests that this residue is close to the enzyme-active site. The W92F mutant was somewhat less effective in the hydrolysis of S-2238 but as effective as wild-type thrombin in platelet aggregation.

TRYPTOPHAN 227 (TRYPTOPHAN 215 IN CHYMOTRYPSIN NUMBERING)

Tryptophan 227 is the other residue that has been addressed by several investigators. Arosio and coworkers[51] showed that W227 mutants had reduced fibrinogen clotting activity and in the hydrolysis of peptide nitroanilide substrates. The extent of activity loss in both fibrinogen clotting and in the hydrolysis of H-D-Phe-Pro-Arg-pNA depended on the mutation, with the greatest loss being shown by W227A and less effect with the W227F mutation. Bell and coworkers[49] showed that W227 was the largest contributor to energy transfer to DAPA. The W227F mutant also had the second largest decrease in activity in hydrolysis, supporting proximity to the enzyme-active site. It is tempting to suggest that the tryptophan residue modified by the generation of 2-hydroxy-5-nitrobenzyl bromide (HNB) by bovine thrombin[72] is either Trp50 or Trp227.

REFERENCES

1. *Protein Engineering*, ed. D.L. Oxender and C.F. Fox, Liss, New York, USA, 1987.
2. Moody, P.C.E. and Wilkinson, A.J., *Protein Engineering*, IRL Press at Oxford University Press, Oxford, UK, 1990.
3. *Protein Engineering Protocols*, ed. K.M. Arndt and K.M. Müller, Humana, Totowa, NJ, USA, 2007.
4. Bachman, J., Site-directed mutagenesis, *Merthods Enzymol.* 529, 241–248, 2013.
5. Russo, G., Gast, A., Schlaeger, E.J., Angiolillo, A., and Pietropaolo, C., Stable expression and purification of a secreted human prethrombin-2 and its activation to thrombin, *Protein Expr. Purific.* 10, 214–225, 1997.

6. DiBella, E.E., Mauer, M.C., and Scheraga, H.A., Expression and folding of recombinant bovine prethrombin-2 and its activation to thrombin, *J. Biol. Chem.* 270, 163–170, 1995.

7. Lövgren, A., Deinum, J., Rosén, S., *et al.*, Characterization of thrombin derived from human recombinant prothrombin, *Blood Coagul. Fibrinolysis* 26, 545–555, 2015.

8. Bishop, P.D., Lewis, K.B., Schultz, J., and Walker, K.M., Comparison of recombinant human thrombin and plasma-derived α-thrombin, *Sem. Thromb. Hemost.* 32(Suppl 2), 86–97, 2006.

9. Soejima, K., Mimura, N., Yonemura, H., *et al.*, An efficient refolding method for the preparation of recombinant human prethrombin-2 and characterization of the recombinant derived α-thrombin, *J. Biochem.* 130, 269–277, 2001.

10. Arcone, R., Pagliuca, M.G., Chinali, A., *et al.*, Thrombin mutants with altered enzymatic activity have an impaired mitogenic effect on mouse fibroblasts and are inefficient modulators of stellation of rat cotical astrocytes, *Biochim. Biophys. Acta* 1451, 173–186, 1999.

11. Russo, G., Gast, A., Shlaeger, E.-J., Angiolillo, A., and Pietropaoli, C., Stable expression and purification of a secreted prothrombin-2 and its activation to thrombin, *Protein Expr. Purific.* 10, 214–229, 1997.

12. Carter, P. and Wells, J.A., Dissecting the catalytic triad of a serine protease, *Nature* 332, 564–568, 1988.

13. Carter, P. and Wells, J.A., Functional interaction among catalytic residues in subtilisin BPN', *Proteins* 7, 335–342, 1990.

14. Krem, M.M. and Di Cera, E., Dissecting substrate recognition by thrombin using the inactive mutant S195A, *Biophys. Chem.* 100, 315–323, 2003.

15. Fenton, J.W., II and Bing, D.H., Thrombin active-site regions, *Sem. Thromb. Hemost.* 12, 200–208, 1986.

16. Bode, W. and Stubbs, M.T., Spatial structure of thrombin as a guide to its multiple sites of interaction, *Sem. Thromb. Hemost.* 19, 321–333, 1993.

17. Tsiang, T., Jain, A.K., Dunn, K.E., *et al.*, Functional mapping of the surface residues of human thrombin, *J. Biol. Chem.* 270, 16854–16863, 1995.

18. Davie, E.W. and Kulman, J.D., An overview of the structure and function of thrombin, *Sem. Thromb. Hemost.* 32(Supp 1), 3–15, 2006.

19. Le Bonniec, B.F., Guinto, E.R., MacGillivray, R.T., Stone, S.R., and Esmon, C.T., The role of thrombin's tyr-pro-pro-trp motif in the interaction with fibrinogen, thrombomodulin, protein C, antithrombin III, and the Kunitz inhibitors, *J. Biol. Chem.* 268, 19055–19061, 1993.

20. Le Bonniec, B.F., Guinto, E.R., and Stone, S.R., Identification of thrombin residues that modulate its interaction with antithrombin III and and α₁-antitrypsin, *Biochemistry* 34, 12241–12248, 1995.

21. Rezaie, A.R., Tryptophan 60-D in the B-insertion loop of thrombin modulates the thrombin-antithombin reaction, *Biochemistry* 35, 1918–1924, 1996.

22. Rezaie, A.R. and Olson, S.T., Contribution of lysine 60f to S1' specificity of thrombin, *Biochemistry* 36, 1026–1033, 1997.

23. Rezaie, A.R. and Yang, L., Deletion of the 60s loop provides new insights into the substrate and inhibitor specificity of thrombin, *Thromb. Heamost.* 93, 1047–1054, 2005.

24. Bode, W., Mayr, I., Baumann, U., *et al.*, The refined 1.9 Å crystal structure of human α-thrombin: interaction with D-Phe-Pro-Arg chloromethylketone and significance of the Tyr-Pro-Pro-Trp insertion segment, *EMBO J.* 8, 3467–3475, 1989.

25. Banner, D.W. and Hadvary, P., Crystallographic analysis at 3.0 Å resolution of the binding to human thrombin of four active site-directed inhitors, *J. Biol. Chem.* 266, 20085–20093, 1991.

26. Wang, S.X., Esmon, C.T., and Fletterick, R.J., Crystal structure of thrombin reveals conformational changes and extended interactions, *Biochemistry* 40, 10038–10046, 3002.

27. Rezaie, A.R. and Esmon, C.T., Molecular basis of residue 192 participation in determination of coagulation protease specificity, *Eur. J. Biochem.* 242, 177–184, 1996.

28. Le Bonniec, B.F. and Esmon, C.T., Glu-192 → Gln substitution in thrombin mimics the catalytic switch induced by thrombomodulin, *Proc. Natl. Acad. Sci USA* 88, 7371–7375, 1991.

29. Marque, P.-E., Spuntarelli, R., Juliano, L., Aiche, M., and Le Bonniec, B.F., The role of glu^{192} in the allosteric control of the S_2' and S_3' subsites of thrombin, *J. Biol. Chem.* 275, 809–816, 2000.

30. Guillin, M.-C. and Bezeaud, A., Thrombin derivatives obtained by autolytic or limited tryptic cleavage, *Sem. Thromb. Hemost.* 18, 224–229, 1992.

31. Dibella, E. and Scheraga, H.A., The role of the insertion loop around tryptophan 148 in the activity of thrombin, *Biochemistry* 35, 4427–4432, 1996.

32. Le Bonniec, B.F., Guinto, E.R., and Esmon, C.T., Interaction of thrombin with des-ETW with antithrombin III, the Kunitz inhibitors, thrombomodulin and protein C. Structural link between the autolysis loop and the Tyr-Pro-Pro-Trp insertion of thrombin, *J. Biol. Chem.* 267, 19341–19348, 1992.

33. Lundblad, R.L., A rapid method for the purification of bovine thrombin and the inhibition of the purified enzyme with phenylmethylsulfonyl fluoride, *Biochemistry* 10, 2501–2506, 1971.

34. Lanchantin, G.F., Friedman, J.A., and Hart, D.W., Interaction of sovbean trypsin inhibitor with thrombin and its effect on prothrombin activation, *J. Biol. Chem.* 244, 865–875, 1969.

35. Ako, H., Foster, R.J., and Ryan, C.A., Mechanism of action of naturally occurring proteinase inhibitors. Studies with anhydrotrypsin and anhydrochymotrypsin purified by affinity chromatography, *Biochemistry* 13, 132–139, 1974.

36. Woodward, C.A. and Ellis, L.M., Hydrogen-exchange kinetics changes on formation of the soybean trypsin inhibitor-trypsin complex, *Biochemistry* 14, 3419–3423, 1975.

37. Yung, B.Y.R. and Trowbridge, C.G., A calorimetric comparison of trypsin and its anhydro modification in complex formation with Kunitz soybean trypsin inhibitor, *J. Biol. Chem.* 255, 9724–9730, 1980.

38. Lundblad, R.L. and Davie, E.W., The activation of Stuart factor (factor X) by activated antihemophilic factor (activated factor VIII), *Biochemistry* 4, 113–120, 1965.

39. Bock, P.E., Craig, P.A., Olson, T., and Singh, P., Isolation of human blood coagulation factor α-Factor Xa by soybean trypsin inhibitor-Sepharose chromatography and its active-site titration with fluorescein mono-*p*-guanidinobenzoate, *Arch. Biochem. Biophys.* 278, 375–288, 1989.

40. Le Bonniec, B.F., Betz, A., Guinto, E.R., Esmon, C.T., and Stone, S.R., Mapping the thrombin des-ETW conformation by site-directed mutants of hirudin. Evidence fo the induction of nonlocal modificcations by mutagenesis, *Biochemistry* 33, 3959–3966, 1994.

41. Goodman, C.A., Deadman, J.J., Le Bonniec, B.E., *et al.*, Heparin enhances the catalytic activity of des-ETW-thrombin, *Biochem. J.* 315, 77–83, 1996.

42. Berliner, L.J., Sugawara, Y., and Fenton, J.W., 2nd, Human α-thrombin binding to non-polymerized fibrin-Sepharose: evidence for an anionic binding region, *Biochemistry* 24, 7005–7009, 1985.

43. Griffith, M.J., Kingdon, H.S., and Lundblad, R.L., The interaction of heparin with human α-thrombin: effect on the hydrolysis of aniline tripeptide substrates, *Arch. Biochem. Biophys.* 195, 378–384, 1979.

44. Dang, G.D., Sabetta, M., and Di Cera, E., Selective loss of fibrinogen clotting in a loopless thrombin, *J. Biol. Chem.* 272, 19649–19651, 1997.

45. Dang, Q.D., Quinto, E.R., and Di Cera, E., Rational engineering of activity and specificity in a serine protease, *Nat. Biotechnol.* 15, 146–149, 1997.

47. Roy, D.B., Rose, I., and Di Cera, E., Replacement of thrombin residue G184 on the activity fails to mimic Na$^+$ binding, *Proteins: Struct. Funct. Genet.* 43, 315–318, 2001.

48. Niu, W., Chen, Z., Bush-Pelc, L.A., *et al.*, Mutant N143P reveals how Na$^+$ activates thrombin, *J. Biol. Chem.* 284, 36175–36185, 2009.

49. Bell, R., Stevens, W.K., Hia, Z., *et al.*, Fluorescence properties and functional roles of tryptophan residues 60d, 96, 148, 207, 215 of thrombin, *J. Biol. Chem.* 275, 29513–29530, 2000.

50. Lottenberg, R., Christensen, U., Jackson, C.M., and Coleman, P.L., Assay of coagulation proteases using peptide chromogenic and fluorogenic substrates, *Methods Enzymol.* 80, 341–361, 1980.

51. Arosio, D., Ayala, Y.M., and Di Cera, E., Mutation of W215 compromises thrombin cleavage of fibrinogen, but not of PAR-1 or protein C, *Biochemistry* 39, 8095–8101, 2000. 237.

52. Peacock, R.B., Davis, J.R., Markwick, P.R.L., and Komives, E.A., Dynamic consequences of mutation of Tryptophan 215 in thrombin, *Biochemistry* 57, 2694–2703, 2018.

53. DiBella, E.T. and Scheraga, H.A., Thrombin specificity: further evidence for the importance of the β-insertion loop and Trp96 and Pro^{60B}Pro^{60C} for the activity of thrombin, *J. Prot. Chem.* 17, 107–208, 1998.

54. Banerjee, D., Rose, T., and Di Cera, E., Replacement of thrombin residue G184 with lys or arg fails to mimic Na$^+$binding, *Prot. Struct. Funct. Genet.* 43, 315–318, 2001.

55. Le Bonniec, B.R., MacGillvray, R.T.A., and Esmon, C.T., Thrombin Glu-39 restricts the P'3 specificity to nonacidic residues, *J. Biol. Chem.* 266, 13796–13803, 1991.

56. Cunningham, B.C. and Wells, J.A., High resolution epitope mapping of hGH receptor interactions by mutagenesis, *Science* 244, 1081–1085, 1989.

57. Weiss, G.A., Watanabe, C.K., Zhang, A., Goddard, A., and Sidhu, S.S., Rapid mapping of protein functional epitopes by combinatorial alanine scanning, *Proc. Natl. Acad. Sci. USA* 97, 8950–8954, 2000.

58. Tang, Q. and Fenton, A.W., Whole-protein alanine-scanning mutagenesis of allostery: a large percentage of a protein can contribute to mechanism, *Hum. Mutat.* 38, 1132–1143, 2017.

59. Tsiang, M., Jain, A.K., Dunn, K.E., *et al.*, Functional mapping of the surface residues of human thrombin, *J. Biol. Chem.* 270, 16854–16863, 1995.

60. Hall, S.W., Gibbs, C.S., and Leung, L.L.K., Identification of critical residues on thrombin mediating its interaction with fibrin, *Thromb. Haemost.* 86, 1466–1474, 2001.

61. Leung, L.K. and Hall, S.W., Dissociation of thrombin's substrate interactions using site-directed mutagenesis, *Trends Cardiovasc. Med.* 10, 89–92, 2000.

62. Gibbs, C.S., Coutre, S.E., Tsiang, M., *et al.*, Conversion of thrombin into an anticoagulant by protein engineering, *Nature* 378, 413–416, 1995.

63. Tsiang, M., Paborsky, L.P., Li, W.X., *et al.*, Protein engineering thrombin for optimal specificity and potency for anticoagulant activity *in vivo*, *Biochemistry* 35, 16449–16457, 1995.

64. Burks, E.A., Chen, G., Georgiou, G., and Iverson, B.L., Burks, E.A., Chen, G., Georgiou, G., and Iverson, B.L., In vitro scanning saturation mutagenesis of an antibody binding pocket, *Proc. Natl. Acad. Sci.* 94, 412–417, 1992.

65. Hall, S.W., Nagashima, M., Zhao, L., Morser, J., and Leung, L.K., Thrombin interacts with thrombomodulin, protein C, and thrombin-activated fibrinolysis inhibitor via specific and distinct domains, *J. Biol. Chem.* 274, 25510–25526, 1999.

66. Myles, T., Church, F.C., Whinna, H.C., Monard, D., and Stone, S.R., Role of thrombin anion binding exosite-1 in the formation of thrombin-serpin complexes, *J. Biol. Chem.* 273, 31203–31208, 1998.

67. Myles, T., Le Bonniec, B.F., and Stone, S.R., The dual role of thrombin's anion-binding exosite-1 in the recognition and cleavage of the protease-activated receptor 1, *Eur. J. Biochem.* 268, 70–77, 2001.

68. Bode, W., Mayr, I., Baumann, U., *et al.*, The refined 1.9 Å crystal structure of human α-thrombin: interaction with D-Phe-Pro-Arg chloromethyl ketone and significance of the Try-Pro-Pro-Trp insertion segment, *EMBO J.* 8, 3467–3475, 1989.

69. Banner, D.W. and Hadvary, P., Crystallographic analysis at 3.0 Å resolution of the binding to human thrombin of four active site-directed inhibitors, *J. Biol. Chem.* 266, 20085–20093, 1991.

70. Guinto, E.R. and Di Cera, E., Critical role of W60d in thrombin allostery, *Biophys. Chem.* 64, 103–109, 1997.

71. Di Bella, E.F. and Scheraga, H.A., Thrombin specificity: further evidence for the importance of the β-insertion looo and Trp[96]. Implications of the hydrophobic interaction between Trp[96] and Pro[60B]Pro[60C] for the activity of thrombin, *J. Prot. Chem.* 17, 197–208, 1998.

72. Uhteg, L.C. and Lundblad, R.L., The modification of tryptophan in bovine thrombin, *Biochim. Biophys. Acta* 491, 551–557, 1971.

5 Chemical Modification of Thrombin

The specific chemical modification of proteins was useful in early studies on the relationship between the structure and function of proteins.[1-3] However, specific chemical modification has been largely supplanted by the site-specific modification of proteins using recombinant DNA technology.[4-6] There is continued use of specific chemical modifications in chemical biology[7,8] and in the preparation of antibody drug conjugates (ADCs).[9,10] I would like to think that some of the early work on the specific chemical modification of lysine residues in thrombin provided early support for the development of the concept of the anion-binding exosite (exosite-1).[11,12] However, I would also to think that I could have been a linebacker in the NFL.

ACTIVE-SITE SERINE IN THROMBIN

Early work on the reaction of DFP with chymotrypsin[13] and trypsin[14] had shown the importance of a serine residue at the enzyme-active site. The reaction/inactivation of thrombin by DFP was reported by Jules Gladner and Kolman Lake in 1956 working at the National Institutes of Health[15,16] and by Kent Miller and Helen Van Vunakis at the New York State Department of Health.[17] These studies were among the first showing that thrombin was a proteolytic enzyme and that the formation of fibrin from fibrinogen was an enzymatic process. Several years later Gladner and coworkers[18] showed the DFP reacted with the active serine, further showing the relationship of thrombin to trypsin and chymotrypsin, although these studies may seem trivial when viewed from 2020 (as if anything is trivial in this particular year).

Many years after the initial studies with DFP, I demonstrated that PMSF inactivated thrombin, presumably reacting with the active-site serine.[19] PMSF had been developed by Fahrey and Gold for the study of enzyme mechanisms.[20] They had noted earlier studies on the use of methanesulfonyl fluorides as insecticides. They subsequently explored the reaction of PMSF with chymotrypsin in further detail, showing a reaction at the enzyme-active site similar to the process of acylation during catalysis.[21] They also observed that the sulfonylated enzyme is reactivated at alkaline pH (pH 8.5). As noted earlier, thrombin modified with PMSF can slowly reactivate; the rate and extent of reactivation are enhanced by the addition such of a nucleophile as hydroxylamine.[22] PMSF has been used to modify thrombin in studies, showing the importance of the enzyme-active site in the reaction of thrombin with cells.[23,24]

Elliott Shaw and coworkers built on the base of information obtained from the study of the reaction of PMSF with the active-site serine in trypsin-like enzymes. Wong and coworkers[25] prepared p-nitrophenyl m-amidinophenylmethyl sulfonate (Figure 5.1) and p-nitrophenyl p-amidinophenylmethylsufonate (Figure 5.1) in an approach to

DOI: 10.1201/b22204-5

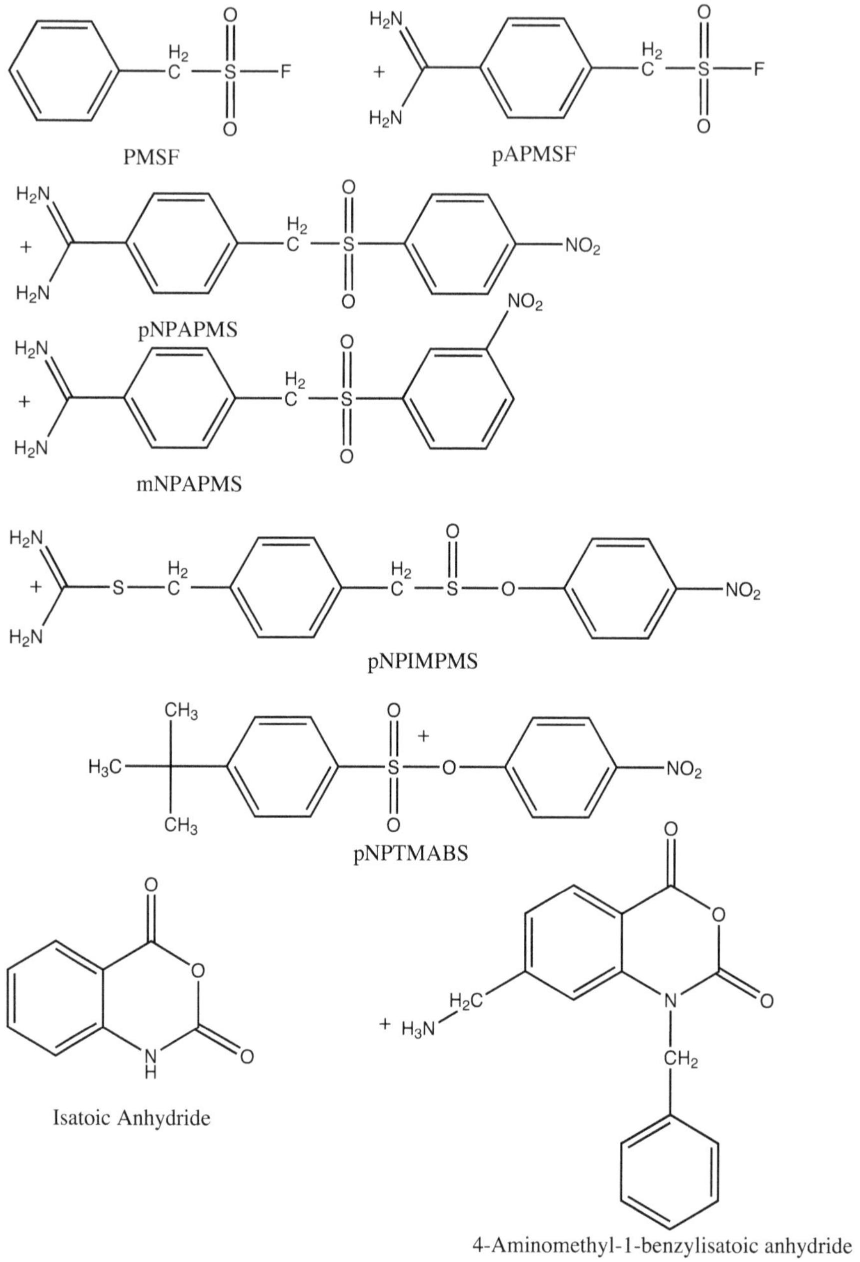

FIGURE 5.1 The structures of various reagents used to modify the active-site serine in thrombin. Shown are phenylmethylsulfonyl fluoride (PMSF), p-amidinophenylmethyl-sulfonyl fluoride (p-APMSF), p-nitrophenlamidinophenylmethyl sulfonate (p-NPAPMS), m-nitrophenylamidinophenylmethyl sulfonate (m-NPAPMS), p-nitrophenylisothiouronium-methylphenylmethyl sulfonate (p-NPIMPMS), and p-nitrophenyltrimehylamonniumbenzene sulfonate (p-NPTMABS).

TABLE 5.1
The Rate of Reaction of Isomers of *p*-Nitrophenyl-*p*-Amidinophenylmethylsu lfonate

Nitrophenol Isomer	pKa	k_2 M^{-1}min^{-1a}			
		Thrombin	Trypsin	Plasma Kallikrein	Plasmin
p-nitrophenol	7.15	4.62×10^2	0	0	0
m-nitrophenol	8.36	0.39×10^2	6.99×10^1	2.5	0
o-nitrophenol	7.23	1.39×10^2	7.3×10^2	7.7×10^1	0

Source: Wong, S.-C. and Shaw E., *Arch. Biochem. Biophys.* 176, 113–118, 1976

[a] Reactions performed in 0.1 M Tris, pH 7.4, at 25°C. In the case of trypsin and thrombin, the reaction mixtures also contained 0.20 M calcium ions (counter ion not specified). Assays for activity were performed in 0.2 M sodium maleate, pH 6.0 with carbobenzoxy-lysine nitrophenyl ester

discriminate between trypsin-like proteases. Both isomers were competitive inhibitors of trypsin, plasmin, kallikrein, chymotrypsin, and thrombin, but only *p*-nitrophenyl *p*-amidinophenylmethylsufonate formed a covalent derivative with thrombin, which was stable for at least 48 hours. Wong and Shaw[26] subsequently showed that the departing group in these pseudosubstrates, nitrophenol, could provide specificity on the reaction of nitrophenyl *p*-amidinophenylmethylsulfonates (NPAPSs) with tryptic-like serine proteases. Previous work[25] had shown that *p*-nitrophenyl-*p*-amidinophenylmethylsulfonate (*p*-NPAP) was a potent inactivator of thrombin but a competitive inhibitor with other tryptic-like serine proteases. The *m*-nitrophenyl (*m*-NPAP) and *o*-nitrophenyl (*o*-NPAP) esters were less specific than the *p*-nitrophenyl ester (Table 5.1). Both of these reacted with trypsin and plasma kallikrein, although slower than thrombin; neither isomer reacted with plasmin. The authors suggest that steric factors in the thrombin-active site are responsible for the difference in reactivity. There is a relationship between the acidity (pKa) of the nitrophenol isomers and the rate of reaction of the particular isomer ester with thrombin.

Shaw and coworkers continued their work on the reaction of sulfonyl esters with thrombin and other tryptic-like serine proteases with derivatives of benzene sulfonic acid and phenylmethylsulfonic acid, which were intended to evaluate reagent length and position of the basic substituent on the aromatic ring.[27] The structures of some of the reagents are shown in Figure 5.1. Another group reported that *p*-amidinophenylmethylsulfonyl fluoride (*p*-APMSF) (Figure 5.1) was a potent inactivator of human α-thrombin.[28] Instability of the reagent precluded accurate determination of reaction rate, but the rate of reaction is likely similar to or faster than that seen with *p*-NPAPMS. *p*-APMSF was also a potent inactivator of trypsin, factor Xa, and plasmin; α-chymotrypsin and acetylcholinesterase were not inactivated by this reagent. Data for the reaction of some of these reagents with thrombin is shown in Table 5.2. Brown and Powers[29] evaluated some acylating agents, carbamylating agents, and carbamylating agents that react with the active site of serine. These reagents did not contain an arginine or other positively charged group, which would provide for binding to the thrombin S_1 site. Isatoic anhydride reacted rapidly with thrombin (4.89 × 10^4 $M^{-1}s^{-1}$); diacylation was very slow (4.4 × 10^{-6} $M^{-1}s^{-1}$). Moorman and Abeles[30] had previously shown that isatoic anhydride was a potent inactivator of chymotrypsin and some other serine proteases; they also demonstrated inactivation of papain. In later work from the Abeles laboratory at Brandeis, Gelb and Abeles[31] showed with studies on thrombin that the specificity of the reaction could be improved by substituents on the aromatic ring. 4-Aminomethyl-1-benzylisatoic anhydride was a potent inhibitor of thrombin, with a lesser effect on trypsin and no effect on the activity of chymotrypsin.

ANHYDROSERINE IN THROMBIN

Another chemical approach to the modification of the active-site serine in thrombin is the conversion to dehydroalanine (2-aminoprop-2-enoic acid).[32] The conversion of the active-site serine to dehydroalanine in chymotrypsin was described by the late Dan Koshland and coworkers in Berkeley,[33,34] demonstrating the importance of this residue in catalysis. There was an argument at the time that modification with DFP introduced a bulky group at the active site, creating steric interference rather than demonstrating the importance of this residue in catalysis. Koshland had an excellent review of the state of the art at that time in the study of the relationship between structure and function in enzymes.[35] Six years later a group at Washington State University at Pullman, Washington, developed a facile method for the preparation of anhydro chymotrypsin[36] and anhydro trypsin[37] based on the base-catalyzed elimination of water from phenylmethylsulfonyl (PMS) enzyme under milder conditions than those used earlier for tosyl-chymotrypsin. The anhydro enzymes bound protein inhibitors, suggesting that no major conformational change occurred on the conversion to anhydro chymotrypsin.

Ashton and Scheraga[32] prepared anhydrothrombin via the β-elimination of water from PMS-thrombin. Their procedure was somewhat more complicated than the studies cited earlier. The base-catalyzed β-elimination reaction was performed in approximately 6 m GuCl and the resulting anhydrothrombin was renatured. It was necessary to inactivate residual thrombin activity, which resulted from incomplete inactivation and reactivation of PMS-thrombin during the process of β-elimination with PPACK. Purification of the renatured anhydrothrombin was accomplished on a column of p-aminobenzamide coupled to agarose.[38] The purified anhydrothrombin maintained structural integrity as judged by mass spectrometric analysis, SDS-PAGE electrophoresis, and sedimentation equilibrium. The purified anhydrothormbin bound to hirudin in close to 1:1 stoichiometry. Human anhydrothrombin was used to identify a binding site for thrombin on the C2 domain of factor VIII.[39] Detail on the preparation of the anhydrothrombin used in this study is not provided other than purification on the p-aminobenzamidine agarose column. Another group showed that anhydrothrombin bound to agarose could bind factor XIII, factor VIII, and protein C.[40] The crystal structure of the complex formed between anhydrothrombin and antithrombin has been reported.[41]

MODIFICATION OF HISTIDINE IN THROMBIN

Histidine is found at the active of site of more than 50% of characterized enzymes.[42,43] Histidine 43 is a member of the catalytic triad in the thrombin-active site, as discussed in Chapter 3. Histidine functions as an acid and a base in catalysis. The imidazole ring has two nitrogen atoms and can exist as a tautomer in solution. Histidine can be alkylated with a reaction preferentially at the N3 ($N\pi$) position. There is little evidence to suggest that the oxidation of histidine in thrombin results in the alteration of activity. Oscar Lucas[44] at the Oregon Health Sciences University observed that oxidized cellulose inactivated thrombin. A more recent study reported that thrombin was inactivated by myeloperoxidase in the presence of hydrogen peroxide and sodium chloride.[45] This system generates hypochlorous acid (HOCl), which is a strong biological oxidizing agent.[46] De Cristofero and Landolfi[45] showed the loss of tryptophan and lysine with a myeloperoxidase–sodium chloride–hydrogen peroxide system, which could account for the loss of thrombin activity. The lack of quantitative amino acid analysis in this study precludes evaluation of the possible loss of histidine.

Acylation of a nitrogen on the imidazole ring is another approach to the chemical modification of histidine. Diethylpyrocarbonate (ethylformic anhydride, ethoxycaronyl ethyl carbonate; DEPC) was introduced by Edith Miles as a reagent for the selective modification of histidine in proteins.[47] This reagent can also modify tyrosine and lysine residues in proteins, but it is possible to distinguish from the various modifications. DEPC does not have any structural features that would provide for selective binding to an enzyme-active site, so it is similar to isatoic anhydride, which modifies the active-site serine of thrombin.[30,31] As such, it is reasonable to assume that the reaction of DEPC with histidine is a measure of the nucleophilicity of the modified ring nitrogen, most like N_2 (N_π).[48,49] The reaction of bovine β-thrombin with DEPC was shown to be slower than the rate of inactivation of bovine α-thrombin (TosArgOMe), suggesting reduced nucleophilicity of the active-site histidine in β-thrombin.[51] This data is shown in Table 5.3 and compared with data obtained for the modification of histidine in other enzymes. The nucleophilicity of the active-site histidine is due to the interaction with Asp99 in the catalytic triad. A similar argument is advanced to explain the enhanced reactivity of His141 in rat liver arginase.[49] It was suggested that the reduced nucleophilicity of the active-site histidine was the cause of reduced fibrinogen clotting activity.[50,51] In looking back at the data from several decades later, it is not possible to exclude the modification of other amino acid residues such as lysine. Bovine β-thrombin does result from a peptide bond cleavage in the exosite-1 domain, which should result in decreased binding of fibrinogen, as well as changes in active-site reactivity (Chapter 3). Chandvekar and Nadkorni[52] studied the effect of gamma irradiation on the activity of human α-thrombin. Fibrinogen clotting activity was lost to a greater extent than esterase (TosArgOMe) activity. It was suggested that the loss of activity was due to modification of tryptophan and tyrosine residues. There was interest in the possible modification of histidine, but no data to support that possibility. These investigators did study the reaction of DEPC with thrombin and showed that the loss of fibrinogen clotting activity and the loss of esterase activity occurred at a similar rate.

It has been possible to modify the active-site histidine in trypsin with iodoacet-amide in the presence of methylguanidine but not in the absence of it.[53] A similar experiment has not been performed with thrombin. As can be seen by the discussion in Chapter 3, there was kinetic and structural evidence to support the importance of a histidine residue in catalysis by serine proteases. There was no chemical evidence unlike that, for example, available for the histidine residues in bovine pancreatic ribonuclease.[54] It remained for Elliott Shaw and his coworkers, first at Tulane University and later in the Biology Department of Brookhaven National Laboratory at Long Island, to develop reagents for the alkylation of histidine residues at the active site of serine proteases. Tosyl-phenylalanine chloromethyl ketone (TPCK, L-1-tosylamido-2-phenylethyl chloromethyl ketone) was shown be a reagent that specifically reacted with the active histidine of chymotrypsin.[55] Subsequent work resulted in the development of tosyl-lysyl chloromethyl ketone (TLCK, 1-chloro-3-tosylamido-2-heptanone).[56] As noted later by this group, there were difficulties in the synthesis of the arginine-containing reagent.[57] While it may seem strange to a generation of scientists living with current technologies, the work of Shaw and coworkers provided the necessary unequivocal evidence for the presence of histidine at the active site of serine proteases. Glover and Shaw[58] subsequently showed that thrombin was also inactivated by TLCK and that the inactivation was associated with alkylation of the N_3 ($N\pi$) position on the imidazole ring of His43 at the enzyme-active site. TLCK and the other peptide chloromethyl ketones, which will be described later, are affinity reagents or affinity labels.[59] As an affinity label, the peptide chloromethyl ketone first binds to the enzyme to form a complex analogous to an enzyme-substrate complex. This binding step is followed by the actual alkylation reaction: a nucleophilic substitution. Other than work cited earlier, there was little additional work on the reaction of TLCK with thrombin. Exner and Koppel[60] reported that cholate enhanced the rate of reaction of TLCK with thrombin. These investigators suggested that the effect was on the alkylation step, which follows the binding step. These investigators also observed that sodium cholate enhanced the rate of hydrolysis of TosArgOMe and TosLysOMe. The reaction of TLCK (and other peptide chloromethyl ketones) is specific for a "hyper-active" histidine. Gilles and Keil[61] showed that TLCK reacted with the active cysteine of α-clostripain and did not react with any histidine residues in the protein.

D-Phe-pro-arg-chloromethyl ketone (PPACK; FPRCH$_2$Cl) was introduced by Kettner and Shaw as specific reagent for the modification of the active-site histidine in thrombin with a minimum second-order rate constant of 6.9×10^8 M^{-1}min^{-1}.[62] As noted by the authors, this is at least three to five orders of magnitude faster than other homologous serine proteases such as human plasma kallikrein, human plasmin, bovine factor Xa, and urokinase. Given its specificity and rapid rate of reaction with thrombin, PPACK has been widely used to inactivate thrombin for various purposes, including crystallographic analysis of thrombin.[63–66] The use of the D-isomer phenylalanine is important, as it allows for interaction of the amine with Gly228.[64] Earlier work[63] had shown the importance of a hydrophobic cage composed of His43, Tyr47, Trp50, Leu96, Ile179, and Trp227 in determining the specificity of the reaction with PPACK. Banner and Hadváry[66] and Tulinsky and coworkers[65] also reported on the crystal structure of PPACK-thrombin, showing movement in the active-site

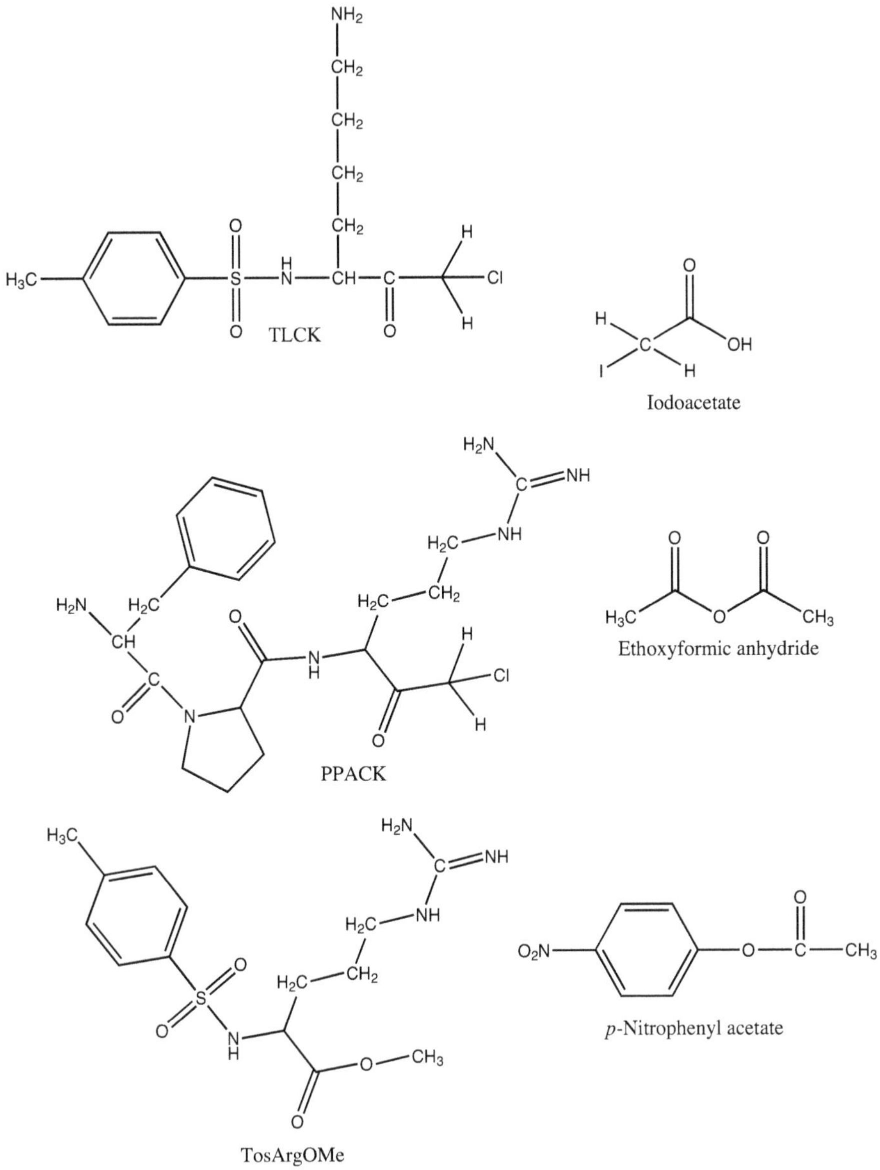

FIGURE 5.2   The structures for reagents which react with histidine residues in enzyme active sites. Shown for comparison are two substrates for thrombin: TosArgOMe is relatively specific, while *p*-nirophenyl acetate is not specific.

region to accommodate the binding of the inhibitors. Bock has prepared fluorescent probes of the thrombin active site via thioester derivatives of PPACK,[67,68] which can be used a reporter group. Data on the rate of reaction of some peptide chloromethyl-ketones (Figure 5.2) are presented in Table 5.3.

TABLE 5.2
Modification of the Active-Site Serine in Thrombin

Enzyme	Rate (M^{-1}min^{-1})	Solvent/Temperature	Inhibitor	Reference
Bovine α-thrombin	348	6 mM Triethanolamine-6 mM imidazole-6 mM acetate, pH 8.0/23°C	PMSF	1
Bovine α-thrombin	522	6 mM Triethanolamine-6 mM imidazole-6 mM acetate, pH 9.0/23°C	PMSF	1
Human α-thrombin	764	10 mM HEPES-10mM Tris-100 mM NaCl-0.1%PEG, pH 7.8/25°C	DFP	2
Human β-thrombin	426	10 mM HEPES-10mM Tris-100 mM NaCl-0.1%PEG, pH 7.8/25°C	DFP	2
Bovine trypsin	271	1,5 mM CaCl$_2$_0.075 M Tris, pH 7.2/25°C	PMSP	3
Bovine trypsin	300	1,5 mM CaCl$_2$_0.075 M Tris, pH 7.2/25°C	DFP	3
Bovine α-thrombin	256	50 mM potassium phosphate, pH 6.0, 24°C	DFP	4
Bovine β-thrombin	89	50 mM potassium phosphate, pH 6.0, 24°C	DFP	4
Bovine α-thrombin	–	50 mM sodium veronal, pH 7.5, 25°C	p-NPAPS[a]	5
Bovine α-thrombin	4620	100 mM Tris, pH 7.4 with 20 mM CaCl$_2$, 25°C	p-NPAPS[a]	6
Bovine trypsin	0	100 mM Tris, pH 7.4 with 20 mM CaCl$_2$, 25°C	p-NPAPS[a]	6
Human plasma kallikrein	0	100 mM Tris, pH 7.4 with 20 mM CaCl$_2$, 25°C	p-NPAPS[a]	6
Bovine α-thrombin	390	100 mM Tris, pH 7.4 with 20 mM CaCl$_2$, 25°C	m-NPAPS[b]	6
Bovine trypsin	69.3	100 mM Tris, pH 7.4 with 20 mM CaCl$_2$, 25°C	m-NPAPS[b]	6
Human plasma kallikrein	2.5	100 mM Tris, pH 7.4 with 20 mM CaCl$_2$, 25°C	m-NPAPS[b]	6
Bovine α-thrombin	1390	100 mM Tris, pH 7.4 with 20 mM CaCl$_2$, 25°C	o-NPAPS[c]	6
Bovine trypsin	730	100 mM Tris, pH 7.4 with 20 mM CaCl$_2$, 25°C	o-NPAPS[c]	6
Human plasma kallikrein	77	100 mM Tris, pH 7.4 with 20 mM CaCl$_2$, 25°C	o-NPAPS[c]	6
Bovine α-thrombin	580	100 mM Tris, pH 7.4, room temperature (assume 23°C)	p-NPIMBS[d]	7
Human plasmin[e]	125	100 mM Tris, pH 7.4, room temperature (assume 23°C)	p-NPIMBS[d]	7

Enzyme	Rate $(M^{-1}min^{-1})$	Solvent/Temperature	Inhibitor	Reference
Bovine α-thrombin	147	100 mM Tris, pH 7.4, room temperature (assume 23°C)	p-NPTMABS[f]	7
Human Plasmin	50	100 mM Tris, pH 7.4, room temperature (assume 23°C)	p-NPTMABS[f]	7
Bovine α-thrombin	0	100 mM Tris, pH 7.4, room temperature (assume 23°C)	p-NPIMPMS[g]	7
Bovine Trypsin	1500	100 mM Tris, pH 7.4, room temperature (assume 23°C)	p-NPIMPMS[g]	7
Human plasmin	7	100 mM Tris, pH 7.4, room temperature (assume 23°C)	p-NPIMPMS[g]	7
Human plasma kallikrein	4	100 mM Tris, pH 7.4, room temperature (assume 23°C)	p-NPIMPMS[g]	7
Human α-thrombin	–	0.75 M NaCl, pH 7.0, 23°C	APMSF[h]	8
Human α-thrombin	2.934×10^6	100 mM HEPES-10 mM CaCl$_2$, pH 7.5, 25°C	isatoic anhydride	9

[a] p-Nitrophenyl p-amidinophenyl-sulfonate
[b] m-Nitrophenyl p-amidinophenyl-sulfonate
[c] o-Nitrophenyl p-amidinophenyl-sulfonate
[d] p-Nitrophenyl p-isothiouroniummethylbenzenesulfonate
[e] Prepared by activation of purified human plasminogen by streptokinase
[f] p-Nitrophenyl p-N(Trimethyl)benzenesulfonate
[g] p-Nitrophenyl p-isothiouroniummethylphenylmethylsulfonate
[h] p-Amidinophenylmethylsulfonyl fluoride [(p-amidinophenyl)methanesulfonyl fluoride]

References
1. Lundblad, R.L., A rapid method for the purification of bovine thrombin and the inhibition of the purified enzyme with phenylmethylsulfonyl fluoride, *Biochemistry* 10, 2501–2506, 1971.
2. Bezeaud, A. and Guillin, M.C., Enzymic and nonenzymic properties of human β-thrombin, *J. Biol. Chem.* 263, 3516–3581, 1988.
3. Fahrney, D.E. and Gold, A.M., Sulfonyl fluorides as inhibitors of esterases I. Rates of reaction with acetylcholinesterase, chymotrypsin, and trypsin, *J. Am. Chem. Soc.* 100, 997–1000,1963.
4. Lundblad, R.L., Nesheim, M.E., Straight, D.L., *et al.*, Bovine α- and β-thrombin. Reduced fibrinogen clotting activity of β-thrombin is not a consequence of reduced affinity for fibrinogen, *J. Biol. Chem.* 259, 6991–6995, 1984.
5. Wong, C.-S. and Shaw, E., Differences in active center reactivity of trypsin homologs. Specific inactivation of thrombin by p-nitrophenyl p-amidinophenylsulfonate, *Arch. Biochem. Biophys.* 161, 536–543, 1974.
6. Wong, C.-S. and Shaw, E., Inactivation of trypsin-like proteases by active-site directed sulfonylation. Ability of the departing group to confer specificity, *Arch. Biochem. Biophys.* 176, 113–118, 1976.
7. Wong, C.-S., Green, G.D.J., and Shaw, E., Inactivation of trypsin-like proteases by sulfonylation. Variation of positively charged group and inhibitor length, *J. Med. Chem.* 21,456–459, 1978.
8. Laura, R., Robison, D.J., and Bing, D.H., (p-amidinophenyl)methanesulfonyl fluoride, an irreversible of serine proteases, *Biochemstry* 19, 4859–4864, 1980.

TABLE 5.3
Chemical Modification of Histidine in Thrombin

Enzyme	Inhibitor	Rate $(M^{-1}min^{-1})$	Solvent/Temperature	Ref
Bovine trypsin	TLCK	336	0.1 M malate, pH 6.0/25°C	1
Bovine α-thrombin	TLCK	10.6	0.05 M ACES-0.04 M Tris, pH 7.0/24°C	2
Bovine β-thrombin	TLCK	3.8	0.05 M ACES-0.04 M Tris, pH 7.0/24°C	2
Bovine α-thrombin	DEP	256	0.05 M potassium phosphate, pH 7.0/24°C	2
Bovine β-thrombin	DEP	86	0.05 M potassium phosphate, pH 7.0/24°C	2
Human α-thrombin	TLCK	40	10 mM HEPES-10mM Tris-100 mM NaCl-0.1%PEG, pH 7.8/25°C	3
Human β-thrombin	TLCK	39	10 mM HEPES-10mM Tris-100 mM NaCl-0.1%PEG, pH 7.8/25°C	3
Bovine α-thrombin	TLCK	23.5	0.05 M sodium phosphate, pH 7.5/22°C	4
Bovine α-thrombin	TLCK	166.4	0.05 M sodium phosphate, pH 7.5 with 2 mM sodium cholate/22°C	4
Human α-thrombin	PPACK	3.6×10^5	0.05 M Tris-0.100 M NaCl-0.1% PEG 6000, pH 7.8/37°C	5
Human β-thrombin	PPACK	1.14×10^5	0.05 M Tris-0.100 M NaCl-0.1% PEG 6000, pH 7.8/37°C	5
Human α-thrombin	PPACK (FPR-CH$_2$Cl)	84×10^7	0.1 M HEPES-0.3 M NaCl-1 mM EDTA-0.1% PEG 8000, pH 7.0/25°C	6
Human α-thrombin	FFR-CH$_2$Cl	90×10^5	0.1 M HEPES-0.3 M NaCl-1 mM EDTA-0.1% PEG 8000, pH 7.0/25°C	6
Human α-thrombin	ATA-FPR-CH$_2$Cl	62×10^4	0.1 M HEPES-0.3 M NaCl-1 mM EDTA-0.1% PEG 8000, pH 7.0/25°C	6
Human α-thrombin	ATA-FFR-CH$_2$Cl	3.6×10^3	0.1 M HEPES-0.3 M NaCl-1 mM EDTA-0.1% PEG 8000, pH 7.0/25°C	6
Bovine α-thrombin	β-AAKCK	72	0.2 M Tris-HCl, pH 7.0, 25°C	7
Bovine α-thrombin	PAKCK	120	0.2 M Tris-HCl, pH 7.0, 25°C	7
Bovine α-thrombin	KAKCK	282	0.2 M Tris-HCl, pH 7.0, 25°C	7
Bovine α-thrombin	Z-KCK	444	0.2 M Tris-HCl, pH 7.0, 25°C	7

Enzyme	Inhibitor	Rate $(M^{-1}min^{-1})$	Solvent/Temperature	Ref
Bovine α-thrombin	EAKCK	720	0.2 M Tris-HCl, pH 7.0, 25°C	7
Bovine α-thrombin	FAKCK	948	0.2 M Tris-HCl, pH 7.0, 25°C	7
Human α-thrombin	dVLKCK	1.38×10^6	100 mM phosphate, pH 7.3, 25°C	8
Human α-thrombin	AFKCK	1.44×10^6	100 mM phosphate, pH 7.3, 25°C	8
Human α-thrombin	pEFKCK	138×10^3	100 mM phosphate, pH 7.3, 25°C	8
Human α-thrombin	dVFKCK	3.24×10^4	100 mM phosphate, pH 7.3, 25°C	8
Human α-thrombin	Ac-GVRCK	3.79×10^4	50 mM Tris-100 mM NaCl, 0.2% PEG 6000, pH 7.8	9
Human α-thrombin	Ac-EGGGVRCK	1.11×10^4	50 mM Tris-100 mM NaCl, 0.2% PEG 6000, pH 7.8	9
Human α-thrombin	Ac-DFLAEGGGVRCK	4.36×10^6	50 mM Tris-100 mM NaCl, 0.2% PEG 6000, pH 7.8	9
Human α-thrombin	Ac-ADSGLEGDF LAGGGVRCK	6.5×10^6	50 mM Tris-100 mM NaCl, 0.2% PEG 6000, pH 7.8	9
Human α-thrombin	FPRCK	1.2×10^9	50 mM Tris-100 mM NaCl, 0.2% PEG 6000, pH 7.8	9

References

1. Shaw, E., Mares-Guia, M., and Cohen, W., Evidence for an active-center histidine in trypsin through use of a specific reagent., 1-chloro-3-tosylamido-7-amino-2-heptanone, the chloromethyl ketone derived from Nα-tosyl-L-lysine, *Biochemistry* 4, 2219–2224, 1965.
2. Lundblad, R.L., Nesheim, M.E., Straight, D.L., *et al.*, Bovine α- and β-thrombin. Reduced fibrinogen-clotting activity of β-thrombin is not a consequence of reduced affinity for fibrinogen, *J. Biol. Chem.* 259, 6991–6995, 1984.
3. Bezeaud, A. and Guillin, M.C., Enzymic and nonenzymic properties of human β-thrombin, *J. Biol. Chem.* 263, 3516–3581, 1988.
4. Exner, T. and Koppel, J.L., Cholate enhancement of interaction between thrombin and tosyl lysine chloromethyl ketone, *Biochim. Biophys. Acta* 329, 233–240, 1973
5. Hofsteenge, J., Braun, P.J., and Stone, S.R., Enzymatic properties of proteolytic derivatives of human α-thrombin, *Biochemistry* 27, 2144–2157, 1988.
6. Bock, P.E., Active-site-selective labeling of blood coagulation proteinases with fluorescence probes by the use of thioester peptide chloromethyl ketones, *J. Biol. Chem.* 267, 14963–14973, 1992.
7. Coggins, J.R., Kray, W., and Shaw, E., Affinity labeling of proteinases with tryptic specificity by peptides with *C*-terminal lysine chloromethyl ketone, *Biochen. J.* 138, 579–585, 1974.
8. Collen, D., Lijnen, H.R., De Cock, F., Durieux, J.P., and Loffet, A., Kinetic properties of tripeptide lysyl chloromethyl ketone and lysine *p*-nitroanilide derivatives towards trypsin-like serine proteinases, *Biochim. Biophys. Acta* 165, 158–166, 1980.
9. Angliker, H., Shaw, E., and Stone, S.R., Synthesis of oligopeptide choromethanes to investigate exending binding regions of proteinases: application to interaction of fibrinogen with thrombin, *Biochem. J.* 292, 261–266, 1993.

TLCK and PPACK both react with the active-site histidine of thrombin, and there are differences in the behavior of the two derivative forms in their interaction with receptors on platelets and other binding partners. While there are crystal structures for FPRCH2-thrombin, none, to the best of my knowledge, exist for TLCK-inactivated thrombin. FPRCH$_2$-thrombin binds to "high affinity" (assumed to be GPIb) sites on blood platelets, while TLCK-inactivated thrombin does not.[69] PPACK-thrombin competes with native thrombin for binding to platelets, while TLCK-thrombin does not bind. A later report from this laboratory[70] suggested that there was incomplete inactivation of thrombin with TLCK. It is not clear that this would make any difference with the earlier results on binding to platelets. A much later study[71] showed that immobilized PPACK-thrombin bound radio-labeled human platelets, while immobilized TLCK-inactivated thrombin did not bind radiolabeled human platelets. Native α-thrombin inhibits PPACK-thrombin binding to platelets and is a more potent inhibitor than FPRCH$_2$-thrombin, suggesting a different mode of binding to blood platelets.[72] Another related study used hydrogen-deuterium exhange[73] to assess changes in solvent exposure in PPACK-thrombin. The researchers observed increased exposure of residues 126–132 and 214–222, exosite-1 had decreased solvent exposure, as did Trp141. It is not prudent to extensively speculate without more information on the structure of TLCK-inactivated thrombin. It is recognized that there is a difference in the binding of an arginine side chain and a lysine side chain in the S$_1$ binding site, and there is something peculiar with the interaction of tosyl derivatives with thrombin, as shown by the substrate activation of the hydrolysis of TosArgOMe. These peculiar observations, at least to me, are those which indicated that exosite-2 function (GpIb) is compromised with TLCK-thrombin. There is, however, no data to support this suggestion.

There are several other observations on the modification of histidine residues in thrombin that are worth discussion. Takeuchi and coworkers[74] reported that thermolysin and human α-thrombin were irreversibly inactivated by a cobalt (III) Schiff base complex. Inactivation of thermolysin was associated with a loss of Zn^{2+}. A axial ligand exchange mechanism was proposed for thermolysin. Such a mechanism has been described in greater detail for the inactivation of transcription factor by cobalt (III) Schiff base complex.[75] The researhers did observe that thrombin was protected from inactivation by leupeptin, which is said to be a competitive inhibitor of thrombin. Unlike thermolysin, thrombin does not have a metal for axial ligand exchange. The authors did not speculate on a mechanism for thrombin inhibition.

N-acetylimidazole was introduced by Riordan and coworkers[76] as a reagent for the modification of "free" or available tyrosine residues in proteins. It was also shown to modify lysine residues and cysteine residues. The relative stabilities of S-acetylcysteine, N^ε-acetyllysine, and O-acetyltyrosine make it possible to distinguish between the several modifications.[76,77] N-acetylimidazole was used most frequently as a reagent to modify tyrosine residues in proteins. Houston and Walsh[78] set off to modify "exposed" tyrosine residues in trypsin. I was in graduate school slightly before Louie Houston and recall some of the early frustration. While there was modification of tyrosine, they discovered a transient modification of serine and perhaps histidine at the enzyme-active site of trypsin. Catalytic activity was mea-sured with BzArgOEt and other ester substrates using a pH Stat. The time course was characterized by an early loss of activity followed by recovery over a period of time. Competitive inhibitors, rather than protecting trypsin, greatly accelerated the rate and extent of inactivation. Benzamidine was the most effective, with a lesser effect with tosylarginine or proflavin. The rate of inactivation was similar at 0°C and 23°C, but the extent of inactivation was greater at 0°C. If the assay was performed with $N\alpha$-Cbz-Lys-pNA, an active-site titrant, complete inactivation was observed. It is suggested that the low pH of the assay would stabilize the O-acetylated serine (or N-acetylated histidine). The stimulatory effect of benzamide is similar to the effect of alkylguandiines on the inactivation of trypsin by iodoacetamide.[53] It is also noted that indole accelerated the carbamylation of the active-site serine in α-chymtrypsin.[79] Since N-acetylimidazole does not have any structural attributes that would contrib-ute to selective active-site binding, it is reasonable that benzamidine is enhancing the nucleophilicity of the active-site serine (or histidine). The total loss of activity when measured at a low pH would be consistent with stabilization of the active-site modification.

Lundblad and others observed that N-acetylation inhibited the clotting activity of thrombin with a minimal effect on esterase activity.[80] Unlike trypsin, inhibition was reduced by the presence of benzamidine and was reversed by hydroxylamine. The results were interpreted as a modification of a tyrosine residue or residues that were important in the action of thrombin on fibrinogen. We were unable to obtain results that would suggest a modification occurred in thrombin similar to that described by Houston and Walsh in trypsin. The inhibition of the proteolytic activity of thrombin by N-acetylimidazole is not unique to fibrinogen. Brown and coworkers[81] showed that modification of thrombin with N-acetylimidazole also inhibited activity in the hydrolysis of protamine sulfate.[81] In this study, the amino terminal group of protamine was blocked by reaction with 2,4-dinitrobenzene and the extent of hydrolysis measured by reaction with fluoresecamine. Later work by Lundblad[82] on the reaction of N-butyrylimidazole with thrombin yielded results analogous to those obtained by Houston and Walsh with trypsin. While benzamidine blocked the loss of activity, the results were consistent with the modification of a serine or histidine residue at the enzyme-active site. There was partial recovery of fibrinogen clotting activity with dilute imidazole, which would provide removal of the butyryl group from the active-site histidine; complete recovery of activity was obtained with hydroxylamine. It was suggested that the treatment with dilute imidazole promoted the release of the butyryl group from the active-site residue (serine or histidine) and yielded an enzyme still modified at tyrosine residues. An examination of the time course of hydrolysis of TosArgOme by the butyrylated enzyme showed a sigmodal time course. The short "lag phase" observed is interpreted as the recovery of active-site function, which would sug-gest that TosArgOMe promotes the removal of the butyryl function; a similar interpretation can be made for the esterase activity of the N-acetylimidazole–treated trypsin.[78] Active-site reactivity (p-nitrophenyl-p'-guanidinobenzoate) was reduced in the enzyme modified with N-butyrylimidazole, consistent with modi-fication by this reagent at the enzyme active site.

PRIMARY AMINO GROUPS (LYSINE, AMINO TERMINAL)

The amino terminal of the B-chain of thrombin, as with other serine proteases, is bound up in an internal salt bridge and is unavailable for chemical modification. Staffan Magnusson and Theo Hofmann were able to modify the amino-terminal iso-leucine of the B-chain of thrombin with nitrous acid at pH 4.35 and 0°C.[83] Fibrinogen clotting activity was lost more rapidly than esterase (BzArgOEt) activity, suggesting that lysine residues and tryptophan residues were also modified under those conditions. The presence of BzArgOEt provided protection of esterase activity but only partial protection of fibrinogen clotting activity. Previous work from the Hofmann laboratory at the University of Toronto[84] had shown that the treatment of trypsin with nitrous acid under similar conditions (pH 4.0, 0°C) resulted in the loss of esterase activity with loss of the amino terminal isoleucine; there was also modification of the lysine residues and tryptophan, which was not associated with the loss of ester-ase activity. BzArgOET protected trypsin both from inactivation and loss of amino-terminal isoleucine by nitrous acid. In the case of thrombin, the loss of the amino terminal isoleucine also paralleled the loss of esterase activity. The loss of fibrinogen clotting activity was likely related to the modification of other amino acid residues such as lysine and tryptophan.

LYSINE

Early work on the chemical modification of proteins was directed at the study of pro-teins such as albumin, gelatin, and ovalbumin, which were available in large quanti-ties in a relatively high state of purification.[85] There were some selective chemical modifications, such as the use of formaldehyde to produce toxoids.[86] Acetic anhy-dride was one of the early reagents used in protein chemistry.[85,87] Landaburu and Seegers were the first to use chemical modification to study thrombin in 1959.[88] Acetic anhydride is used to modify lysine residue proteins in the presence of satu-rated sodium acetate.[89,90] The initial pH is 9 and decreases to 6 during the course of the reaction unless controlled by a pH state. Landaburu and Seegers[88] observed a complete loss of fibrinogen clotting activity with a decrease in pH to 5.6, while esterase (TosArgOMe) was observed to increase. If the reaction with acetic anhy-dride was performed in 15% potassium carbonate, there was a 90% loss of esterase activity (TosArgOMe). There was a greater loss of amino groups and modification of tyrosine in the presence of potassium carbonate; O-acetyltyrosine was formed in the presence of sodium acetate but was unstable in that solvent. The acetylated thrombin is referred to as thrombin-E (esterase thrombin) in analogy to the degraded form of thrombin, which has greatly reduced fibrinogen clotting activity but essentially unchanged esterase activity. The observations on acetylated bovine thrombin were extended by Seegers and coworkers[91] to included physical characterization of the acetylated protein. This work also showed that bovine α-thrombin was composed of two polypeptide chains. Later work by Tangen and Bygdeman[92] demonstrated that, contrary to earlier observations,[93] acetylated thrombin did not aggregate platelets. Tangen and Bygdeman[92] also showed that Reptilase, the coagulant protein isolated from *Bothrops atrox* venom, did not aggregate platelets. This is likely a reflection of the absence of an exosite-2–like domain in the snake venom enzyme. Nakagami and coworkers[94] showed that acetylated bovine thrombin retained activity in the hydrolysis of a tripeptide nitroanilide substrate but had greatly reduced activity in the clotting of fibrinogen or the activation of protein C in the presence or absence of thrombomodulin. The loss of activity in protein C activation was somewhat less than that in the presence of thrombomodulin but still greater than 95% in either situation. Inhibition of acetylated thrombin with either antithrombin or hirudin was reduced with a greater effect on the interaction with hirudin. Later work from this laboratory[95] confirmed some of these early observations and extended the work to include succinylthrombin and maleylthrombin. Maleylthrombin was shown have a smaller loss of activity in fibrinogen clotting and protein C activation; there was about 25% inactivation of amidase activity with maleylthrombin

Pyridoxal phosphate [PLP, pyridoxal-5′-phosphate; (4-formyl-5-hydroxy-6-methylpyridin-yl)methyl dihydrogen phosphate] is the active form of vitamin B_6. Pyridoxal has been shown to be a relatively specific reagent for the modification of lysine residues in a basic region of a protein.[96] This specificity for a basic region is based on the presence of a methylene phosphate group *ortho* to the reactive carbonyl function, which forms a Schiff base with an unprotonated ε-amino group of lysine which, while relatively stable, can be converted a stable carbon-nitrogen bond by reduction with sodium borohydride. Mike Griffith was clever enough to use PLP to modify thrombin in the hopes of identifying the heparin binding site.[11] Griffith observed that PLP reacted with thrombin with the incorporation of approximately 2 moles of PLP per mole of thrombin with loss of fibrinogen clotting activity and loss of heparin acceleration of reaction with antithrombin. There was a minimal effect on amidase activity with tripeptide nitroanilide substrates (TosGlyProArgpNA; BzPheValArgpNA) but a greater effect of the hydrolysis of BzArgpNA with a large decrease in K_M (50 μM vs. 350 μM) and a somewhat smaller decrease in V_{max} (0.007 μmol min^{-1} vs. 0.019 μmol min^{-1}). A study of reaction kinetics indicated modification occurred at two sites on thrombin, and modification at the second site was blocked by heparin. Modification at the second site was also inhibited by heparan sulfate and dermatan sulfate but not by other sulfated mucopolysaccharides such as chondroitin sulfate. White and coworkers[97] subsequently evaluated the effect of phosphopyridoxylation of human α-thrombin on binding and activation of human blood platelets. Phosphopyridoxylation of thrombin decreased the ability to activate platelets as judged by both aggregation and release of serotonin; there was also loss of both high-affinity and low-affinity binding. If the reaction with PLP was performed in the presence of heparin, the ability to activate platelets was partially retained and high-affinity binding restored; there was still loss of low-affinity binding. This is consistent with heparin protecting exosite-2, which is critical for binding to GpIb. Heparin does block the interaction of thrombin with GpIb.[98] Bauer and coworkers[99] reported on the effect of the phosphopyridoxylation of thrombin on binding to mini-pig aortic endothelial cells. The binding of thrombin to endothelial cells is complex, with as many as three binding sites proposed.[100] In results similar to platelets, Bauer and coworkers[99] observed that the phosphopyridoxylation of thrombin in the absence of heparin eliminated high-affinity and reduced low-affinity binding to the endothelial cells. High-affinity binding was protected when reaction with PLP was performed in the presence of heparin. This is consistent with the observation that membrane-bound heparan sulfate is important for the binding of heparin to endothelial cells[101] and the ability of heparin to displace PPACK-thrombin from endothelial cells.[102] Kaminski and McDonagh[103] showed the phosphopyridoxylated human α-thrombin did not bind to fibrin monomer, while modification of thrombin with active site–directed reagents such as PMSF and PPACK did not influence binding to the fibrin-monomer. These experiments used fibrin bound to agarose prepared by the thrombin hydrolysis of fibrinogen that had been covalent coupled to agarose. These investigators also showed that native thrombin bound to fibrin monomer had the same activity as unbound enzyme in the hydrolysis of a tripeptide nitroanilide substrate, S-2238. Church and coworkers were able to identify the lysine residues in thrombin that were modified by PLP.[12] Lysine residues 21, 65, 174, and 252 in thrombin were modified by PLP in the absence of heparin, while lysine-21 and lysine-65 were modified in the presence of heparin. We now know that residue lysine-65 is in exosite-1, which does not bind heparin, while lysine residues 174 and 252 are in exosite-2, which is known to bind heparin.[104] Yamaguchi and coworkers[105] modified human thrombin with PLP. In results similar to those obtained by Griffith,[11] incorporation of approximately 2 moles of PLP per mole of thrombin resulted in the loss of the ability to bind heparin but there was retention of amidase activity. They did observe a decreased rate of reaction of the modified thrombin with heparin cofactor II in the presence of heparin or dermatan sulfate. Pyridoxylation of human α-thrombin also reduced the ability of thrombin to bind thrombomodulin, as well as fibrinogen, suggesting that thrombo-modulin binds to exosite-1.[106] Modification of thrombin with PLP did not influence binding to p-chlorobenzylamido-agarose.[107]

The studies cited here showed that heparin interfered with the "high-affinity" binding of thrombin to blood platelets, and other studies suggested the importance of glycocalicin/GP1b in the binding of thrombin to platelets. Two studies several years apart showed that the phosphopyridoxylation of thrombin inhibited the binding of thrombin to GpIb[108,109] Tenisien and coworkers[108] showed that modification of throm-bin with PLP inhibited binding to GpIb but binding to another high-molecular-weight protein and platelet protease nexin was retained. Binding was measured by use of a covalent crosslinking reagent [bis(sulfosuccinimidyl)suberate]. When the modifica-tion was performed in the presence of heparin, binding to Gp1b was retained. Later work from De Candia and coworkers[109] showed that phosphopyridoxylation of throm-bin reduced binding to purified platelet glycocalicin. These investigators also showed that heparin interfered with the binding of thrombin to isolated glycocalicin.

There are several other studies on the modification of lysine residues in throm-bin which should be briefly discussed. Kang used ethyl acetimidate to modify lysine residues in human thrombin; modification of 78% of the lysine residue was achieved.[110] There was a 90% decrease in k_{cat} and a small change in K_M for a tri-peptide nitroanilide substrate (BzPheValArgpNA); there was also a 90% decrease in fibrinogen clotting activity. Amidation is a modification of amino groups which maintains positive charge rather than neutralization or a charge change. Another example of the effect of amidation on the activity of thrombin is provided by the work of Silverberg.[111] Silverberg used methyl acetimidate to modify prothrombin; approximately 80% modification of lysine was achieved in this study. There was little, if any, conformational change in prothrombin as result of the modification. Thrombin obtained from the activation of the acetimidated prothrombin had little activity toward fibrinogen, factor V, or the peptide bond prothrombin between fragment 1 and prethrombin 1; differing from the Kang study, activity was maintained in the hydrolysis of D-Phe-Pip-Arg-pNA. While it is difficult to work with two somewhat different studies, the results would suggest that amidation does modify a residue near the active site in thrombin, which is not available in prothrombin or more likely that in the absence of conformational analysis in the study on thrombin, there is considerable conformational change resulting from the extensive modification of lysine in the thrombin study. Modification of four to five lysine residues in bovine thrombin (likely a mixture of α- and β-thrombin) with 2,4,6-trinitrobenzene sulfonic acid resulted in a 60%–70% loss of fibrinogen clotting activity but only a slight reduction in the rate of the reaction with antithrombin.[112] There was, however, a greater than 50% decrease in the effect of heparin on the rate of reaction of thrombin with antithrombin. Fluoroscein-5′-isothiocyanate was used to label human α-thrombin for binding studies with murine bone marrow–derived mast cells.[113] Fluoroscein-5-isothiocynate reacts with lysine residue in proteins.[114,115] Fluorescein 5′-isothiocyanate was also used to identify the DNA aptamer binding site in human α-thrombin.[116] Lys-21 and Lys-65 in exosite-1 were identified as being protected by the DNA aptamer.

Reaction of partially purified citrate-activated equine thrombin with β-naphthoquinone 4,6-disulfonic acid resulted in the loss of both esterase (TosArgOMe) and clotting activity; esterase activity was somewhat more resistant to inactivation than clotting activity.[117]

TYROSINE

Tyrosine has been an attractive target for chemical modification in proteins,[118] with several reagents introduced in the 1960s for such work. Bert Vallee and his colleagues at Harvard introduced tetranitromethane (TNM) for the modification of tyrosine residues in proteins.[119,120] There was earlier work,[107] but the reagent had been ignored until the work of the Harvard group. Tage Astrup, working at the Carlsberg Institute, reported the first study on the reaction of TNM, with thrombin showing the loss of fibrinogen clotting activity.[121] Some 20+ years later, Sokolovsky and Riordan[122] showed that reaction of thrombin with TNM greatly decreased fibrinogen clotting activity with little effect on esterase activity. Several years later Lundblad and Harrison used TNM to modify tyrosine residues in bovine thrombin (a mixture of α- and β-thrombin, approximately 70% α-thrombin) in 30 mM sodium phosphate, pH 8.0.[123] Consistent with the earlier findings of Sokolovsky and Riordan,[122] there was loss of fibrinogen clotting activity without loss of esterase activity (TosArgOMe). The esterase activity was performed at a single concentration of substrate, which was above K_M. The presence of 4 mM benzamidine did not prevent the loss of fibrinogen clotting activity. Amino acid analysis showed the nitration of approximately 5 moles of tyrosine per mole of thrombin. Gel filtration did not indicate the presence of aggregates. Later work from this laboratory[124] showed that the rate of inactivation of thrombin by TNM was two to three times more rapid in a Tris buffer (50 mM

Tris, pH 8.0) than in 50 mM sodium phosphate-100 mM NaCl, pH 8.0. A plot of the first-order rate constant against TNM concentration in either solvent suggested that the inactivation was due to the modification of a single tyrosine residue. Structural analysis showed nitration of Tyr71 and Tyr85 with minor modification at Tyr154. Modification of the human proteins showed modification at Tyr71 and Tyr86 with lesser modification at Tyr134 and Tyr148. The structural analysis was performed in 50 mM sodium phosphate, pH 8.0; similar results were obtained with a protein modified in a Tris buffer. There was a 30% decrease in k_{cat} for the hydrolysis of either TosGlyProArgpNA or H-D-PhePipArgpNA but a 70% decrease in k_{cat} for the release of fibrinopeptide A. Try71 is a component of exosite-1, which might explain the large effect on the hydrolysis of the peptide bond in fibrinogen. This work was performed with a more highly purified preparation of bovine α-thrombin.[125]

Different ultraviolet (UV) spectroscopy of nitrated thrombin in the presence of sodium did not show the same differences as native α-thrombin in the presence of sodium ions.[126] The results showed that the modification of a single tyrosine residue was sufficient to abolish the difference spectra observed in the presence of sodium ions. Sonder and Fenton[127] showed that the modification of four tyrosine residues with tetranitromethane resulted in a 10-fold decrease in the binding of proflavine. Fenton and coworkers[128] also reported that nitrated human thrombin did not bind to the cation-exchange resin, Amberlite CG-50, and showed reduced affinity for a fibrin-agarose matrix. The extent of modification (5 moles of nitrotyrosine per mole of thrombin) is similar to that reported in work with the bovine enzyme. These investigators did note aggregation of nitrated thrombin, as well as a decreased active site (ca 60%), as determined by titration with NPGB. Workman and coworkers[129] reported that nitration of bovine α-thrombin eliminated high-affinity binding, and the ability to cause aggregation and release was at least 10-fold less that of unmodified thrombin. Nitrated thrombin also showed reduced binding to cultured fibroblasts, as well as reduced ability to stimulate mitogenesis.[130] The results were complicated by reduced active-site reactivity (32% vs. 85% in unmodified) as assessed by titration with *p*-nitophenyl-*p*'-guanidinobenzoate.

N-acetylimidazole is another reagent that was used for the modification of tyrosyl residues in proteins.[131] The original work proposed that reactivity of tyrosyl residues could differentiate between "free" and "buried" tyrosyl residues, but other work[132] suggested that the interpretation of results can be complicated. Kotoku and coworkers[133] were the first to report the use of *N*-acetylimidazole to modify tyrosine residues in thrombin. They reported the modification of 4 moles of tyrosine per mole of equine thrombin. Other work suggests that there were 4 tyrosine residues, which had a pKa of 10.2; other tyrosine residues had higher pKa values (two with pKa 11.1, and three were not available for titration in the native protein). Since the reaction with *N*-acetylimidazole involves tyrosine as nucleophile, it is reasonable to assume that the four tyrosine residues with the lower pKa values are the residues modified by *N*-acetylimidazole. These investigators reported little loss of fibrinogen clotting activity but a substantial loss of esterase (TosArgOMe). The loss of activity is reversed during the time course of the reaction. While *O*-acetyltyrosine is unstable, it is difficult to rationalize this data with other studies. Some years later Lundblad and coworkers[80] showed that *N*-acetylimidazole reacted with bovine thrombin (70%

α, 30% β) resulting in approximately 90% loss of fibrinogen clotting activity at pH 7.5 (50 mM sodium phosphate). Reaction at pH 8.0 resulted in a more rapid initial rate of inactivation but a lesser extent (70%) of inactivation, reflecting the instability of the reagent. The loss of fibrinogen clotting activity was fully reversed by neutral hydroxylamine. There was no effect on esterase (TosArgOMe) activity. The enzyme was protected from inactivation by benzamidine by TosArgOMe, but not by TosArg. If the purified thrombin was allowed to stand in 0.05 M NaCl for 24 hours, 30% of the fibrinogen clotting activity was lost. Reaction of this "aged" thrombin with N-acetylimidazole resulted in 50% inhibition of fibrinogen clotting activity with the modification of six to seven tyrosine residues compared to four to five residues in the freshly thawed enzyme preparation. The relative amounts of α-thrombin and β-thrombin did not change during the process of "aging." Unlike the human protein, we were never able to demonstrate the autolysis of bovine α-thrombin; we only observed the formation of bovine β-thrombin during citrate activation. Later work from this laboratory showed that reaction of bovine α-thrombin with N-butyrylimidazole resulted in the modification of one mole of tyrosine per mole of thrombin.[82] There was loss of both fibrinogen clotting activity (greater than 90%) and esterase (TosArgOme) activity. Interpretation of the data is complicated by a concomitant modification at the enzyme-active site. Treatment of the modified enzyme with dilute imidazole restored the fibrinogen clotting activity to approximately 30% of the unmodified enzyme, a value similar to that observed with N-acetylimidazole.[80]

TRYPTOPHAN

Some of the chemical modification studies were performed before the availability of structural information. For example, the concept of the tryptophan loop: residues Try47 (bovine V), Pro48 (bovine Pro), Pro49 (bovine Pro), and Trp50 (bovine Trp,) have been suggested as having a role in thrombin specificity.[66] Other work showed that Trp50 (bovine Trp) and Trp227 (bovine Trp) were important in binding a fibrinogen peptide.[134] There are a number of studies using site-specific mutagenesis to support a role for Trp50, Trp148, and Trp227, which are discussed in Chapter 4.

There are few studies on the chemical modification of tryptophan in thrombin. Kotoku and coworkers[133] used hydrogen peroxide in dioxane to modify tryptophan residues in equine thrombin. One to two moles of tryptophan were oxidized per mole of thrombin with a 40% loss of esterase (TosArgOMe) activity and an 80% loss of fibrinogen clotting activity. Uhteg and Lundblad[135] used both N-bromosuccimide (NBS) and 2-hydroxy-5-nitrobenzyl bromide (HNB) to modify tryptophan residues in bovine thrombin. Reaction of NBS (2-fold molar excess) with bovine α-thrombin (23°C, 0.1 M sodium acetate, pH 4.0) resulted in the oxidation of 1 mole of tryptophan per mole of thrombin; there was 90% loss of fibrinogen clotting activity and a 20% loss of esterase (TosArgOMe) activity. Use of a 50-fold molar excess of NBS resulted in oxidation of 5 moles of tryptophan per mole of thrombin and the total loss of fibrinogen clotting activity. Reaction of HNB with bovine α-thrombin resulted in the incorporation of 1 mole of hydroxylbenzyl per mole of thrombin. The reaction of thrombin with HNB resulted in 80% loss of fibrinogen clotting activity and 20% loss of esterase (TosArgOMe) activity. Prior oxidation of thrombin with NBS (0.77 mole Trp oxidized) reduced HNB incorporation from 1.04 to 0.28 moles per mole of thrombin. Bob Horton at North Carolina State University developed a novel reagent, 2-acetoxy-5-nitrobenzyl bromide, for the modification of tryptophan in chymotrypsin. [136,137] 2-Acetoxy-5-nitrobenzyl bromide is not active for the modification of tryptophan but bears a structural resemblance to *p*-nitrophenyl acetate. Hydrolysis of the acetyl function from 2-acetoxy-5-nitrobenzyl bromide generates HNB at the enzyme-active site, where it will either modify tryptophan or undergo hydrolysis, which results in inactivation.[128] Uhteg and Lundblad[135] showed that the reaction of bovine α-thrombin with 2-acetoxy-5-nitrobenzyl bromide incorporated 0.84 moles of hydroxynitrobenzyl per mole of thrombin with a 75% loss of fibrinogen clotting activity and a 25% loss of esterase (TosArgOMe) activity. Prior reaction of thrombin with PMSF or TLCK resulted in a small decrease in the extent of modification with HNB. These results supported the function of a tryptophanyl residue in thrombin function but did not identify the residue. Somewhat later, Stevens and Nesheim[137] reported on the reaction of NBS with bovine α-thrombin and prethrombin-2. These investigators reported more extensive modification of tryptophan in thrombin than that observed by Uhteg and Lundblad.[135] Stevens and Nesheim[137] performed their modification reactions in 0.05 M sodium acetate, pH 4.5, while Uhteg and Lundblad[135] performed their studies in 0.1 M sodium acetate, pH 4.0. Spande and coworkers[138] reported more extensive modification of chymotrypsin with NBS at 0.05 M sodium acetate. Stevens and Nesheim[137] also used a large excess of NBS. These investigators concluded that Trp50, Trp148, and Trp227 are most likely important for thrombin function. As expected, there were differences between α-thrombin and prethrombin-2 in susceptibility to modification of residues near the substrate binding site and enzyme-active site. These results were interpreted as supporting a role for a tryptophan residue in the function of thrombin.

REFERENCES

1. Stark, G.R., Recent developments in chemical modification and sequential degradation of proteins, *Adv. Prot. Chem.* 24, 261–308, 1970.
2. Means, G.E. and Feeney, R.E., Chemical modification of proteins: history and applications, *Bioconjugate Chem.* 1, 2–12, 1998.
3. Lundblad, R.L., *Chemical Modification of Biological Polymers*, CRC Press, Boca Raton, FL, USA, 2012.
4. De Simone, G., Di Fiore, A., Truppo, E., *et al.*, Exploration of the residues modulating the catalytic features of human carbonic anhydrase XIII by site-specific mutagenesis approach, *J. Enzyme Inhib. Med. Chem.* 34, 1506–1510, 2015.
5. Xuan, W., Collins, D., Koh, M., *et al.*, Site-specific incorporation of a thioester containing amino acid into proteins, *ACS Chem. Biol.* 13, 578–581, 2018.
6. Pomés, A., Mueller, G.A., and Chruszcz, M., Structural aspects of the allergen-antibody interaction, *Front. Immunol.* 11, 2067, 2020.
7. Thirumuragan, P., Matasuk, D., and Jozwiak, K., Click chemistry for drug development and diverse chemical-biology applications, *Chem. Rev.* 113, 4905–4979, 2013.
8. Yang, A., Ho, S., Ahm, J., *et al.*, A chemical biology route to site-specific authentic protein modifications, *Science* 354, 623–626, 2016.

9. Chudasama, V., Marvani, A., and Caddick, S., Recent advances in the construction of antibody-drug conjugates, *Nat. Chem.* 8, 114–119, 2016.

10. Poudel, Y.B., Chowdari, N.S., Chang, H., *et al.*, Chemical modification of linkers provides stable linker-payload for the generation of antibody-drug conjugates, *Med. Chem. Lett.* 11, 2190–2194, 2020.

11. Griffith, M.J., Covalent modification of human α-thrombin with pyridoxal-5-phosphate. Effect of phosphopyridoxylation on the interaction with heparin, *J. Biol. Chem.* 254, 3401–3406, 1979.

12. Church, F.C., Pratt, C.W., Noyes, C.M., *et al.*, Structural and functional properties of human α-thrombin, phosphopyridoxylated α-thrombin, and γ$_T$-thrombin. Identification of lysyl residues in α-thrombin that are critical for heparin and fibrinogen interactions, *J. Biol. Chem.* 254, 15419–15425, 1989.

13. Schaffer, N.Y., Ay, S.C., Jr., and Summerson, W.H., Serine phosphoric acid from diisopropylphorpphoryl chymotrypsin, *J. Biol. Chem.* 202, 67–76, 1953.

14. Dixon, G.H. and Neurath, H., The reaction of DFP with trypsin, *Biochim. Biophys. Acta* 20, 572–574, 1956.

15. Glader, J.A. and Laki, K., The inhibition of thrombin by diisopropylphosphorofluoridate, *Arch. Biochem. Biophys.* 62, 501–503, 1956.

16. Gladner, J.A., Laki, K., and Stohlman, I., Labeled DIP thrombin, *Biochim. Biophys. Acta* 27, 218, 1958.

17. Miller, K.D. and Van Vunakis, H., The effect of diisopropyl fluorophosphate on the proteinase and esterase activity of thrombin and on prothrombin and its activation, *J. Biol. Chem.* 223, 227–237, 1956.

18. Gladner, J.A. and Laki, K., The active site of thrombin, *J. Am. Chem. Soc.* 80, 1203–1204, 1958.

19. Lundblad, R.L., A rapid method for the purification of bovine thrombin on the inhibition of the purified enzyme with phenylmethylsulfonyl fluoride, *Biochemistry* 10, 2501–2505, 1971.

20. Fahrney, D.E. and Gold, A.M., Sulfonyl fluorides as inhibitors of esterases. I. Rates of reaction with acetylcholinesterase, α-chymotrypsin, and trypsin, *J. Am. Chem. Soc.* 85, 977–1000, 1963.

21. Gold, A.M. and Fahrney, D., Sulfonyl fluorides as inhibitors of esterases. II. Formation and reactions of phenylmethanesulfonyl α-chymotrypsin, *Biochemistry* 3, 783–791, 1964.

22. Workman, E.F., Jr., White, G.C., II., and Lundblad, R.L., Structure-function relationships In the interaction of thrombin with blood platelets, *J. Biol. Chem.* 252, 7118–7123, 1977.

23. Hayakawa, Y., Tazawa, S., Ischikawa, T., *et al.*, Thrombin regulation of tissue-type plasminogen activator synthesis in cultured human fetal lung fibroblasts, *Thromb. Res.* 71, 457–465m, 1993.

24. Uemitsu, N., Sugiyama, M., and Matsumiya, H., The reaction of phenylmethanesulfonyl-subtilisin, *Biochim. Biophys. Acta* 258, 562–565, 1972.

25. Wong, C.-S. and Shaw, E., Differences in active center reactivity of trypsin homologs. Specific inactivation of thrombin with nitrophenyl-*p*-amidinophenylmethanesulfonate, *Arch. Biochem. Biophys* 161, 536–543, 1974.

26. Wong, S.-C. and Shaw, E., Inactivation of trypsin-like proteases by active-site directed sulfonation. Ability of the departing group to confer selectivity, *Arch. Biochem. Biophys.* 76, 113–118, 1976.

27. Wong, C.-S., Green, G.D.J., and Shaw, E., Inactivation of trypsin-like proteases by sulfonylation. Variation of positively group and inhibitor length, *J. Med. Chem.* 21, 456–459, 1978.

28. Laura, R., Robison, D.J., and Bing, D.H., (*p*-Amidinophenyl)methanesulfonyl fluoride, an irreversible inhibitor of serine proteases, *Biochemistry* 19, 4859–4864, 1980.
29. Brown, A.D. and Powers, J.C., Rates of thrombin acylation and diacylation upon reaction with low molecular weight acylating agents, carbamylatinga agents and carbonylating agents, *Bioorg. Med. Chem.* 3, 1091–1097, 1995.
30. Moorman, A.R. and Abeles, R.H., A new class of serine protease inactivators based on isatoic anhydride, *J. Am. Chem. Soc.* 104, 6785–6786, 1982.
31. Gelb, M.H. and Abeles, R.H., Substituted isatoic anhydrides: selective inactivators of trypsin-like serine proteaes, *J. Med. Chem.* 29, 585–589, 1986.
32. Ashton, R.W. and Scheraga, H.A., Preparation and characterization of anhydrothrombin, *Biochemistry* 34, 6454–6463, 1995.
33. Strumeyer, D.H., White, W.N., and Koshland, D.E., Jr., Role of serine in chymotrypsin action. Conversion of the active serine to dehydroalanine, *Proc. Natl. Acad. Sci. USA* 50, 931–935, 1963.
34. Weiner, H., White, W.N., Hoare, D.G., and Koshland, D.E., Jr., The formation of anhy-drochymotrypsin by removing the elements of water from the serine at the active site, *J. Am. Chem. Soc.* 88, 3851–3859, 1966.
35. Koshland, D.E., Jr., Correlation of structure and function in enzyme activity, *Science* 142, 1533–1541, 1964.
36. Ako, H., Ryan, C.A., and Foster, R.J., The purification by affinity chromatography of proteinase inhibitor binding specific of anhydro chymotrypsin, *Biochem. Biophys. Res. Commun.* 46, 1639–1645, 1972.
37. Ako, H., Foster, R.J., and Ryan, C.A., The preparation of anhydro-trypsin and its reactivity with naturally occurring proteinase inhibitors, *Biochem. Biophys. Res. Commun.* 47, 1402–1407, 1972.
38. Hixson, H.F., Jr. and Nishikawa, A.M., Affinity chromatograpy of bovine trypsin and thrombin, *Arch. Biochem. Biophys.* 154, 501–510, 1973.
39. Nogami, K., Shima, M., Hosokawa, K., *et al.*, Factor VIII C2 domain contains the throm-bin binding site responsible for thrombin-catalyzed cleavage of Arg[1689], *J. Biol. Chem.* 275, 25774–15780, 2000.
40. Hosokawa, K., Ohnishi, T., Shima, M., Nogata, M., and Koide, T., Preparation of Anhydrothrombin and characterization of its interaction with normal thrombin sub-strates, *Biochem. J.* 354, 309–313, 2001.
41. Dementiev, A., Petitou, M., Herbert, J.-M., and Gettins, G.W., The ternary complex of antithrombin-anhydrothrombin-heparin reveals the basis of inhibitor specificity, *Nat. Struc. Mol. Biol.* 11, 863–867, 2004.
42. Scheider, F., Histidine in enzyme active centers, *Angew. Chemie* 17, 583–592, 1978.
43. Rebek, J., Jr., On the structure of histidine and its role in enzyme active sites, *Struct. Chem.* 1, 129–131, 1990.
44. Lucas, O.N., Inactivation of thrombin by oxidized cellulose, *J. Oral. Therapeut. Pharmacol.* 3, 262–268, 1967.
45. De Cristofero, R. and Landolfi, R., Oxidation of human α-thrombin by the myeloperox-idase-H_2O_2-chloride system: structural and functional effects, *Thromb. Haemost.* 83, 253–261, 2000.
46. Davies, M.J., Myeloperoxidase: mechanism, reactions, and inhibition as a therapeutic strategy in inflammatory diseases, *Pharmacol. Ther.* 218, 107685, 2021, doi: 10.1016/j.pharmther.2020.107685.
47. Miles, E.W., Modification of histidyl residues in proteins by diethylpyrocarbonate, *Methods Enzymol.* 47, 431–449, 1977.

48. Osterman-Golkar, S., Ehrenberg, I., and Solynosy, F., Reaction of diethylpyrocarbonate with nucleophiles, *Acta Chem. Scand B* 28, 215–228, 1974.

49. Colleluori, D.M., Reczkowski, R.S., Emig, F.A., *et al.*, Probing the active site of hyper-reactive histidine residue of arginase, *Arch. Biochem. Biophys.* 444, 15–26, 2005.

50. Lundblad, R.L., Neshem, M.E., Straight, D.L., *et al.*, Bovine α- and β-thrombin. Reduced fibrinogen-clotting activity of β-thrombin is not a consequence of reduced affinity for fibrinogen, *J. Biol. Chem.* 259, 6991–6995, 1984.

51. Lundblad, R.L., Noyes, C.M., Mann, K.G., and Kingdon, H.S., The covalent differences between bovine α and β-thrombin. A structural explanation for the changes in catalytic activity, *J. Biol. Chem.* 254, 8524–8528, 1979.

52. Chandvekar, L.P. and Nadkorni, G.P., Radiation-induced change in purified prothrombin and thrombin, *Biochim. Biophys. Acta* 706, 1–8, 1982.

53. Inagami, T., The alkylation of the active site o trypsin with iodoacetamide in the presence of alkylguanidines, *J. Biol. Chem.* 240, PC3453–PC3455, 1965.

54. Lin, M.C., Stein, W.H., and Moore, S., Further studies on the alkylation of the histidine residues at the active site of pancreatic ribonuclease, *J. Biol. Chem.* 243, 6167–6701, 1968.

55. Schoellmann, G. and Shaw, E., Direct evidence for the presence of histidine at the active center of chymotrypsin, *Biochemistry* 2, 252–235, 1963.

56. Shaw, E., Mares-Guia, M., and Cohen, W., Evidence for an active-center histidine in trypsin through use of a specific reagent, 1-chloro-3-tosylamido-7-amino-2-heptanone, the chloromethyl ketone derived from Nα-tosyl-L-lysine, *Biochemistry* 4, 2219–2224, 1965.

57. Shaw, E. and Glover, G., Further observations o substrate-derived chloromethyl ketones that inactivate trypsin, *Arch. Biochem. Biophys.* 139, 298–305, 1970.

58. Glover, G. and Shaw, E., The purification of thrombin and isolation of a peptide containing the active center histidine, *J. Biol. Chem.* 246, 4594–4601, 1971.

59. Plapp, B.V., Application of affinity labeling for the studying structure and function in enzymes, *Methods Enzymol.* 87, 469–499, 1982.

60. Exner, T. and Koppel, J.L., Cholate enhancement of interaction between thrombin and tosyl lysine chloromethyl ketone, *Biochim. Biophys. Acta* 329, 233–240, 1973.

61. Gilles, A.-M. and Keil, B., Evidence for an active-center cysteine in the SH proteinase α-chlostripain through the use of N-tosyl-lysine chloromethyl ketone, *FEBS Lett.* 173, 58–62, 1984.

62. Kettner, C. and Shaw, E., D-Phe-Pro-ArgCH$_2$Cl-A selective affinity label for thrombin, *Thrombosis Res.* 14, 969–973, 1979.

63. Bode, W., Mayr, J., Baumann, U., *et al.*, The refined 1.9Å crystal structure of human α-thrombin: interaction with D-Phe-Pro-Arg-chloromethylketone and significance of the Tyr-Pro-Pro-Trp insertion segment, *EMBO J.* 8, 3467–3475, 1989.

64. Bode, W., Turk, D., and Karshikov, A., The refined 1.9-Å x-ray crystal structure of D-Phe-Pro-Arg chloromethylketone-inhibited human α-thrombin: structure analysis, overall structure, electrostatic properties, detailed active site geometry, and structure-function relationship, *Protein Sci.* 1, 426–471, 1992.

65. Vijayalakshmi, J., Padwanabhan, K.P., Mann, K.G., and Tulinsky, A., The isomorphous structures of prethrombin-2, hirugen- and PPACK-thrombin. Changes occurring accompanying activation and exosite binding to thrombin, *Protein Sci.* 3, 2254–2271, 1971.

66. Banner, D. and Hadváry, P., Crystallographic analysis at 3.0-Å resolution of the binding to human thrombin of four active site-directed inhibitors, *J. Biol. Chem.* 266, 20085–20093, 1991.

67. Bock, P.E., Active-site-selective labeling of blood coagulation proteinases with fluorescence probes by the use of thioester peptide chloromehyl ketones. I Specificity of thrombin labeling, *J. Biol. Chem.* 267, 14963–14973, 1992.

68. Bock, P.E., Active-site-selective labeling of blood coagulation proteinases with fluorescence probes by the use of thioester peptide chloromehyl ketones. II Properties of thrombin derivatives as reporters of prothrombin fragment 2 binding and specificity of the labeling approach for other proteinases, *J. Biol. Chem.* 267, 14974–14981, 1982.

69. Harmon, J.T. and Jamieson, G.A., Activation of platelets by α-thrombin is a receptor-mediated vente. D-phenylalanyl-L-prolyl-L-arginine chloromethyl ketone thrombin but not Nα-tosyl-L-lysine chloromethyl ketone thrombin binds to the high affinity thrombin receptor, *J. Biol. Chem.* 261, 15928–15933, 1986.

70. Greco, N.J., Tenner, T.E., Jr., Tandon, N.H., and Jamieson, G.A., PPACK-thrombin inhibits thrombin-induced platelet aggregation and cytoplasmic acidification but does not inhibit platelet shape change, *Blood* 75, 1983–1990, 1990.

71. Wu, H.-f., White, G.C., II., Workman, E.F., Jrl., Jenzano, J.W., and Lundblad, R.L., Affinity chromatography of platelets on immobilized thrombin: retention of catalytic activity by platelet-bound thrombin, *Thromb. Res.* 67, 419–427, 1992.

72. Schmaier, A.H., Meloni, F.J., Nawarawong, W., and Jiang, Y.P., PPACK-thrombin Is a noncompetitive inhibitor of thrombin binding to human platelets, *Thromb. Res.* 67, 479–489, 1992.

73. Croy, C.H., Koeppe, J.R., Bergquist, S., and Komives, E.A., Allosteric changes in solvent accessibility observed upon active site occupation, *Biochemistry* 43, 5246–5356, 2004.

74. Takeuchi, T., Böttcher, A., Quezada, C.M., Meade, T.J., and Gray, H.B., Inhibition of thermolysin and thrombin by Cobalt (III) Schiff base complexes, *Bioorg. Med. Chem.* 7, 815–819, 1999.

75. Matosziuk, L.M., Holbrook, R.J., Marius, L.M., *et al.*, Rational design of [Co(acacen)L$_2$] inhibitor of protein function, *Dalton Trans.* 42, 4002–4012, 2013.

76. Riordan, J.F., Wacker, W.E.C., and Vallee, B.L., N-Acetylimidazole: a reagent for determination of 'free" tyrosine residues in proteins, *Biochemistry* 4, 1758–1765, 1965.

77. Kong, K.-H., NIshida, N., Inoue, H., and Takahashi, K., Tyrosine-7 is an essential residue for the catalytic activity of human class F1-glutathione transferase: chemical modification and site-directed mutagenesis, *Biochem. Biophys. Res. Commun.* 182, 1127–1135, 1972.

78. Houston, L.L. and Walsh, K.A., The transient inactivation of trypsin by mild acetylation with N-acetylimidazole, *Biochemstry* 9, 156–166, 1970.

79. Shaw, D., Stein, W.C., and Moore, S., Inactivation of chymotrypsin by cyanate, *J. Biol. Chem.* 239, PC671–PC673, 1964.

80. Lundblad, R.L., Harrison, J.H., and Mann, K.G., On the reaction of purified bovine thrombin with N-acetylimidazole, *Biochemistry* 12, 409–413, 1973.

81. Brown, F., Freedman, M.L., and Troll, W., Sensitive fluorescent determination of trypsin-like proteases, *Biochem. Biophys. Res. Commun.* 53, 75–81, 1973.

82. Lundblad, R.L., The reaction of bovine thrombin with N-butyrylimidazole. Two different reactions resulting in the inhibition of catalytic activity, *Biochemistry* 14, 1033–1037, 1975.

83. Magnusson, S. and Hofmann, T., Inactivation of bovine thrombin by nitrous acid, *Can. J. Biochem.* 48, 432–437, 1970.

84. Scrimger, S.T. and Hofmann, T., The involvement of the amino terminal amino acid in the activity of the pancreatic proteases. II. The effect of nitrous acid on trypsin, *J. Biol. Chem.* 247, 2528–2533, 1967.

85. Kenchington, A.W., Chemical modification of the side chains of gelatin, *Biochem. J.* 68, 458–460, 1958.

86. Eaton, M.D., Chemical modification of diphtheria toxin. I. The mechanism of detoxification by formaldehyde, *J. Immunol.* 22, 419–436, 1937.
87. Olcott, H.S. and Fraenkel-Conrat, H., Specific group reagents for proteins, *Chem. Rev.* 41, 151–197, 1947.
88. Landaburu, R.H. and Seegers, W.H., The acetylation of thrombin, *Can. J. Biochem.* 37, 1361–1366, 1959.
89. Riordan, J.F. and Vallee, B.L., Acetylation, *Methods Enzymol.* 11, 565–570, 1967.
90. Riordan, J.F. and Vallee, B.L., Acetylation, *Methods Enzymol.* 25, 494–499, 1972.
91. Seegers, W.H., McCoy, L., Kipfer, R.K., and Murano, G., Preparation and properties of thrombin, *Arch. Biochem. Biophys.* 128, 194–201, 1968.
92. Tangen, O. and Bygdeman, C., Study of the clotting, esterase and platelet aggregating activities of thrombin, acetylated thrombin and Reptilase®, *Scand. J. Haematol.* 9, 333–338, 1972.
93. Seegers, W.H., Diekamp, U., and McCoy, L.E., Induction of platelet aggregation with acetylated thrombin, in *Platelet Adhesion and Aggregation in Thrombosis: Countermeasures*, ed. E.F. Mammen, G.F. Andersnn, and M.B. Barnhart, pp. 115–124, F.K. Schatauer Verlag, Stuttgart, Germany, 1970.
94. Nakagomi, K., Ajisaka, K., and Yokota, I., Effects of acetylthrombin on protein activation and fibrinogen clotting, *Thromb. Res.* 59, 713–722, 1990.
95. Nakagami, K. and Ajiasaka, K., Chemical modification of alpha-thrombin and *in vitro* characterization of anticoagulant activity, *Biochem. Int.* 22, 75–84, 1990.
96. Lundblad, R.L., *Chemical Reagents for Protein Modification*, 4th edn, Chapter 13, pp. 523–580, CRC Press/Taylor & Francis, Boca Raton, FL, USA, 2014.
97. White, G.C., Lundblad, R.L., and Griffith, M.J., Structure-function relatiions in platelet-thrombin reactions. Inhibition of platelet-thrombin interactions by lysine modifications, *J. Biol. Chem.* 256, 1763–1766, 1981.
98. De Candia, E., De Cristofaro, R., and Landolfi, R., Thrombin-induced platelet activation is inhibited by high- and low-molecular weight heparin, *Circulation* 99, 3308–3314, 1999.
99. Bauer, P.I., Machovich, R., Arányi, P., *et al.*, Mechanism of thrombin binding to endothelial cells, *Blood* 61, 368–372, 1987.
100. Carney, D.H., Redin, W., and McCroskey, L., Role of high-affinity thrombin receptors in postclotting cellular effects of thrombin, *Sem. Thromb. Hemostasis* 18, 91–103, 1992.
101. Shimada, K. and Ozawa, T., Evidence that cell surface heparan sulfate is involved in the high-affinity thrombin binding to cultured porcine aortic endothelial cells, *J. Clin. Invest.* 75, 1308–1316, 1985.
102. Hatton, M.W.C. and Moar, S.L., Comparative behavior of thrombin and an inactive derivative, FPR-thrombin, toward the rabbit vascular endothelium, *Circulation Res.* 67, 221–229, 1990.
103. Kaminski, M. and McDonagh, J., Inhibited thrombins. Interaction with fibrinogen and fibrin, *Biochem. J.* 242, 881–887, 1983.
104. Chen, K., Stafford, A.R., Wu, C., *et al.*, Exosite-2 directed ligands attenuated protein C activation by thrombin-thrombomodulin complex, *Biochemistry* 56, 3119–3128, 2017.
105. Yamaguchi, R., Koide, T., and Sakuragawa, N., Binding of heparin or dermatan sulfate to thrombin is essential for the sulfated polysaccharide-accelerated inhibition of thrombin by heparin cofactor II, *FEBS Lett.* 225, 109–112, 1987.
106. Jakobowski, H.V. and Owen, W.G., Macromolecular determinants on thrombin for fibrinogen and thrombomodulin, *J. Biol. Chem.* 264, 11117–11121, 1989.
107. Lundblad, R.L., Tsai, J., Wu, H.-F., *et al.*, Hydrophobic affinity chromatography of human thrombin, *Arch. Biochem. Biophys. I* 302, 109–112, 1993.

108. Tenisien, C., Jandrot-Perrus, M., Huisse, M.-G., and Guillin, M.-C., Effect of phospho-pyridoxylation on thrombin interaction with platelet glycoprotein GpIb, *Blood Coagul. Fibrin.* 2, 521–528, 1991.

109. De Candia, E., De Cristofaro, R., De Marco, L., *et al.*, Thrombin interaction with plate-let GpIb: role of the heparin binding domain, *Thromb. Haemost.* 77, 735–740, 1997.

110. Kang, E.P., Amidinated thrombin: preparation and peptidase activity, *Thromb. Res.* 12, 177–180, 1977.

111. Silverberg, S.A., Chemically modified bovine prothrombin as a substrate in studies on activation kinetics and fluorescence changes during thrombin formation, *J. Biol. Chem.* 255, 8550–8559, 1980.

112. Machovich, R., Regoeczi, E., and Hatton, M.W.C., Altered inactivation kinetics of trini-trohenylated thrombin by antithrombin III in the presence of heparin, *Thromb. Res.* 17, 383–391, 1980.

113. Razin, E., Baranes, D., and Marx, G., Thrombin-mast cell interactions Bind and cell activation, *Expt. Cell Res.* 160, 380–386, 1985.

114. Fredy, J.W., Cutolo, G., Poret, B., *et al.*, Diverted natural Lossen-type rearrangement for bioconjugation through in situ myosinase-triggered isothiocyanate synthesis, *Bioconjug. Chem.* 30, 1385–1394, 2019.

115. Zhang, X., Hemar, Y., Ly, L., Molecular characteristics of the β-lactoglobulin conju-gated with fluorescein isothiocynate. Binding sites and structural changes as a function of pH, *Int. J. Biol. Macrol.* 140, 377–383, 2019.

116. Paborsky, L.R., McCurdy, S.N., Griffin, L.C., Toole, J.J., and Leung, L.L.K., The single =-stranded DNA aptamer-binding site of human thrombin, *J. Biol. Chem.* 268, 20808–20811, 1993.

117. Kotoku, I., Matsushima, A., Bando, M., and Inada, Y., Tyrosine and tryptophan residues and amino groups in thrombin related to enzymic activities, *Biochim. Biophys. Acta* 214, 490–497, 1970.

118. Lundblad, R.L., *Chemical Reagents for Protein Modification*, Chapter 11, pp. 435–487, CRC Press/Taylor & Francis, Boca Raton, FL, USA, 2014.

119. Sokolovsky, M., Riordan, J.F., and Vallee, B.L., Tetranitromethane. A reagent for nitra-tion of tyrosine residues in proteins, *Biochemistry* 5, 3582–3589, 1966.

120. Riordan, J.F., Sokolovsky, M., and Vallee, B.L., Environmentally sensitive tyrosyl resi-dues. Nitration with tetranitromethane, *Biochemistry* 6, 358–361, 1967.

121. Astrup, T., Inactivation of thrombin by means of tetranitromethane, *Acta Chem. Scand.* 7, 744–748, 1947.

122. Sokolovsky, M. and Riordan, J.F., Nitration of pepsin, thrombin, and carboxypeptidase B with tetraniromethane, *Israel J. Chem.* 7, 575–561, 1969.

123. Lundblad, R.L. and Harrison, J.H., The differential effect of tetranitromethane on the proteinase and esterase activity of bovine thrombin, *Biochem. Biophys. Res. Commun.* 45, 1344–1349, 1971.

124. Lundblad, R.L., Noyes, C.M., Featherstone, G.L., Harrison, J.H., and Jenzano, J.W., The reaction of bovine α-thrombin with tetranitromethane. Characterization of the modified protein, *J. Biol. Chem.* 263, 3729–3734, 1988.

125. Workman, E.F., Jr. and Lundblad, R.L., On the preparation of bovine α-thrombin, *Thromb. Haemostas.* 39, 193–200, 1978.

126. Villanueva, G.B. and Perret, V., Effect of sodium and lithium salts on the conformation of human α-thrombin, *Thromb. Res.* 29, 489–498, 1983.

127. Sonder, S.A. and Fenton, J.W., II, Proflavin binding within the fibrinopeptide groove adjacent to the catalytic site of human α-thrombin, *Biochemistry* 23, 1818–1823, 1984.

128. Fenton, J.W., II., Olson, T.A., Zabinski, M.P., and Wilner, G.D., Anion-binding exosite of human α-thrombin and fibrinogen recognition, *Biochemistry* 27, 7106–7112, 1988.

129. Workman, E.F., Jr., White, G.C., II, and Lundblad, R.L., High affinity binding of thrombin to platelets. Inhibition by tetranitromethane and heparin, *Biochem. Biophys. Res. Commun.* 75, 925–932, 1977.

130. Glenn, K.C., Carney, D.H., Fenton, J.W., II., and Cunningham, D.D., Thrombin active site regions required for fibroblast receptor binding and initiation of cell division, *J. Biol. Chem.* 255, 6609–6616, 1980.

131. Riordan, J.F., Wacker, W.E.C., and Vallee, B.L., *N*-acetylimidazole: a reagent for the determination of "free" tyrosyl residues in proteins, *Biochemistry* 4, 1758–1765, 1965.

132. Gorbornoff, M.J., Exposure of tyrosine residues in proteins. III. The reaction of cyanuric fluoride and *N*-acetylimidazole with ovalbumin, chymotrypsinogen and trypsinogen, *Biochemistry* 8, 2591–2598, 1969.

133. Kotoku, I., Matsushima, A., Bando, M., and Inada, Y., Tyrosine and tryptophan residues and amino groups in thrombin related to enzymatic activity, *Biochim. Biophys. Acta* 214, 400–407, 1970.

134. Martin, P.D., Robertson, W., Turk, D., *et al.*, The structures of residues 7–17 of the A(α) chain of human fibrinogen bound to bovine thrombin at 2.3 Å resolution, *J. Biol. Chem.* 267, 7911–7924, 1992.

135. Uhteg, L.C. and Lundblad, R.L., The modification of tryptophan in bovine thrombin, *Biochim. Biophys. Acta* 491, 551–557, 1977.

136. Horton, H.R. and Young, G., 2-Acetoxy-5-nitrobenzyl chloride, a reagent designed to introduce a reporter group near the active site of chymotrypsin, *Biochim. Biophys. Acta* 194, 272–278, 1969.

137. Stevens, W.K. and Nesheim, M.E., Structural changes in the protease domain of prothrombin upon activation as assessed by *N*-bromosuccinimide modification of tryptophan residues in prethrombin-2 and thrombin, *Biochemistry* 32, 2787–2794, 1993.

138. Spande, T.F., Green, N.M., and Witkop, B., The reactivity toward *N*-bromosuccinimide of tryptophan in enzymes, zymogens, and inhibited enzymes, *Biochemistry* 5, 1926–1933, 1966.

6 The Interaction of Thrombin with Fibrinogen and Fibrin

THROMBIN ADSORPTION TO FIBRIN

It is clear from early work on thrombin that the interaction with fibrinogen is complex. Beck has presented an excellent discussion of the early work on blood coagulation, including the work on the interaction of thrombin and fibrinogen.[1] As he notes, the concept of a factor intrinsic to blood coagulation was difficult to accept. The affinity of thrombin for fibrin led early investigators to question the enzymatic action of thrombin on fibrinogen.[2,3] These early studies presented data supporting a stoichiometric relationship between thrombin and fibrinogen. However, these studies followed much earlier studies, such as those of Buchanan[4] and Gamgee,[5] who argued that thrombin activity was a ferment. Buchanan was able to isolate a clot-promoting material from fibrin formed on the dilution of blood with water. This fibrin could be added to blood or fibriniferous "liquids" (exudates from hydroceles),[3,6] resulting in coagulation. In retrospect, this was a significant observation that a "fibriniferous" fluid did not clot on standing. I was not able to find fibriniferous in another citation. It is a combination word formed from "fibrin" and "iferous," which would mean having or bearing fibrin. Gamgee[5] repeated Buchanan's work using both hydrocele fluid and "salted plasma" ($MgSO_4$ added to blood; blood allowed to settle and the supernatant plasma diluted with water). Gamgee's data would suggest that the coagulant material remained bound to fibrin in water but could be eluted with a sodium chloride ("common salt") solution (8%; ~1.4 M). While these early studies are not definitive, it is not unreasonable to assume that the coagulant material obtained by Buchanan and Gamgee was thrombin that bound to fibrin with reasonable affinity. Several studies used fibrin as a source of thrombin.[3,6] Lea and Green[7] observed that their fibrin ferment eluted from fibrin was identical to that prepared from blood by alcohol precipitation.[7] Howell, in 1910,[8] was able to reproduce these early studies purifying thrombin by elution from fibrin with sodium chloride (8% assumed W/V; ca. 1.4 M). It could be argued that these early studies might be the first examples of affinity purification of an enzyme. It is also of interest that Howell's studies were prompted by another study,[9] also from the Physiology Laboratory at Johns Hopkins. Rettger,[9] as with Lea and Green,[6] argued that thrombin was not a ferment (enzyme) and that the reaction between thrombin and fibrinogen was stoichiometric. The fact that enzymes were proteins was not settled until Sumner's work on urease in 1927,[10,11] and even then, the battle over the nature of thrombin continued—but that is discussed in greater detail in Chapter 1. As a personal example, the protein nature of my old friend, blood coagulation factor VIII, was still in question as late as 1966.[12] While I am not sure, this paper may have caused Don Hanahan to ask me whether factor VIII was a protein as the first question on my oral exam. My stumbling answer haunts me to the current day. Regardless, despite the misgivings of the physical chemist who was the graduate school "wild card" on my thesis committee, I was permitted to proceed. I do regret not having a stronger background in thermodynamics. Despite the "hallowed" image of scientific research, we scientists are human and can fall in love with our hypotheses, and leaving any relationship is hard. However, this devotion is to our detriment, and we would do better with multiple hypothesis and not regard the demise of a favored hypothesis as a personal affront. J. Scott Armstrong at the Wharton School (University of Pennsylvania) published a useful paper[13] discussing the relative merits of a single hypothesis (advocacy) and multiple hypotheses. Regardless of Rettger's mistaken (in hindsight) interpretation of the data, his paper is a good review of the state of the art in coagulation at that time. Howell[8] first attempted to obtain thrombin from canine fibrin placed in 8% NaCl/40°C/24 hours. Under these conditions, the fibrin clot was dissolved, resulting in a large amount of protein in a solution that had thrombin activity. A more satisfactory product was obtained from fibrin obtained from porcine blood (8% NaCl/40°C/48–72 hours). Under these conditions a viscous solution was obtained, which was processed by repeated chloroform extraction to obtain thrombin. It is clear that the fibrin obtained from canine blood and porcine blood contained plasmin/plasminogen, which permitted "solubilization" of the fibrin clot with release of thrombin. Some years later with purified proteins, the binding of thrombin to fibrin was confirmed.[14]

Fibrinogen is one of several proteins in blood that are substrates for thrombin.[15] The late Dave Aronson, one of the most clever individuals who I have ever met and who includes some real stars, published a paper in 1976 on the relative sensitivity of some blood coagulation proteins to the action of thrombin. This work is compared with some more recent studies in Table 2.1. It is clear that Dave's intuition in 1976 was incredibly accurate. He is indeed missed. It is of interest that the three best substrates in blood for thrombin all contain sulfated tyrosine residues, which are thought to bind exosite-2 (see Chapter 3 and later).

The early work on thrombin and fibrin(ogen) was briefly discussed earlier and in more detail in Chapter 1. The concept of thrombin as a proteolytic enzyme was still being debated into the 1940s. There was no question as to the fact that thrombin could convert fibrinogen to fibrin, an insoluble protein similar to protein denaturation. Thus, it was reasonable that Wöhlisch and Jöhling[16] described thrombin as an enzyme that denatures protein, a denaturase. Even Erwin Chargaff and his student, Aaron Bendich, two distinguished biochemists, noted that they could not distinguish between fibrinogen denatured by heat or organic chemicals and fibrin produced by the action of thrombin.[17] They concluded that thrombin combined with fibrinogen to form fibrin by an unknown mechanism. This work was performed during WWII when a majority of the scientific community were involved in the war effort. John Ferry described the state of the understanding of thrombin and fibrinogen at that time in an excellent review article in 1983.[18]

As of 1943, there was no scientific consensus on the mechanism of fibrin formation from fibrinogen in the presence of thrombin. This is despite the observation that papain, a recognized proteolytic enzyme, was reported to clot fibrinogen.[19,20] There was little advance in the understanding, as judged by scientific publications, of this process until 1949 despite the clinical use of thrombin with or without fibrinogen during WWII.[18,21] It is difficult for me to define the entire process by which thrombin was established as a protease that converted fibrinogen to fibrin—eventually described as limited proteolysis by a unique protease.[22] Limited proteolysis is a process by which only a few of several susceptible bonds are cleaved in a substrates and is frequently associated with the formation of a protein with biological activity. Laki and Mommaerts[23] were the first to suggest that the formation of the fibrin clot from fibrinogen was a two-step process. They combined thrombin and fibrinogen at pH 5.1 and observed no physical change in the solution. When the pH was raised, the formation of a fibrin aggregate was observed. The longer the incubation time at pH 5.1, the shorter the time for fibrin aggregate formation. When the pH was raised to pH 6.1, it was possible to inhibit the formation of a fibrin aggregate by inclusion of a neutral salt. Laki later described these experiments in greater detail.[24] Further support for a two-step process was provided by Laki and Mihalyi,[25] who showed that thrombin caused a physical change in iodinated fibrinogen without the formation of a fibrin clot. At the same time, two groups in England, one at Cambridge and the other at Leeds, described proteolysis occurring during the conversion of fibrinogen to fibrin in the presence of thrombin. The key observation was the presence of a new amino-terminal amino acid in fibrin. This was the first demonstration of a chemical change in fibrinogen caused by thrombin. One member of the Leeds group, the late Lazlo Lorand, was at the same time describing his early experiments on the role of fibrin-stabilizing factor (factor XIII) in producing urea-insoluble fibrin (crosslinked fibrin).[27] This contribution was a single paper in *Nature* from investigators at two different institutions. It would be nice if such collaboration occurred in today's environment. Work slowly progressed on the understanding of the thrombin-catalyzed conversion of fibrinogen to fibrin. Both Lorand[28] and Kowarzyk[29] reported on the appearance of nonprotein nitrogen (Kjeldahl nitrogen not precipitated by trichloroacetic acid) resulting from the action of thrombin on fibrinogen. Although not stated by these authors, this was the first suggestion of the release of fibrinopeptides from fibrinogen by thrombin. Lorand continued to advance the concept of fibrinopeptide.[28,30,31] Lorand and Middlebrook[32] subsequently demonstrated that fibrinogen was composed of three different polypeptide chains. Two of the chains had *N*-terminal tyrosine and one chain had *N*-terminal glutamic acid. Fibrin was shown to have two chains with *N*-terminal glycine and one *N*-terminal tyrosine. The identification of new *N*-terminal glycine residues in fibrin is consistent with earlier observations.[26] Lorand summarized his early observations on thrombin, fibrinogen, and fibrin-stabilizing factor with a review article in 1954.[33] Ehrenpreis and Scheraga[34] reviewed the work and suggested that thrombin converted fibrinogen to fibrin (now known to be fibrin I) in a reversible reaction. They described the inaction of fibrinogen and fibrin by thrombin. It is likely that their bovine thrombin, "citrate thrombin," contained plasmin. Lazlo Loran, now in the Department of Chemistry at Northwestern University, continued his work on the action of thrombin.[35] In previous work, Lorand and Middlebrook[36] had suggested that thrombin cleaved fibrinogen at an arginylglycine peptide bond. Lorand and Yudkin[35] showed that while neither tosylarginine nor an arginylglycine dipeptide inhibited the action of thrombin on fibrinogen, tosylarginylglycine did inhibit the clotting of fibrinogen by thrombin. It was also shown that thrombin did not cleave the peptide bond in either arginylglycine or tosylarginylglycine. These results provided support for the thrombin cleavage of an arginylglycine peptide bond in fibrinogen during conversion to fibrin. Gladner and coworkers[37] reported the isolation and characterization of fibrinopeptide A and fibrinopeptide B in 1959. Laki and Gladner[38] contributed an excellent review of the status of thrombin and fibrinogen in 1964. This work and two other excellent review articles[39,40] summarized the work on the action of thrombin on fibrinogen at that time. Laki and Gladner[38] also reported on a method for the purification of thrombin on XE-64 ion exchange resin (XE-64 ion exchange resin is a carboxylic acid–based cation exchange resin also known as IRC-50; as a carboxylic acid–based resin, XE-64 would be considered to a weak ion exchange resin) with phosphate buffers. These investigators also provided information on the hydrolysis of a number of synthetic substrates by thrombin, as well as on the inactivation of thrombin by diisopropylfluorophosphate; DFP. They also reported on the composition of a peptide containing the serine residue modified by DFP, showing homology with chymotrypsin and trypsin and supporting the classification of thrombin as a proteolytic enzyme. Finally, Lake and Gladner[38] summarized work on the structure of fibrinogen and confirmed that cleavage of four arginylglycine peptide bonds by thrombin was responsible for the conversion of fibrinogen to fibrin.

This discussion summarized the state of understanding of the fibrinogen-fibrin transition, and today it is known to be a complex process.[39–41] While the current work is focused on the action of thrombin on fibrinogen to form fibrin I and subsequently to form fibrin II and not on the subsequent polymerization steps, it is necessary to briefly discuss factors that influence the quality of the fibrin clot, as this influences both the binding of thrombin and the expression of fibrin-bound thrombin activity. α-Thrombin preferentially forms fibrin I, which forms fibrinprotofibrils (linear polymers of fibrin I).[42–45] Thrombin subsequently cleaves fibrin I to yield fibrin II with the release of fibrinopeptide B.[46,47] Fibrin II can undergo lateral polymerization. Calcium ions have been shown to affect the formation of observed fibrin clots from fibrinogen by the action of thrombin on fibrinogen,[48–54] but this effect has to do with the binding of calcium ions to fibrinogen[55] and not an effect on the catalytic activity of thrombin. Ratnoff and Potts[49] reported that strontium could replace calcium ions in supporting fibrin monomer polymerization, while Carr and Powers[53] observed that magnesium ions could replace calcium ions in the polymerization process.

Fibrin I undergoes linear polymerization (end-to-end polymerization), resulting in thin fibers with lower mass to length ratios than those observed with the polymerization of fibrin II, which undergoes lateral polymerization resulting in thick fibers with higher mass to length ratios.[47,55] Thin fibers are the product of fibrin formation at high concentrations of thrombin.[56–60] The observations of Shen and coworkers[60] link the release of fibrinopeptide B (FPB) to lateral polymerization. Alisa Wolberg[60] has an excellent review of the effect of thrombin on fibrin polymer formation. In subsequent work, Wolberg and coworkers[61] also observed the elevated blood prothrombin concentration resulted in the formation of thin fibrin fibers, which are thought to be more resistant to fibrinolysis.[57,62] The issue of elevated prothrombin and thrombosis is not new. I recall some discussion of the role of elevated prothrombin in the development of thrombosis with prothrombin complex concentrates[63] resulting from the longer half-life of prothrombin compared to factor IX.[64]

Other factors influence the polymerization of fibrin monomers, and early researchers were careful to distinguish between an effect on fibrin polymerization and an effect on the activity of thrombin. It has been known for some time that dextran accelerated the rate of clotting of fibrinogen by thrombin.[65] Laurell[65] reported that dextran accelerated the clotting of fibrinogen by either thrombin or staphylocoagulase. Subsequently Abildgaard[66] observed that both dextran and ficoll accelerated the thrombin clotting of citrated human plasma. Abildgaard also reported that neither dextran nor ficoll affected the rate of the appearance of N-terminal glycine during the action of thrombin on fibrinogen, showing that these components did not have a direct effect on thrombin. Much later Carr and Gabriel[67] showed that dextran accelerated the rate of lateral polymerization at low concentrations, while higher concentrations inhibited polymerization, and it was incorporated into the fibrin clot. Lower-molecular-weight polyols were shown to be mostly ineffective.[68] Fenton and Fasco[69] showed that incorporation of poly(ethylene)glycol also accelerated the rate of fibrinogen clotting and was a useful addition to mechanical and visual assay systems for thrombin. Other factors such as γ-globulin[70,71] and albumin have been shown to affect fibrin structure.[72–76]

THE BINDING OF THROMBIN TO FIBRIN

The =fibrinogen molecule is composed of three pairs of polypeptide chains, a dimer of a heterotrimer.[55,77] The majority of plasma fibrinogen consisted of 2Aα, 2Bβ, and 2 γ(γA) chains. A small portion of plasma fibrinogen (15%–30%) is composed of a variant that contains a γ-chain extended at the carboxyl terminal, designated γ′ (gamma prime).[78,79] This extension of 20 amino acids is relatively acidic and contains two sulfated tyrosine residues[80] and is considered to be the high-affinity binding site for thrombin.[77] Fibrinogen γ′ has been shown to form fibrin with decreased fiber mass (diameter) and decreased protofibril packing, similar to fibrin formed with high concentrations of thrombin.[80]

The adsorption of thrombin to fibrin has been a subject of study for some time.[3,4,6] Work by Liu and coworkers[81] can be considered seminal for the subsequent work in the modern era on the binding of thrombin to fibrin. These researchers were the first to suggest that there were two classes of binding sites of thrombin on fibrin: a high-affinity site (0.39 moles thrombin/mole fibrinogen) and a low-affinity site (1.6 moles thrombin/mole fibrinogen). Binding of thrombin at the low-affinity site is freely reversible but considerably tighter at the high-affinity site. These researchers observed that the binding of thrombin to fibrin did not require the enzyme-active site, as modification with PMSF did not affect binding. Thrombin bound to fibrin equilibrated with thrombin solution in a slow process. These investigators also reported that thrombin bound to fibrin was partially protected from inactivation by antithrombin and that thrombin activity could be obtained from the fibrin clot by treatment with plasmin, which was also partially protected from inactivation by antithrombin. Work from several other laboratories showed that thrombin could bind to a fibrin monomer bound to agarose,[82–84] as well as fibrin bound to a surface.[85] Vali and Scheraga[86] reported on the binding of PMSF-inactivated bovine thrombin to various fragments bound to cyanogen bromide-activated agarose. They found that PMSF-thrombin bound to fragments from the D domain and the E domain. Later work[87,88] showed that the high-affinity binding site was located on the D domain of the γ' chain (1.05 moles thrombin/mole fibrinogen; $k_a = 4.5 \times 10^6$ M^{-1}), while the low-affinity site was located in the fibrin E domain (1.69 moles thrombin/mole thrombin; $k_a = 0.28 \times 10^6$ M^{-1}). Siebenlist and coworkers[88] also suggested that polymerization of the fibrin monomer influences thrombin binding. Other work[89] from Mike Mosesson's laboratory showed that while PPACK-thrombin bound to fibrin, hirudin did block binding. These results were consistent with the earlier observations showing that the active site of thrombin was not important in binding to fibrin, but the anion-binding site (exosite-1), where hirudin binds, was important. This work did present evidence suggesting that PPACK-thrombin enhanced fibrin polymerization. Later work[91] suggested that thrombin formed a dimer that would bridge the low-affinity binding sites on the fibrin monomer. Mosesson and coworkers[90] also showed that the polymerization of γA/γA fibrinogen was also enhanced by PPACK-thrombin, suggesting that the high-affinity site was not involved in the enhancement of fibrin polymerization by PPACK-thrombin. PPACK-γ-thrombin, which lacks a fully functional exosite-1, did not enhance polymerization, and the HD-1 DNA aptamer, which binds to exosite-1, blocked polymerization. These two observations suggested the importance of a functional exosite-1 in the PPACK-thrombin enhancement of fibrin polymerization. In subsequent work Mosesson[91] showed that the HD-22 aptamer, which binds to exosite-2, also blocked the enhancement of fibrin polymerization by PPACK-thrombin. It was suggested a thrombin dimer bound two fibrin monomers, with binding between the low-affinity site and exosite-1. A dimeric thrombin has been observed in the crystal structure of a thrombin-heparin fragment complex.[92] The asymmetric unit contained two thrombin dimers, with the individual dimer sharing binding to a sulfated octasaccharide. A dimeric human or bovine thrombin in complex with low-molecular-weight heparin (LMWH) in solution has been demonstrated using analytical ultracentrifugation.[93] These investigators also reported the crystal structure of a bovine thrombin–sucrose octasulfate complex. The addition of PPACK-thrombin to either fibrinogen or fibrin did not affect the thermal stability of either protein, suggesting a lack of effect of PPACK-thrombin on the stability of these proteins.[94] The same study showed that the presence of calcium ions markedly increased the thermal stability of fibrinogen or fibrin. The tetrapeptide, Gly-Pro-Arg-Gly (GPRG), increased the stability of fibrinogen but not fibrin. It was suggested the binding of GPRG of fibrinogen created a fibrin-like structure. Finally, PPACK-thrombin was much more stable than native thrombin.

Bänninger and coworkers[95] suggested that the binding of thrombin to fibrin depended on the structure of the fibrin clot. These investigators studied the binding of PPACK-thrombin to a batroxobin clot where only fibrinopeptide A (FPA) is released from des-AA fibrin and raised concern about the existence of two different binding sites, as well as the importance of the physical entrapment of thrombin in the fibrin clot. Gugerell and coworkers[96] observed that fibrin matrices obtained at a low thrombin concentration (4U/mL) were more effective in wound healing (rat excisional wound healing model) than those obtained at high thrombin concentrations (800–1200 U/mL). Matrices obtained at low thrombin concentrations would be thicker than those obtained at high thrombin concentrations and would be expected to degrade faster than those obtained at low thrombin concentrations. More effective wound healing was observed at lower thrombin concentrations. It was also observed that keratinocytes were degraded at higher thrombin concentrations, which may be related to the release of thrombin from newly formed clots during contraction. The importance of physical entrapment of thrombin in a fibrin clot continues to receive support[97] and is consistent with the burst release of thrombin from a clot during fibrinolysis,[98] as discussed later in more detail. The fibrin clot entraps a variety of macromolecules found in blood, which can be released by physical compression of the clot. Burnouf and coworkers[99] reported that the contents expressed from a fibrin clot could be used as a replacement for fetal bovine serum in cell culture. Fibrinolysis (plasmin) has been shown to cause the release of latent transforming growth factor beta (TGF-β) from a fibrin clot.[100]

With the binding of thrombin to fibrin established, studies now refined the nature of the binding. Pechnik and coworkers[101] extended the understanding of low-affinity binding. They used x-ray crystallography to the study the binding of a site in the E domain of α and β chains to the exosite-1 region of thrombin. There was also continued characterization of the high-affinity binding site in the D domain of the γ′ chain of fibrinogen.[102] Meh and coworkers[102] observed that the binding site for thrombin on the γ′ chain was an inherent domain found in both fibrinogen and fibrin. These researchers also suggested that the low-affinity sites for binding thrombin are located on each half of the dimeric E domain in fibrin. They also suggested that the low-affinity binding sites are recognized by the anion-binding exosite (exosite-1) in thrombin.[103] Fredenburgh and coworkers[104] suggested that the binding of thrombin to fibrin could involve binding both exosites in γA/γ′ fibrinogen. Thrombin would bind to the low-affinity site on fibrin via exosite-1 and to the high-affinity site via exosite-2.[105] Haynes and coworkers[106] proposed a model for thrombin binding to fibrin that involved initial binding to the low-affinity sites with subsequent movement to the high-affinity site on the γ′ chain. Zhu and coworkers[107] showed that thrombin binds tightly to the fibrinogen γ′-chain under conditions of venous flow. These researchers suggest that thrombin activity can be regulated by binding to the fibrinogen γ′ chain during *in vivo* coagulation. This work suggested that thrombin bound in fibrin generated a small amount of fibrin, possibly suggesting a short half-life for thrombin bound to fibrin. They did show that fluorophore-labeled PPACK did react with thrombin bound to fibrin. They note that the sequestration of thrombin from bulk solution is consistent with the role of fibrin as antithrombin I. Several constituents of blood can neutralize thrombin activity and, as such, have been designated as anthrombins.[108] Further work from the University of Pennsylvania laboratory[109] showed the intrinsic tenase activity (factors XIa/IXa/VIIIa) continues to support thrombin generation in the fibrin clot, consistent with earlier observations from Hemker and coworkers on the feedback action of thrombin.[110,111] As briefly noted earlier, the γ′-chain extension contains sulfated tyrosine residues.[112] Sulfated tyrosine residues are also found in GP1bα[113] and in factor VIII[114] and are thought to be critical for the interaction of these proteins with thrombin.

Many other proteins can be specifically or nonspecifically associated with the fibrin clot.[115] Notable among these is histidine-rich glycoprotein, which binds specifically to fibrin(ogen) and can compete with thrombin for binding to the high-affinity binding site on the γ'-chain as reported by Vu and workers.[116] The binding of histidine-rich glycoprotein to the γ'-chain is dependent on the presence of Zn^{2+}. Vu, Fredenburgh, and Weitz have reviewed the functions of zinc ions in blood coagulation.[117] Szuldyrzyński and coworkers[118] observed that the bleeding time was lower in coronary artery disease (CAD) patients; the concentration of histidine-rich glycoprotein was also lower. It is possible that the lower histidine-rich glycoprotein concentration would suggest more fibrin-bound thrombin, which could accelerate coagulation.

Regardless of the mechanism of the binding of thrombin to fibrin, thrombin is released from the fibrin clot either as free thrombin, which was physically entrapped in the clot, or as thrombin bound to fragments of fibrin resulting from fibrinolysis. In the majority of situations, free thrombin would be rapidly inactivated by plasma proteinase inhibitors.[119,120] Thrombin bound to fibrin fragments[121–123] is more resistant to inactivation by these plasma proteinase inhibitors.[124–127] Francis and coworkers[121] showed that most of the thrombin bound to fibrin was released by incubation in buffer or manual compression with 10%–15% remaining bound to the fibrin. These investigators also established that bound thrombin retained catalytic activity. Consistent with earlier studies, the bound thrombin could be solubilized with digestion by plasmin. There is variability in the ability of plasma proteinase inhibitors to inactivate thrombin bound to fibrin or fibrin fragments, but thrombin would be susceptible to inactivation by low-molecular-weight inhibitors such as PPACK,[124] dabigatran (Pradaxa),[128–132] or other potential direct thrombin inhibitors.[133,134] As an example, Deinum and coworkers[133] characterized an experimental drug, AZD0837, as a novel oral anticoagulant (NOAC). The investigators showed that active drug, ARHO67637, was as effective in the inhibition of fibrin bound-thrombin as free thrombin. Like many NOACs, AZD0837 is a prodrug which is converted *in vivo* to the active drug, ARH067637. It has been shown that a covalent complex of heparin and antithrombin is a potent inhibitor of thrombin bound to fibrin.[135] Smith and coworkers[136] showed that a covalent antithrombin-heparin complex that was bound to fibrin-bound thrombin was a potent anticoagulant. In their model, the heparin-antithrombin-thrombin complex bound to fibrin effectively presents heparin to promote activation of thrombin or other coagulation proteases.

CATALYTIC ACTIVITY OF FIBRIN-BOUND THROMBIN

One consequence of the binding of thrombin by fibrin is the "inactivation" of thrombin by removing it from the bulk solution. Fibrin was known as antithrombin I,[108,137] antithrombin II is known as heparin cofactor II, and antithrombin III is now known as antithrombin. Other antithrombin activities also were identified. These included antithrombin IV, an ether-extractable activity from plasma; antithrombin V, which was an activity associated with hypergammaglobulinemia in rheumatoid arthritis; and antithrombin VI, which was likely a fibrin-degradation product (FDP). When removed from solution, thrombin bound to fibrin retains catalytic activity. One example briefly noted earlier is the role of thrombin bound to fibrin I in the formation of fibrin II by removal of FPB. Hogg and Jackson[138] reported that a complex of thrombin, heparin, and antithrombin resulted in an increase in K_M and decrease in k_{cat} in the hydrolysis of H-D-IleProArgpNA and Cbz-GlyProArgpNA. They also reported a decrease in the catalytic efficiency in the hydrolysis of prothrombin to form prethrombin I and prothrombin fragment 1, but there was no effect in the release of FPA and FPB. Subsequently Chen and coworkers[139] reported that zinc ions (Zn^{2+}) increased the affinity of heparin for fibrin, promoting the formation of the heparin-fibrin-thrombin complex and enhancing the protection of thrombin from inactivation by antithrombin. The importance of Zn^{2+} in coagulation has been reviewed,[117] and the importance of Zn^{2+} in binding histidine-rich glycoprotein to fibrin was discussed earlier.

Thrombin bound to fibrin can still promote the feedback reactions of thrombin in the activation of factor V and factor VIII in coagulation.[110] Chen and Diamond[109] emphasized the *in situ* activation of factor XI in the fibrin clot in the activation of the intrinsic coagulation pathway. Further work by Hemker and coworkers[111] showed that the binding of thrombin by fibrin(ogen) enhanced thrombin generation by reducing inhibition by antithrombin and α_2-macroglobulin; γ' fibrinogen was more effective than γ fibrinogen. Earlier work[140] had shown that fibrin-1 enhanced prothrombin activation, while fibrinogen inhibited prothrombin activation. The use of ionic contrast agents is suggested to present a lower risk of thrombosis than nonionic agents. Ionic contrast agents compete with thrombin by binding to the high-affinity exosite.[141] As a result, there would be less fibrin-bound thrombin to participate in the growth of thrombi. This would be the same effect observed with reduced histidine-rich glycoprotein; increased binding of thrombin to fibrin promotes coagulation. There is also the well-known role of fibrin in the activation of factor XIII by thrombin.[142–146] Factor XIII is the precursor of factor XIIIa (fibrin-stabilizing factor). Factor XIIIa is a transglutaminase which catalyzed the formation of crosslinks between fibrin chains, yielding a clot that is not soluble in a chaotropic solvent such as urea. Fibrin participates as a cofactor in the reaction. While it is likely that the fibrin-bound thrombin catalyzes the activation of factor XIII, I could not find definitive experimental evidence to support this function.

While these observations are important for the function of thrombin in the humoral phase of blood coagulation, it is my sense that fibrin-bound thrombin is also of significance in interaction with cells. Thrombin bound to fibrin does activate blood platelets.[147,148] Phillips and coworkers[149] showed that fibrin-bound thrombin increases the expression of binding sites on blood platelets for factor VIIIa, potentially increasing the rate of intrinsic coagulation and thrombosis. I have long felt that the IXa/VIIIa interaction is the key to thrombosis. Phillips and coworkers[148] also demonstrated that a mutant γ fibrinogen chain was ineffective, which suggested the importance of high-affinity binding of thrombin to fibrin in the activation of blood platelets. As noted by Lovely and coworkers,[105] the binding of thrombin to fibrin via exosite-2 leaves the enzyme-active site available for functioning. Platelets do adhere to fibrin under flow conditions, but adhesion is markedly increased if exogenous thrombin is added to the fibrin[149] (I am assuming the thrombin was present on the preformed fibrin surface). Binding was dependent on the exosite-2 domain. Binding of platelets to thrombin was also observed with thrombin alone bound to a surface. The binding of platelets to immobilized thrombin was mediated through GP1bα.

Fibrin-bound thrombin can affect cells outside of the circulatory system. Some of this interest has been driven by the increasing use of fibrin sealant and the possible effect of fibrin-bound thrombin in wound healing. There is considerable literature on the interaction of fibrin with cells,[150–152] but the possibility that these effects might be due to the presence of thrombin is frequently not considered. Early studies showed that fibrin could stimulate fibroblasts independent of thrombin.[153] Fibrinogen bound to a surface can exhibit epitopes considered unique to fibrin[154] and could affect cell behavior.[155,156] However, the binding of fibrinogen to a surface depends on the quality of the surface,[157] and the bound fibrinogen can bind in different orientations,[158] which likely influences conformational change. While acknowledging that there is considerable literature on the interaction of thrombin with fibrinogen and other extracellular matrix proteins not cited, it is possible to speculate that the binding of thrombin in the extracellular matrix[159] resulted in the modulation of thrombin activity in a manner similar to the binding to fibrin discussed earlier. Thrombin bound to the extracellular matrix does possess mitogenic activity toward vascular smooth muscle cells.[160]

Based on studies cited earlier,[138] it is likely that the binding of thrombin of fibrin does modulate activity much in the same way as observed in the binding of thrombin to glycocalicin (GP1b).[161] Since thrombin can function in the interstitial space,[162] it is important to consider the potential role of fibrin-bound thrombin in modulating cell function in the extravascular space.[163–166] Smadia and coworkers[166] observed that thrombin bound to fibrin in plasma enhanced the expression of endothelial protein C receptor and thrombomodulin on endothelial progenitor cells and enhanced both progenitor endothelial cell proliferation and migration. It is suggested that this effect is mediated through PAR-1 receptors on endothelial progenitor cells.[167] Oh and coworkers[168] suggest that fibrin-bound thrombin can enhance fibroblast proliferation by enhancing the ability of fibrin to bind fibronectin. Proliferation of CCL39 cells (Chinese hamster lung fibroblasts) induced by fibrin clots is suggested to be due to thrombin released from the clot.[169] Others suggest that fibrin alone is responsible for the migration of smooth muscle cells into a fibrin gel.[170,171]

REFERENCES

1. Beck, E.A., Historical development of the prothrombin concept, in *Prothrombin*, ed. H.C. Hemker and J.J. Veltkamp, pp. 15–24, Leiden University Press, Leiden, Netherlands, 1975.
2. Collingwood, B.J. and MacMahon, M.T., The nature of thrombin and anti-thrombin, *J. Physiol.* 47, 44–53, 1913.
3. Howell, W.H., The preparation and properties of thrombin, together with observations on thrombn and prothrombin, *Am. J. Physiol.* 26, 453–473, 1910.
4. Buchanan, A., On the coagulation of the blood and other fibriniferous liquids, *London Med. Gazette* Vol 1 (new series), 617–621, 1845.
5. Gamgee, A., Some old and new experiments on the fibrin ferment, *J. Physiol.* 2, 145–163, 1878.
6. Lea, S. and Green, J.R., Some notes on the fibrin ferment, *J. Physiol.* 4, 380–386, 1884.

7. Schmidt, A., Neue Untersuchungen über die Fasterstoffgerinnung, *Pflüger's Arch. Ges. Physiol.* 6, 413–, 1872.

8. Howell, W.H., The preparation and properties of thrombin together with observations on antithrombin and prothrombin, *Am. J. Physiol.* 26, 453–473, 1910.

9. Rettger, L.J., The coagulation of blood, *Am. J. Physiol.* 24, 406–435, 1909.

10. Simoni, R.D., Hill, R.H., and Vaughn, M., Urease, the first crystalline enzyme and the proof that enzymes are proteins: the work of James B. Sumner, *J. Biol. Chem.* 277, 23e, 2002.

11. Manchester, K.L., The crystallization of enzymes and virus proteins: laying to rest the colloidal concept of living systems, *Endeavor* 28, 25–29, 2004.

12. Veder, H.A., Is the antihaemophilic globulin a protein?, *Nature* 209, 202, 1966.

13. Armstrong, J.S., Advocacy and objectivity in science, *Manag. Sci.* 25, 423–428, 1979.

14. Seegers, W.H., Nieft, M.L., and Loomis, F.C., Note on the adsorption of thrombin on fibrin, *Science* 101, 520–521, 1945.

15. Aronson, D.L., Comparison of the actions of thrombin and the thrombin-like venom enzymes ancrod and batroxobin, *Thromb. Haemost.* 36, 9–13, 1976.

16. Wöhlisch, E. and Jöhling, L., Das thrombi als Fibrinogendenaturase und seine Bezieungen zum papain, *Biochim. Zeitschrift* 297, 353–368, 1938.

17. Charaff, H.E. and Bendich, A., On the coagulation of fibrinogen, *J. Biol. Chem.* 149, 93–110, 1943.

18. Ferry, J.D., The conversion of fibrinogen to fibrin: events and recollections from 1942 to 1983, *Ann. N. Y. Acad. Sci.* 408, 1–10, 1983.

19. Eagle, H. and Harris, T.N., Studies in blood coagulation V> The coagulation of blood by proteolytic enzymes (trypsin, papain), *J. Gen. Physiol.* 20, 543–560, 1937.

20. Ferguson, J.H. and Ralph, P.H., Ninhydrin, crystalline papain, and fibrin clots, *Am. J. Physiol.* 138, 648–651, 1943.

21. Ferry, J.D. and Morrison, P.R., Chemical, clinical, and immunological studies on the products of human plasma fractionation XVI. Fibrin clots, fibrin films, and fibrinogen plastics, *J. Clin. Invest.* 23, 566–572, 1944.

22. Scheraga, H.A. and Laskowski, M., Jr., The fibrinogen-fibrin conversion, *Adv. Prot. Chem.* 12, 1–131, 1957.

23. Laki, K. and Mommaerts, W.F.H.M., Transition of fibrinogen to fibrin is a two-step reaction, *Nature* 156, 664, 1945.

24. Laki, K., The polymerization of proteins: the action of thrombin on fibrinogen, A5h, *Biochem. Biophys.* 32, 317–324, 1951.

25. Laki, K. and Mihalyi, E., Action of thrombin on iodinated fibrinogen, *Nature* 163, 66, 1949.

26. Bailey, K., Bettelheim, F.R., Lorand, L., and MIddlebrook, W.R., Action of thrombin in the clotting of fibrinogen, *Nature* 167, 614–615, 1952.

27. Lorand, L., Fibrin clots, *Nature* 166, 694–695, 1950.

28. Lorand, L., 'Fibrino-peptide': aspects of the fibrinogen-fibrin transformation, *Nature* 167, 992–993, 1952.

29. Kowarzyk, H., Thrombin and thrombin-protease, *Nature* 169, 614–615, 1952.

30. Lorand, L. and Middlebrook, W.R., Studies on fibrino-peptide, *Biochim. Biophys. Acta* 9, 581–582, 1952.

31. Lorand, L., Fibrino-peptide, *Blochem. J.* 52, 200–203, 1952.

32. Lorand, L. and Middlebrook, W.R., The action of thrombin on fibrinogen, *Biochem. J.* 52, 196–199, 1952.

33. Lorand, L., Interaction of thrombin and fibrinogen, *Physiol. Rev.* 34, 742–752, 1954.

34. Ehrenpreis, S. and Scheraga, H.A., Observations on the analysis for thrombin and the inactivation of fibrin monomer, *J. Biol. Chem.* 227, 1043–1061, 1957.

35. Lorand, L. and Yudkin, E.P., The effect of arginyl peptides on the clotting of fibrinogen with thrombin, *Biochim. Biophys. Acta* 25, 29–30, 1957.

36. Lorand, L. and Middlebrook, W.R., Specific specificity of fibrinogen as revealed by end group studies, *Science* 118, 515–516, 1953.

37. Gladner, J.A., Folk, J.E., Laki, K., and Carroll, W.C., Thrombin-induced formation of co-fibrin. I. Isolation, purification, and characterization of co-fibrin, *J. Biol. Chem.* 234, 63–66, 1959.

38. Laki, K. and Gladner, J.A., Chemistry and physiology of the fibrinogen-fibrin transition, *Physiol. Rev.* 44, 127–160, 1964.

39. Mosesson, M.W., Thrombin interaction with fibrinogen and fibrin, *Sem. Thromb. Hemost.* 19, 362–367, 1993.

40. Doolittle, R.F., The conversion of fibrinogen to fibrin: a brief history of some key events, *Matrix Biol.* 60–61, 5–7, 2017.

41. Siedelmann, J.J., Gran, J., Jespersen, J., and Kluft, C., FIbrin clot formation and lysis: basic mechanisms, *Sem. Thromb. Hemost.* 26, 605–618, 2000.

42. Blömbäck, B., Hessel, B., Hogg, D., and Therkildsen, L., A two-step fibrinogen-fibrin transition In blood coagulation, *Nature* 275, 501–505, 1978.

43. Blömbäck, B., The fibrinopeptide, *Thromb. Haemost.* Suppl 20, 201–210, 1978.

44. Lewis, S.D., Shields, P.P., and Shafer, J.A., Characterization of the kinetic pathway for liberation of fibrino-peptides during the assembly of fibrin, *J. Biol. Chem.* 265, 10192–10199, 1985.

45. Siebenlist, K.R., Di Orio, J.P., Budzynski, A.Z., and Mosesson, M.W., The polymerization and thrombin-binding properties of des-(Bβ1–42)-fibrin, *J. Biol. Chem.* 265, 18650–18655, 1990.

46. Naski, M.C., and Shafer, J.A., α-Thrombin catalyzed hydrolysis of fibrin I. Alternative binding modes and accessibility of active site in fibrin I-bound thrombin, *J. Biol. Chem.* 265, 1401–1407, 1995.

47. Blombäck, B. ad Bark, N., Fibrinopeptides and fibrin gel structure, *Biophys. Chem.* 112, 147–151, 2004.

48. Robbins, K.C., A study on the conversion of fibrinogen to fibrin, *Am. J. Physiol.* 142, 581–588, 1944.

49. Ratnoff, O.D. and Potts, A.M., The accelerating effect of calcium and other cations on the conversion of fibrinogen to fibrin, *J. Clin. Invest.* 33, 206–210, 1954.

50. Shen, L.L., McDonagh, R.P., McDonagh, J., and Hermans, J., Jr., Fibrin gel structure: influence of calcium and covalent cross-linking on the elasticity, *Biochem. Biophys. Res. Commun.* 56, 793–798, 1974.

51. Okada, M. and Blö==mbäck, B., Calcium and fibrin gel structure, *Thromb. Res.* 29, 269–280, 1983.

52. Perizzolo, K.E., Sullivan, S., and Waugh, D.F., Effect of calcium binding and of EDTA and CaEDTA on the clotting of bovine fibrinogen by thrombin, *Arch. Biochem. Biophys.* 237, 520–534, 1985.

53. Carr, M.E. and Powers, P.L., Differential effects of divalent cations on fibrin structure, *Blood Coag. Fibriolysis* 2, 741–747, 1991.

54. Profuma, A., Turci, M., Damonte, G., *et al.*, Kinetics of fibrinopeptide release by thrombin as a function of $CaCl_2$ concentration. Different susceptibility of FPA and FPB and evidence for a fibrinogen isoform-specific effect at physiological Ca^{2+} concentrations, *Biochemistry* 42, 12335–12348, 2003.

55. Weisel, J.W., Fibrinogen and fibrin, *Adv. Prot. Chem.* 70, 247–299, 2005.

56. Blomback, B., Hessel, B., Hogg, D., and Therkildsen, L., A two-step fibrinogen-fibrin transition in blood coagulation, *Nature* 275, 501–505, 1978.

57. Gabriel, D.A., Muga, K., and Boothroyd, E.M., The effect of fibrin structure on fibrino-lysis, *J. Biol. Chem.* 267, 24259–24263, 1992.

58. Rowe, S.L., Lee, S., and Stegeman, J.P., Influence of thrombin concentration on the mechanical and morphological properties of cell-seeded fibrin hydrogels, *Acta Biomater.* 2, 59–67, 2007.

59. Shen, L.L., Hermans, J., McDonagh, J., and McDonagh, R.P., Role of fibrinopeptide B release: comparison of fibrin formed by thrombin and ancrod, *Am. J. Physiol.* 232, H628–H633, 1977.

60. Wolberg, A.S., Thrombin generation and clot structure, *Blood Rev.* 21, 131–142, 2007.

61. Wolberg, A.S., Monroe, D.M., Roberts, H.R., and Hoffman, M., Elevated prothrombin results in clot with altered fibrin structure: a possible mechanism for increased thrombotic risk, *Blood* 101, 3008–3013, 2013.

62. Torbet, J., The thrombin activation pathway modulates the assembly, structure, and lysis of human plasma clots *in vitro*, *Thromb. Haemost.* 73, 785–793, 1995.

63. Kim, H.C., McMillan, C.W., White, G.C., *et al.*, Purified factor IX using monoclonal immunoaffinity techniques: clinical trials in hemophilia B and comparison in prothrombin complex concentrates, *Blood* 79, 568–575, 1992.

64. Franchini, M. and Lippi, G., Prothrombin complex concentrates: an update, *Blood Transfus.* 8, 149–154, 2010.

65. Laurell, A.B., Influence of dextran on the conversion of fibrinogen to fibrin, *Scand. J. Clin. Lab. Invest.* 3, 262–266, 1951.

66. Abildgaard, U., Acceleration of fibrin polymerization by dextran and ficoll. Interaction with calcium and plasma proteins, *Scand. J. Clin. Lab. Invest.* 18, 518–524, 1966.

67. Carr, M.E. and Gabriel, D.A., Dextran-induced changes in fibrin fiber size and density based on wavelength dependence of gel turbidity, *Macromolecules* 13, 1473–1477, 1981.

68. Ferry, J.D., and Shuman, S., The conversion of fibrinogen to fibrin. I. The influence of hydroxyl compounds on clotting time and clot opacity, *J. Am. Chem. Soc.* 71, 3198–3204, 1949.

69. Fenton, J.W. and Fasco, M.J., Polyethylene glycol enhancement of clotting of fibrinogen in visual and mechanical assays, *Thromb. Res.* 4, 809–817, 1974.

70. Coleman, M., Vigliano, E.M., Weksler, M.E., and Nachman, R.L., Inhibition of fibrin monomer polymerization by lambda myeloma globulin, *Blood* 39, 210–223, 1972.

71. Davey, F.R., Gordon, G.B., Boral, L.I., and Gottlieb, A.J., Gamma globulin inhibition of fibrin clot formation, *Ann. Clin. Lab. Sci.* 6, 72–76, 1976.

72. Carr, M.E., Fibrin formed in plasma is composed of fibers more massive than those formed from purified fibrinogen, *Thromb. Haemost.* 59, 535–539, 1988.

73. Galanakis, D.K., Lane, B.P., and Simmons, S.R., Albumin modulates lateral assembly of fibrin monomers. Evidence of enhanced fine fibrils formed of unique synergy with fibrinogen, *Biochemistry* 26, 2389–2400, 1986.

74. Torbet, J., Fibrin assembly in human plasma and fibrinogen/albumin mixures, *Biochemistry* 25, 5309–5314, 1986.

75. Marx, G. and Harari, N., Albumin indirectly modulates fibrin and protofibril ultrastructure, *Biochemistry* 28, 8242–8248, 1989.

76. Levy, G.K., Ong, J., Birch, M.A., Justin, A.W., and Markaki, A.E., Albumin-enriched fibrin hydrogel embedded in active ferromagnetic networks improves osteoblast differentiation and vascular self-organization, *Polymers* 11, 1743, 2019.

77. Mosesson, M., Fibrinogen and fibrin polymerization: appraisal of the binding events events that accompany fibrin generation and fibrin clot assembly, *Blood Coagul. Fibrin.* 8, 257–267, 1997.

78. Chung, D.W. and Davie, E.W., γ and γ' chains of human fibrinogen are produced by alternative mRNA splicing, *Biochemistry* 23, 4232–4236, 1984.

79. Doolittle, R.F., Hong, S., and Wilcox, D., Evolution of the fibrinogen γ chain: implications for the binding of factor XIII, thrombin, and platelets, *J. Thromb. Haamost.* 7, 1431–1433, 2009.
80. Dominguez, M.M., Macrae, F.L., Duval, C., *et al.*, Thrombin and fibrinogen γ' impact clot structure by marked effect on intrafibrillar structure and protofibril packing, *Blood* 127, 487–495, 2016.
81. Liu, C.Y., Nossel, H.L., and Kaplan, K.L., The binding of thrombin by fibrin, *J. Biol. Chem.* 254, 10421–10425, 1979.
82. Fenton, J.W., II, Olson, T.A., Zabinski, M.P., and Wilner, G.D., Anion-binding exosite pf human α-thrombin and fibrin(ogen) recognition, *Biochemistry* 27, 7106–7112, 1988.
83. Kaminski, M.P. and McDonagh, J., Studies on the mechanism of thrombin. Interaction with fibrin, *J. Bol. Chem.* 258, 10530–10535, 1983.
84. Berliner, L.J. and Sugawara, Y., Human α-thrombin binding to nonpolymerized fibrin-Sepharose: evidence for an anion-binding site, *Biochemstry* 24, 7005–7009, 1985.
85. Wilner, G.D., Danitz, M.P., Mudd, M.S., Hsieh, K.H., and Fenton, J.W., Selective immobilization of α-thrombin by surface-bound fibrin. *J. Lab. Clin. Med.* 97, 403–411, 1981.
86. Vali, Z., and Scheraga, H.A., Localization of the binding site for fibrin for the secondary binding site of thrombin, *Biochemistry* 27, 1956–1963, 1988.
87. Meh, D.A., Siebinllist, K.R., and Mosesson, M.W., Identification and characterization of the thrombin binding site on fibrin. *U. Biol. Chem.* 271, 23121–23125, 1996.
88. Siebenlist, K.R., Di Orio, J.P., Budzynski, A.Z., and Mosesson, M.W., The polymerization and thrombin-binding properties of des(B1–42)-fibrin, *J. Biol. Chem.* 265, 18650–18655, 1990.
89. Kaminski, M., Siebenlist, K.R., and Mosesson, M.W., Evidence for thrombin enhancement of fibrin polymerization that is independent of catalytic activity, *J. Lab. Clin. Med.* 117, 718–725, 1991.
90. Mosesson, M.W., Hernandez, J., and Siebenlist, K.R., Evidence that catalytically-inactivated thrombin formed non-covalently linked dimers the bridge between fibrin/fibrinogen fibers and enhance fibrin polymerization, *Biophys. Chem.* 110, 93–100, 2004.
91. Mosesson, M.W., The effect of a thrombin exosite-2 binding DNA aptamer (HD-22) on non-catalytic thrombin enhanced polymerization, *Thromb. Haemost.* 97, 327–328, 2007.
92. Carter, W. J., Cama, E., and Huntington, J.A., Crystal structure of thrombin heparin bound to heparin, *J. Biol. Chem.* 280, 2745–2749, 2005.
93. Desai, B.J., Boothello, R.S., Mehta, A.Y., *et al.*, Interaction of thrombin with sucrose octasulfate, *Biochemistry* 50, 6973–6982, 2011.
94. Crossen, J. and Diamond, S.L., Thermal shift assay to probe melting of thrombin, fibrinogen, fibrn monomer: gly-pro-arg-gly induces a fibrin monomer state in fibrinogen, *Biochim. Biophys. Acta* 1865, 129805, 2021.
95. Bänninger, H., Lämmle, B., and Furlan, M., Binding of α-thrombin to fibrin depends on the quality of the fibrin network, *Biochem. J.* 298, 151–163, 1994.
96. Gurgerell, A., Pasteiner, W., Nürnberger, S., *et al.*, Thrombin as important factor for cutaneous would healing: comparison of fibrin matrices in vitro and in a rat excisional wound healing model, *Wound Repair Regen.* 72, 740–748, 2014.
97. Kelly, M. and Leiderman, K., A mathematical model of bivalent binding suggests physical trapping within fibrin fibers, *Biophys. J.* 117, 1442–1455, 2019.
98. Bloom, A.L., The release of thrombin from fibrin by fibrinolysis, *Br. J. Haematol.* 8, 129–133, 1962.
99. Burnouf, T., Lee, C.-Y., Loo, C.-W., *et al.*, Human blood-derived fibrin releasates: composition and use of cell lines and human primary cells, *Biologicals* 40, 21–30, 2012.

100. Grainger, D.J., Wakefield, L., Bethell, H.N., Farndale, R.W., and Metcalfe, J.C., Release and activation of platelet latent TGF-β in blood clots during dissolution with plasmin, *Nat. Med.* 1, 932–937, 1995.

101. Pechnik, I., Modrazo, J., Mosesson, M.W., *et al.*, Crystal structure of the complex between thrombin and the central "E" region of fibrin, *Proc. Natl. Acad. Sci.* 101, 2718–2723, 2004.

102. Meh, D.A., Siebenlist, K.R., and Mosesson, M.W., Identification and characterization of the thrombin binding site on fibrin, *J. Biol. Chem.* 274, 23121–23125, 1996.

103. Fenton, J.W., II., Olson, T.A., Zabinski, M.P., and Wilner, G.D., Anion-binding exosite of human α-thrombin and fibrin(ogen) recognition, *Biochemistry* 27, 7106–7113, 1988.

104. Fredenburgh, J.C., Stafford, A.F., Leslie, B.A., and Weitz, J.I., Bivalent binding to γAγ' fibrinogen engages both exosites of thrombin and can protect it from inhibition by the antithrombin-heparin complex, *J. Biol. Chem.* 283, 2470–2477, 2008.

105. Lovely, R.S., Moodell, M., and Farrell, D.H., Fibrinogen γ' chain binds thrombin exosite II, *J. Thromb. Haemost.* 1, 124–131, 2003.

106. Haynes, L.M., Orfeo, T., Mann, K.G., Everse, J.T., and Brummel-Ziedins, K.E., Probing the dynamics of clot-bound thrombin and venous shear rate, *Biophys. J.* 112, 1634–1644, 2017.

107. Zhu, S., Chen, J., and Diamond, S.L., Establishing the transient mass balance of thrombosis from tissue factor to thrombin to fibrin under venous flow, *Arterioscler. Thromb. Vasc. Biol.* 38, 1528–1536, 2018.

108. Mosesson, M.W., Update on antithrombin I (fibrin), *Thromb. Haemost.* 98, 105–108, 2007.

109. Chen, J. and Diamond, S.L., Reduced model to predict thrombin and fibrin during thrombosis on collagen/tissue factor under flow conditions, *PLoS One* 15(8), e107266, 2019.

110. Kumar, R., Béguin, S., and Hemker, H.C., The influence of fibrinogen and fibrin on thrombin generation—Evidence for feedback activation of the clotting system by clot bound thrombin, *Thromb. Haemost.* 72, 713–721, 1994.

111. Kremers, R.M.W., Wagenvoord, R.J., and Hemker, H.C., The effect of fibrin(ogen) on thrombin generation and decay, *Thromb. Haemost.* 112, 486–494, 2014.

112. Meh, D.A., Siebenlist, K.R., Brennen, S.O., Holyst, T., and Mossenson, M.W., The amino acid sequence in fibrin responsible for high affinity thrombin binding, *Thromb. Haemost.* 85, 470–474, 2001.

113. De Cristofero, R. and De Flippes, V., Interaction of the 1 68–182 region of glycoprotein Ibalpha with the heparin binding site of thrombin inhibits the enzyme activation of factor VIII, *Biochem. J.* 373, 593–601, 2003.

114. Michnick, D.A., Pittman, D.R., Wise, R.J., and Kaufman, R.J., Identification of individual tyrosine sulfation sites within factor VIIIi required for optimal activity and efficient cleavage, *J. Biol. Chem.* 265, 20095–20102, 1994.

115. Bryk, A.H., Natorska, J., Zabczyk, M., *et al.*, Plasma fibrin clot proteomics in patients with acute pulmonary embolism: association with clot properties, *J. Proteomics* 229, 103946, 2020.

116. Vu, T.T., Stafford, A.R., Leslie, B.A., *et al.*, Histidine-rich glycoprotein binds fibrin(ogen) with high affinity and competes with thrombin for binding to the γ' chain, *J. Biol. Chem.* 286, 30314–30323, 2011.

117. Vu, T.T., Fredenburgh, J.C., and Weitz, J.I., Zinc; An important cofactor in hemostasis and thrombosis, *Thromb. Haemost.* 109, 421–430, 2013.

118. Szuldyrzyński, K., Jankowski, M., Potaczek, D.P., and Undas, A., Plasma fibrin clot properties as determinants of bleeding times in human subjects: association with histidine-rich glycoprotein, *Dis. Markers* 2020, 719828, 2020.

119. Lundblad, R.L., Bradshaw, R.A., Gabriel, D., *et al.*, A review of the therapeutic uses of thrombin, *Thromb. Haemost.* 91, 851–860, 2004.

120. Naski, M.C. and Shafer, J.A., α-Thrombin with fibrin clots: inactivation by antithrombin III, *Thromb. Res.* 69, 453–461, 1993.

121. Francis, C.W., Markham, R.E., Jr., Barrow, G.H., *et al.*, Thrombin activity of fibrin thrombi and soluble plasmic derivatives, *J. Lab. Clin. Med.* 102, 220–230, 1983.

122. Goodwin, C.A., Kakker, V.V., and Scully, M.F., Generation of forms fragment E with differing thrombin binding properties during digestion of fibrinogen by plasmin, *Biochem. J.* 281, 613–618, 1992.

123. Weitz, J.I., Leslie, B., and Hudoba, M., Thrombin binding to soluble fibrin degradation products where it is protected from inactivation by heparin-antithrombin but susceptible to inactivation by antithrombin-independent inhibitors, *Circulation* 97, 544–553, 1998.

124. Weitz, J.I., Leslie, B., and Hudoba, M., Thrombin bound to soluble fibrin degradation products where is it protected from inhibition by heparin-antithrombin but is susceptible to inactivation by antithrombin-independent inhibitors, *Circulation* 97, 544–552, 1998.

125. Kinjoh, K., Nakamura, M., Sunagawa, M., and Kosugi, T., Isolation of bound thrombin consisting of thrombin and *N*-terminal fragment from clot lysate, *Hematologica* 32, 457–465, 2002.

126. Komatsu, Y. and Hayashi, H., Most clot-bound thrombin activity is inhibited by plasma antithrombin during clot aging but a very small fraction survives, *Biol. Pharm. Bull.* 23, 502–505, 2000.

127. Meddahi, S. and Samama, M.M., Is the inhibition of both clot-associated thrombin and factor Xa more clinically relevant than either one alone?, *Blood Clog. Fibrinol.* 20, 207–214, 2009.

128. Scagliane, F. New oral anticoagulants: comparative pharmacology with vitamin K antagonists, *Clin. Pharmacokinet.* 52, 69–82, 2013.

129. Hankey, G.J. and Eikelboorm, J.W., Dabigatran etexilatez: a new oral thrombin inhibitor, *Circulation* 123, 1436–1450, 2008.

130. Stangier, J., Clinical pharmacokinetics and pharmacodynamics of the oral direct thrombin inhibitor, dabigatran etexilate, *Clin. Pharmacokinet.* 47, 285–295, 2008.

131. Satra, S., Papageorgiou, L., Larsen, A.K., *et al.*, Comparison of antithrombin-dependent and direct inhibitors of factors Xa or thrombin on the kinetics and qualitative characteristics of blood clots, *Res. Practice Thromb. Haemost.* 2, 696–707, 2018.

132. Narayanan, V. and Bharagava, A., Efficacy and safety of dabigatran in prevention of venous thromboembolism in patients undergoing major orthopedic surgeries: a review, *Int. J. Res. Orthoped.* 5, 196–200, 2019.

133. Deinum, J., Mattsson, C., Inghardt, T., and Elg, M., Biochemical and pharmacological effects of the direct thrombin inhibitor, AR-H067637, *Thromb. Haemost.* 101, 1051–1059, 2009.

134. Lip, G.Y.H., Rasmussen, L.H., Olsson, S.B., *et al.*, Exposure-response for biomarkers of anticoagulant effect by the direct thrombin inhibitor, AZD)837, in patients with atrial fibrillation, *Brit. J. Haematol.* 80, 1362–1373, 2015.

135. Berry, L.R., Becker, D.L., and Chan, A.K.C., Inhibition of fibrin-bound thrombin by a covalent antithrombin-heparin complex, *J. Biochem.* 132, 167–176, 2002.

136. Smith, L.J., Mewhart-Buist, T.A., Berry, L.R., and Chan, A.K.C., Antithrombin-heparin complex increases the anticoagulant activity of a fibrin clot, *Res. Lett. Biochem.* 2008, 639829, 2008.

137. Biggs, R. and Denson, K.W.E., Natural and pathological inhibitors of blood coagulation, in *Blood Coagulation, Haemostasis and Thrombosis*, 2nd edn, ed. R. Biggs, Chapter 7, pp. 133–158, Blackwell Scientific, Oxford, UK, 1972.

138. Hogg, P.J. and Jackson, C.M., Formation of a ternary complex between thrombin, fibrin monomer and heparin influences the action of thrombin of its substrates, *J. Biol. Chem.* 265, 248–255, 1990.

139. Chen, H.H., Leslie, B.A., Stafford, A., *et al.*, By increasing the affinity of heparin for fibrin, Zn^{2+} promotes the formation of a ternary heparin-thrombin-fibrin complex that protects thrombin from inactivation by antithrombin, *Biochemistry* 51, 7965–7973, 2012.

140. Okwusidi, J., Anvari, N., and Ofosu, F.A., Modulation of intrinsic prothrombin activation by fibrinogen and fibrin 1, *J. Lab. Clin. Med.* 121, 64–70, 1993.

141. Fay, W.P. and Parker, A.C., Effect of radiographic contrast agents in thrombin formation and activity, *Thromb. Haemost.* 80, 266–272, 1998.

142. Janus, J.J., Lewis, S.D., Lorand, L., and Shafer, J.A., Promotion of thrombin-catalyzed activation of factor XIII by fibrinogen, *Biochemistry* 22, 6269–6272, 1983.

143. Naski, M.C., Lorand, L., and Shafer, J.A., Characterization of the kinetic pathway for fibrin promotion of α-thrombin-catalyzed activation of plasma factor XIII, *Biochemistry* 30, 934–941, 1991.

144. Maurer, M.C., Trumbo, T.A., Isetti, G., and Turner, B.T.J., Probing interaction between coagulants thrombin, factor XIII, and fibrin(ogen), *Archs. Biochem. Biophys.* 445, 36–45, 2006.

145. Hornyak, T.J. and Shafer, J.A., Interaction of factor XIII with fibrin as substrate and cofactor, *Biochemistry* 31, 423–429, 1992.

146. Bagoly, Z. and Muszbek, L., Factor XIII: what does it look like?, *J. Thromb. Haemost.* 17, 714–716, 2019.

147. Kumar, R., Béguin, S., and Hemker, H.C., The effect of fibrin clots and clot-bound thrombin on the development of platelet procoagulant activity, *Thromb. Haemost.* 79, 962–966, 1994.

148. Phillips, J.E., Lord, S.T., and Gilbert, G.E., Fibrin stimulates platelets to increase factor VIII binding site expression, *J. Thromb. Haemost.* 2, 1806–1815m, 2004.

149. Weeterings, C., Adelmeijer, J., Myles, T., de Groot, P.G., and Lisman, T., Glycooprotein Ibα-mediated platelet adhesion and aggregation to immobilized thrombin under conditions of flow, *Arteriosler. Thromb. Vasc. Biol.* 26, 670–675, 2006.

150. Geer, D.J., Schwartz, D.D., and Andreadis, S.T., Fibrin promotes migration in a three-dimensional *in vitro* model of wound healing, *Tissue Engineer.* 8, 787–798, 2002.

151. Laurens, N., Kodwijk, P., andDe Moat, M.P.M., Fibrin structure and wound healing, *J. Thromb. Haemost.* 4, 932–939, 2006.

152. Sese, N., Cole, M., and Tawik, B., Proliferation of human keratinocytes and fibroblasts in three-dimensional constructs, *Tissue Eng. Pt. A.* 129, 1022–1027, 2000.

153. Sporn, L.A., Bunce, L.A., and Francis, C.W., Cell proliferation on fibrin: modulation by fibrinopeptide cleavage, *Blood* 86, 1802–1810, 1995.

154. Zamarron, C., Ginsburg, M.W., and Plow, E.F., Monoclonal antibodies specific for a conformationally altered site on fibrinogen, *Thromb. Haemost.* 64, 41–46, 1990.

155. Guadiz, G., Sporn, L.A., and Simpson-Haidaris, P.J., Thrombin cleavage independent deposit of fibrinogen in the extracellular matrices, *Blood* 90, 2644–2653, 1997.

156. Rybarczyk, B.J., Lawrence, S.O., and Simpson-Haidaris, P.J., Matrix-fibrinogen enhanced wound closure by increasing both cell proliferation and migration, *Blood* 102, 4035–4043, 2003.

157. Geer, C.B., Rus, I.A., Lord, S.T., and Schoenfisch, M.H., Surface-dependent fibrinopeptide A accessibility to thrombin, *Acta Biomater.* 3, 663–668, 2007.

158. Dyr, J.E., Tichý, I., Juroušková, M., *et al.*, Molecular arrangement of adsorbed fibrinogen molecules characterized by specific monoclonal antibodies and a surface plasmon resonance sensor, *Sens. Actuators* B 51, 268–272, 1998.

159. Bar-Shavit, R., Eldor, A., and Vlodavsky, I., Binding of thrombin to subendothelial extracellular matrix. Protection and expression of functional properties, *J. Clin. Invest.* 84, 1096–1104, 1989.

160. Bar-Shavit, R., Benezra, M., Eldor, A., *et al.*, Thrombin immobilized to extracellular matrix is a potent mitogen for vascular smooth muscle cells: nonezymatic mode of action, *Cell Regul.* 1, 453–463, 1990.

161. Jandrot-Perrus, M., Clemetson, K.J., Huisse, M.G., and Guillin, M.-C., Thrombin interaction with glycoprotein Ib: effect of glycocalicin on thrombin specificity, *Blood* 80, 2781–2786, 1992.

162. de Ridder, G.G., Lundblad, R.L., and Pizzo, S.V., Actions of thrombin in the interstitium, *J. Thromb. Haemost.*, 14, 40–47, 2016.

163. Gandoss, E., Lunven, C., and Berry, C.N., Role of clot-associated (-derived) thrombin in cell proliferation induced by fibrin clots *in vitro*, *Brit. J. Pharmacol.* 129, 1021–1027, 2000.

164. Macasev, D., Di Orio, J.P., Gugercell, A., *et al.*, Cell compatibility of fibrin sealant: in vitro study with cells involved in soft tissue repair, *J. Biomater. Appl.* 26, 129–140, 2011.

165. Gugerell, A., Schlossleitnen, K., Wolbank, S., *et al.*, High thrombin cncentrations in fibrin sealants induce apoptosis in human keratinocytes, *J. Biomed. Mater. Res. Part A* 100A, 1239–1247, 2012.

166. Smadia, D.M., Basire, A., Amelot, A., *et al.*, Thrombin bound to a fibrin clot confers angiogenic and hemostatic properties on endothelial progenitor cells, *J. Cell. Mol. Med.* 12, 975–986, 2008.

167. d'Audigier, A., Cochain, C., Rossi, E., *et al.*, Thrombin receptor PAR-1 activation on endothelial cells enhanced chemotaxis-associated genes expression and leukocyte recruitment by a COX-2 dependent mechanism, *Angiogenesis* 18, 347–359, 2015.

168. Oh, J.H., Kim, H.J., Kim, T.J., *et al.*, The effects of the modulation of the fibronectin-binding capacity of fibrin thrombin on osteoblast differentiation, *Biomaterials* 33, 4089–4099, 2012.

169. Gandossi, E., Lunven, C., and Berry, C.N., Role of clot-associated (-derived) thrombin in cell proliferation induced by fibrin clots *in vitro*, *Brit. J. Pharmacol.* 129, 1021–1027, 2000.

170. Nomura, H., Naito, M., Igudi, A., Thrompson, W.D., and Smith, E.B., Fibrin gel induces the migration of smooth muscle cells from rabbit aortic explants, *Thromb. Haemost.* 82, 1347–1352, 1999.

171. Kodama, M., Naito, M., Nomura, H., *et al.*, Role of D and E domains in the migration of vascular smooth muscle cells into fibrin gels, *Life Sci.* 71, 1139–1148, 2002.

7 Thrombin and Platelets

Thrombin "activates" platelets, causing adherence, aggregation, the release of bioactive materials,[1-5] and the formation of platelet procoagulant activity, known earlier as platelet factor 3.[6-10] Wright and Minot[1] published one of the very early studies on platelet shape change/aggregation but failed to identify a role for thrombin in these processes. It is my sense, at the risk of being parochial, that Shermer and coworkers[2] were to the first to describe a discrete role for thrombin in the activation of platelets, observing that there was little consensus by early studies on the role of thrombin. This study showed that bovine, canine, or human thrombin could aggregate canine platelets or human platelets in a reaction that required calcium ions. Davey and Luscher[4] made several seminal observations of the action of thrombin on platelets. First was the observation that the action of thrombin on fibrinogen did not play a role on the action of thrombin on platelets. Second, thrombin-like factors, including enzymes from snake venoms, could cause the release of ADP from platelets. Several snake venom enzymes clotting fibrinogen in a reaction inhibited by DFP also aggregated platelets in a reaction inhibited by DFP. Third, the thrombin-stimulated release of ADP from platelets was inhibited by DFP. Finally, trypsin and papain, which had specificity similar to thrombin, also stimulated ADP release from platelets, whereas chymotrypsin, with a specificity different from trypsin, did not stimulate ADP release from platelets.

The binding of thrombin to blood platelets has been known for some time,[11-24] and it was mostly assumed that the specific binding of thrombin to a receptor on the platelet surface was necessary for the activation of the platelet. This early assumption stemmed in large part from the early work of Detwiler and Feinman[11] in the Department of Biochemistry at SUNY Downstate Medical Center in Brooklyn, who reported on the kinetics of the release of calcium ions from platelets in the presence of thrombin. This was a seminal paper that proposed a process for platelet activation, where thrombin bound to the platelet, forming a thrombin-platelet complex, followed by transformation of the complex resulting in protein activation. These investigators proposed that thrombin did not turn over; in other words, thrombin bound to a platelet did not dissociate to react with another platelet. They also reported that hirudin blocked the stimulation of platelets by thrombin but had no effect if added seconds after the addition of thrombin. Another early study that provided additional support for the assumption of the necessity of thrombin binding to platelets prior to activation came from Washington University in St. Louis a year later. Tollefsen and coworkers[12] reported data that supported the presence of two classes of binding sites on the platelet surface: one class consisting of 50,000 sites per platelet with a $K_D = 30$ nM and a second class of 500 binding sites with a $K_D = 0.2$ nM. While Davey and Luscher[4] had shown that DFP inhibited the action of thrombin on platelets, Tollefson and coworkers[12] showed that DIP-thrombin bound to platelets with the same characteristics of native α-thrombin, but, as observed by Davey Luscher,[4] did not result in platelet

DOI: 10.1201/b22204-7

aggregation. DIP-thrombin did enhance the release of ^{14}C-serotonin at a low thrombin concentration. They suggested that the two classes of binding resulted from a single binding site with different affinities. Ganguly[13] observed the binding of thrombin to platelets was not affected by aspirin, DFP, or PMSF. The observations with DFP and PMSF were consistent with other studies showing that the active site of thrombin is not required for binding to platelets. Aspirin inhibits the cyclooxygenase enzyme important in the synthesis of prostaglandins, showing this process is not necessary for the binding of thrombin to platelets. Ganguly also showed that binding of thrombin to platelets was associated with formation of a complex with a molecular weight in excess of 200 kDa, as determined by native gel electrophoresis or gel filtration. The samples for these studies were obtained from platelets solubilized with Triton X-100. Triton X-100 is a nonionic detergent, one of several that can be used for the isolation of membrane protein complexes.[14] A crosslinking agent was not used in these studies. This study suggests that the putative thrombin-membrane complex is behaving as an integral membrane protein complex. Saturation binding was observed by Ganguly in this study[13] when the binding of thrombin to platelets was studied in platelet-rich plasma. It was stated that the thrombin concentration was insufficient to cause the clotting of plasma under these conditions. As noted in a later study,[15] saturation binding was observed at lower thrombin concentrations. It is possible that the proteinase inhibitors in plasma, as noted later in a study from another laboratory, substantially reduced the effective thrombin concentration in the plasma experiments. The later study from the Ganguly laboratory[15] reproduced the lack of saturation binding at high thrombin concentrations but showed saturation binding at lower thrombin concentrations. These studies at lower thrombin concentrations suggested high affinity ($K_D = 0.02$ U/mL) with 500 sites per platelet consistent with observations from other laboratories. The Detwiler laboratory followed with a study[16] showing that trypsin and papain, as observed earlier by Davey and Luscher,[4] could also activate platelets but at higher concentrations, and the process of platelet activation also took a longer period of time. It was of interest that the investigators were brave enough to apply Michaelis-Menton kinetics to the reaction of thrombin with platelets using the release of adenosine triphosphate (ATP) as a measure of reaction velocity. The k_{cat} for the reaction of bovine trypsin (1.1 sec^{-1}) was similar to that for bovine thrombin (0.5 sec^{-1}). However, there was a large difference in K_M for the reaction: 7.6 µM for bovine trypsin compared to 0.04 µM for bovine thrombin. While we generally think of substrate affinity when consider K_M values, a more dramatic change in K_M is seen when phospholipid is included with factor IXa in the assembly of the tenase complex.[17] The concept here is concentration of the reactants—in this case, thrombin (enzyme) and a platelet receptor (substrate) from bulk solution, increasing the observed affinity of the substrate for the enzyme. Martin and coworkers[17] did not observe the stimulation of platelet activation by DIP-thrombin at low native thrombin concentrations reported by Tollefsen and coworkers.[12] In a subsequent study, Martin and coworkers[18] showed that maximal calcium ion secretion was obtained with the occupancy of 500–600 sites per platelet, a number consistent with the earlier observations.[12,13] In another study published in 1976,[19] Tollefsen and Majerus studied the binding of native thrombin and DIP-thrombin to platelets. These studies demonstrated high-affinity binding and low-affinity binding to human blood platelets. They used glutaraldehyde[20] to crosslink thrombin to the binding site on the platelet surface. SDS-PAGE analysis showed the presence of a single 200-kDa product at all concentrations of thrombin, suggesting the presence of a single binding site for thrombin on the platelet surface. These investigators did not observe the formation of this product in the absence of glutaraldehyde. However, the electrophoretic analysis in this study was performed in the presence of SDS (sodium dodecyl sulfate) (SDS-PAGE), which would dissociate most protein-protein complexes. Mohammed and coworkers,[21] in the Department of Pathology at the University of North Carolina at Chapel Hill, studied the action of the multiple forms of human thrombin, α-thrombin, β-thrombin, and γ-thrombin. There was a direct correlation between fibrinogen clotting activity and the ability to activate platelets. Static studies on the binding of the various forms of thrombin to platelets did not show significant differences. It was observed that TLCK-thrombin bound to platelets to a greater extent than native thrombin. This study was complicated, as with the Ganguly study,[13] by failure to observe saturation binding. It was also noted that thrombin was not removed by extensive washing of the platelets. Work in my laboratory in Chapel Hill consistent with high affinity andl low affinity binding sites for thrombin on plateltets.[22,23] Ganguly and Gould[24] isolated a glycoprotein from solubilized platelets that bound thrombin. This protein also inhibited the clotting of fibrinogen by thrombin. Characterization of the protein suggested that the isolated protein was glycoprotein I with a molecular weight of 150 kDa.[25,26] At that time three platelet surface glycoproteins had been identified and numbered according to molecular weight.[27] The inhibition of fibrinogen clotting activity may represent the antithrombin activity of platelets described by Tullis and coworkers.[28] Thrombin bound to platelets does retain catalytic activity toward low-molecular-weight substrates.[29] We biochemists are sometime deluded into thinking that Tris or HEPES buffers are physiological. Occasionally we will add albumin or use a physiological solution such as Hanks. In a study which, to the best of my knowledge, has only been cited once, that by Gil White and me in our review of some years ago,[30] Rotoli and Majerus[31] showed that citrated, defibrinated plasma inhibited the binding of DIP-thrombin to platelets; dialyzed plasma was somewhat less effective. These investigators showed that while antithrombin or α_2-macroglobulin inhibited the binding of thrombin to platelets, there were other plasma proteins that also inhibited binding. For example, I did not find a study that determined if von Willebrand factor (vWF) affected the binding of thrombin to platelets. It has been reported that thrombin does stimulate the binding of vWF to human platelets.[32] Ganguly,[13] in a study discussed earlier, was able to show saturation binding of thrombin to platelets in the presence, but not absence, of plasma.

Thrombin has been reported to bind to other platelet proteins which are not associated with platelet activation. Bennett and Glenn[33] reported that a portion of thrombin that initially bound to platelets was found to form a stable complex with a 40-kDa protein after platelet activation (shape change). The amount of thrombin bound in this complex increased after shape change was induced by concanavalin A. Yeo and Detwiler,[34] in 1985, reported that there were four different forms of thrombin-bound to platelets: an early equilibrating state, a second form which dissociated slowly, a stable form which did not dissociate, and a large SDS-stable complex. The formation of the stable complexes formed at 37°C but only slightly at 22°C or with inhibited thrombin. They suggested that the rapidly equilibrating form was associated with platelet activation and the other complexes formed only after platelet activation. In 1986, McGowan and Detwiler[35] suggested that there were two types of receptors for thrombin that could modulate platelet function. One receptor is sensitive to chymotrypsin and not responsive to γ-thrombin. This site was also observed to be coupled with the inhibition of adenyl cyclase and the activation of prostaglandins. Secretion of ATP and phosphorylation of a 40-kDa protein and 20-kDa protein were delayed, as was aggregation in thrombin action on chymotrypsin-treated platelets or with γ-thrombin. McGowan and Detwiler[35] proposed that there were two receptors to account for their results, suggesting a receptor separate from PAR receptors to account for the effect of thrombin on adenyl cyclase and phospholipase A. Detwiler and coworkers[36] subsequently reviewed the various complexes formed between thrombin and platelets. There is a product of 77 kDa, which likely represents a complex between thrombin and protease nexin-1. There is a second complex of >450 kDa, which is suggested to be the product composed of thrombin-protease nexin-1, with thrombospondin arising from disulfide exchange. Another study[37] reported on the reaction of thrombin with cytoplasmic antiprotease (CAP) forming a complex of 58 kDa and 70 kDa. The 58-kDa complex was converted to a 70-kDa complex upon heating. It was suggested the 58-kDa product was a partially unfolded SDS-protein complex. Thrombin has also been shown to form a complex with protease nexin-1 expressed during platelet activation.[38,39] Gronke and coworkers[38] suggested that the platelet protease nexin-1 had a higher affinity for thrombin than other platelet membrane receptors. The SDS-stable 77-kDa thrombin-platelet nexin-1 complex accounts for most of the thrombin bound after platelet activation. A subsequent paper from this laboratory[39] showed that platelet protease nexin-1 was released from platelets after activation. Hagen and coworkers,[40] using immunoelectrophoresis, showed that thrombin formed a complex that was stable in Triton X-100; the complex was not formed with PMSF. Charles Knupp, who trained with Gil White and Harold Roberts in the Department of Medicine and then moved on the East Carolina School of Medicine in Greenville, has also worked on the reaction of thrombin with protease nexin-1.[41] He showed that the thrombin-protease nexin-1 complex is only formed after platelet activation in either normal or chymotrypsin-treated platelets. Hirudin or DAPA could block the formation of the thrombin-protease nexin complex if added within 1 minute after the addition of thrombin. Appearance of the thrombin-protease nexin-1 complex occurs in the supernatant fraction after formation of the membrane protease nexin-1-thrombin complex. The various data on the interaction of thrombin and protease nexin-1 do not appear to be directly associated with the process of platelet activation.

These early studies suggested that a platelet surface component such as glycoprotein Ib (glycocalicin) is responsible for the binding of thrombin prior to activation. This would be consistent with the rapid response required for the action of thrombin on platelets during the hemostatic response. Gil White and I summarized some of these studies in a review published in 2005.[30] Sixteen years later it is my sense that the proposal of a single binding site for thrombin on platelets, as proposed by Tollefsen and Majerus in 1976,[19] seems even more reasonable today than it did 45 years ago. Tollefesen and Majarus[19] made several important points in this paper. First, it is possible to show high- and low-affinity binding to platelets. Second, equilibrium binding studies may be misleading because of slow dissociation of thrombin from platelets. Third, DIP-thrombin binds the same as native α-thrombin. Fourth, thrombin binds to the same receptor with either high-affinity binding or low-affinity binding. These investigators and others[42–47] also used chemical crosslinking to identify the site of thrombin binding to platelets (Table 7.1). The reagents used would establish crosslinks between lysine residues on thrombin and lysine residues on a putative receptor. These studies yielded a product of approximately 200 kDa, consistent with a covalent complex between thrombin (Mr = 35,000[48]) and glycoprotein Ibα (MW = 150,000[49,50]). Jung and Moroi[42] also observed a 200-kDa product on crosslinking thrombin to platelets with either N-succinimidyl(4-azidophenyldithio)propionate (SADP; a hydrophobic reagent capable of penetrating a cell membrane) or 3,3′-dithiobis(sulfosuccinimi dyl)propionate (DTSSP; a hydrophilic reagent incapable of entering a cell membrane). More product was obtained with DTSSP than with SADP. A 200-kDa product was obtained with both native and TLCK-thrombin. Jandrot-Perrus and coworkers[43] used bis(sulfosuccinimidyl)sububerate in their studies and obtained derivatives of M_r >400 kDa, 240 kDa, and 79 kDa on SDS electrophoresis on crosslinking thrombin with platelets. The 240-kDa product was also obtained with TLCK-thrombin, but the 78-kDa product was not observed. It is notable that Jandrot-Perrus and coworkers[43] failed to obtain a similar complex with γ-thrombin and platelets. They also reported a complex stable in 0.1% Triton X-100 between labeled thrombin and GpIb on immuno-electrophoresis. Such a complex was also reported by Ganguly.[13] It was further noted that the 78-kDa complex was formed in the absence of crosslinking reagents. Larsen and coworkers[44] observed that the inactivation of thrombin by TLCK eliminated the ability of thrombin to activate platelets, as well as the ability to cause the change in membrane polarization as assessed by the change in fluorescence of a cyanine dye.[51] However, preincubation of TLCK-thrombin with platelets did decrease, but did not eliminate, the change in fluorescence caused by native thrombin. Native thrombin was not able to displace TLCK-thrombin a 10:1 molar ratio. The authors suggested the data supported two binding sites for thrombin on platelets. Larsen and Simons[45] prepared a conjugate of thrombin consisting of a diazonitrophenyl moiety coupled to a carbohydrate side chain. Photolysis of the derivative noncovalently bound to thrombin in the dark yielded products characterized by gel filtration. The was a difference in the elution position of the products obtained with native or TLCK-thrombin. There was also a difference in the products on SDS-PAGE; the product obtained with native thrombin had an M_r = 120 kDa, while the product obtained with TLCK-thrombin had a M_r = 200 kDa. Danishefsky and Detwiler[47] also used a photoaffinity label (ethyl-N-5-azido-2-nitrobenzylaminoacetimidate) bound to ^{125}I-thrombin to identify platelet surface proteins that bind thrombin. These investigators identified four radiolabeled products ranging in size from 125 kDa to 210 kDa (Table 7.1). As with other investigators, the formation of these products was blocked by hirudin. It is likely that at least one of these products was a complex formed between thrombin and GpIbα. They also observed a supernatant complex consistent with a complex of thrombin and thrombospondin. Tollefsen and Majerus[19] had shown that a 200-kDa complex was formed with DIP-thrombin. Simons and coworkers[32] reviewed the binding of thrombin to platelets, including the crosslinking studied. Knupp and White[52] showed the TLCK-thrombin would inhibit the binding of thrombin to human platelets. They measured the binding

TABLE 7.1
Crosslinking Thrombin to Platelets Using Chemical Methods

Reagent and Conditions	Products	Reference
Glutaraldehyde in PBS, pH 6.5 for 30 minutes at 23°C with DIP-thrombin, analysis with SDS-PAGE on 5% acrylamide gels	A 200-kDa product was formed over a range of thrombin concentrations.[a] There was higher-molecular-weight material that did not enter the gel, which increased with higher concentrations of glutaraldehyde.[b]	1
Native gel electrophoresis of solubilized (Triton-X-100) platelet ^{125}I-thrombin after centrifugation[c]	Radioactive material a gel origin suggests a complex with M_r greater than 200 kDa. Similar results with G-200 gel filtration. This apparent complex is stable to a nonionic detergent.	2
^{125}I-TLCK thrombin in pH 7.4 PBS and SADP[c]—photolytic activation	200-kDa product at low thrombin concentration (6 nM). 200-kDa product and 167-kDa product at high thrombin concentration (100 nm). A complex was not formed with SADP and glycocalicin. It was suggested the 200-kDa product is a complex of thrombin and GpIb.	3
^{125}I-TLCK thrombin in pH 7.4 PBS and DTSSP[d] in pH 7.4 PBS	200-kDa product obtained with additional radioactivity (thrombin) present in a high-molecular-weight band which was suggested to be higher molecular cross-linked products of thrombin and platelet proteins. It could be the high-molecular material described by other investigators. As with SADP, the 200-kDa product is suggested to a heterodimer or thrombin and GpIb.	3
[125]I-Human α-thrombin labeled with ANBAI[e] in 60 mM sodium acetate-20 mM EDTA, pH 7.0, containing 100 mM sucrose	Products with M_r 210, 185, 155, and 125 kDa. Formation of the complexes was blocked by hirudin. A 490-kDa product was observed in the supernatant fractions following platelet activation.	4

Source: Richards, F.M. and Knowles, J.R., *J. Mol. Biol.* 37, 231–233, 1968; Migneault, I., Dartiguenave, C., Bertrand., M.J., and Waldrop, K.C., *Biotechniques* 37[5], 790–802, 2004

[a] The results are consistent with a single high-affinity binding site for thrombin on blood platelets

[b] The radioactivity that did not enter the gel may represent the high-molecular-weight (>450 kDa) material observed by other investigators. The increase of material at a higher glutaraldehyde concentration may represent oligomeric material with a molecular weight higher than the 200 kDa found for the likely heterodimer formed under these reaction conditions. The chemistry of glutaraldehyde is complex, and there are higher molecular forms of the reagent observed under different reaction conditions

[c] SADP, N-succinimidyl(4-azidophenyldithio)propionate. SADP is also a hydrophobic (lipophilic) compound and would be able to penetrate the membrane and react with proteins in the membrane, as well as the extracellular space. It was noted that SDAP did crosslink Gp1b and other proteins in experiments performed in the absence of thrombin

[d] DTSSP, 3,3'-dithiobis(sulfosuccinimidyl propionate. DTSSP is a hydrophilic reagent that cannot penetrate the membrane

[e] ANBAI, ethyl-N-5-azido-2-nitrobenzoylaminoacetamide

References
1. Tollefsen, D.M. and Majerus, P.W., Evidence for a single class of thrombin-binding sites on human platelets, *Biochemistry* 15, 2144–2149, 1976.
2. Ganguly, P., Binding of thrombin to human platelets, *Nature* 247, 306–307, 1974.
3. Jung, S.M. and Moroi, M., Crosslinking of platelet glycoprotein Ib by *N*-succinimidyl(4-azidophenyl dithiopropionate and 3,3'-dithiobissulfosuccinimidyl propionate, *Biochim. Biophys. Acta* 761, 152–162, 1983.
4. Danishefsky, K.J. and Detwiler, T.C., Photoaffinity labeling of platelet thrombin-binding proteins, *Biochim. Biophys. Acta* 801, 48–57, 1984.

of ^{125}I-thrombin to platelets that had been treated with neuraminidase, galactose oxidases, and sodium borohydride. The data suggested a complete inhibition of the high-affinity binding of ^{125}I-thrombin by TLCK-thrombin, with a lesser effect on low-affinity binding. They also observed that TLCK-thrombin did not inhibit the hydrolysis of glycoprotein V (GPV) by native thrombin. As noted earlier, Tollesfsen and Majerus[14] showed with glutaraldehyde crosslinking that thrombin binds to the same receptors at low-affinity binding as at high-affinity binding. While this data supported the proposition that GpIbα was the primary receptor for thrombin on the platelet surface, data was lacking to show a relationship between receptor occupancy and platelet activation.

GLYCOPROTEIN 1Bα(GP1Bα) AS THE HIGH-AFFINITY RECEPTOR FOR THROMBIN ON BLOOD PLATELETS AND ITS ROLE IN PLATELET ACTIVATION

Some of the early work was reviewed earlier; what was lacking was a coupling of receptor occupancy to physiological response. Early work had shown that platelets obtained from patients with Bernard-Soulier syndrome were defective in response to either vWF or thrombin.[53–55] Nurden and Caen[56] reported that a protein with a molecular weight of 155 kDa was missing from Bernard -Soulier platelets; this would be consistent with Gp1bα. Cooper and coworkers[57] showed that treatment of platelets with *Serratia marcescens* protease resulted in cleavage of Gp1bα and reduced sensitivity to thrombin activation; loss of response to vWF and ristocetin was more pronounced.

Glycoprotein Gp1bα is a glycoprotein on the platelet surface,[50,58] existing as a complex known as Gp1b-IX-V.[58–64] There was early work showing the association of Gp1b and Gp IX.[59,60] Modderman and coworkers[61] were the first to show the presence of GPV in association with Gp1b and GpIX. Gp1b is composed of two membrane-bound polypeptide chains, Gp1bα and Gp1bβ, which are linked by a disulfide bond.[62] The association of GpIb and GPX is well understood. The role of GPV in the GP *bb-V-IX* complex is less well understood.[63] The Gp1bα chain is a receptor for both vWF and thrombin. While the signaling in response to vWF binding is well understood,[64] the mechanism by which thrombin can initiate a signal in addition via the GpIb-IX-V complex independent of PAR-1 cleavage is poorly understood. There is a soluble form of Gp1bα known as glycocalcicin.[65,66] Glycocalicin is large fragment (Mr 110,000) derived from membrane-bound Gp1bα by limited proteolysis.[66] Glycocalicin is highly glycosylated (more than 50% carbohydrate) and contains the binding site for thrombin.[47] Jamieson and coworkers[65] suggested that there was a single binding site for thrombin on platelets, since glycocalicin seemed to contain all the binding sites for thrombin, consistent with the earlier suggestion of Tollefsen and Majerus.[19]

Tom Detwiler, whose laboratory was the first to study the binding of thrombin to platelets, presented a model for the interaction of thrombin with platelets in 1981.[67] He proposed a model where there was reversible binding of thrombin to a receptor on platelets, which leads to the modification (proteolysis) resulting in activation of the platelets. This is still a reasonable model if one considers the Gp1bα and PAR-1 to be a single receptor. In the same issue of *The Annals of the New York Academy of Sciences*, Berndt and Phillips[68] showed that thrombin did cleave GPV and suggested that this was associated with platelet activation. These investigators detected a large fragment (Mr ≈ 69,000) on 10% SDS-PAGE resulting from the putative cleavage of GPV. It was shown that the fragment derived from the thrombic proteolysis of platelets was identical to a fragment obtained from the proteolysis of purified platelet GPV. They did not detect any material derived from the thrombic proteolysis. It would have been unlikely that they would have detected the fragment derived from the proteolysis of PAR-1; that fragment, 41 amino acids, would have a molecular weight of approximately 4.5 kDa and would be difficult to detect on the SDS-PAGE system used by Berndt and Phillips.[68]

It is fair to say that much of the previous work on the binding of thrombin to a receptor on platelets got knocked into a cocked hat by Shaun Coughlin and colleagues in 1991.[69,70] I don't intend to spend much time on the PAR receptors except to say that the work at UCSF was very clever. The PAR receptor is a classical G-protein coupled receptor (GPCR), except the ligand and receptor are in the same transmembrane protein and the ligand is exposed by proteolysis of the PAR protein. The coupling to the G-protein and subsequent intracellular pathways is beyond the scope of the current work and is covered elsewhere.[71,72] It is my intent to focus on the interaction of thrombin with the Gp1b–IX-V complex. De Candia has an excellent discussion on the role of the PAR proteins.[73]

It is generally accepted that the exosite-2 domain of thrombin binds to a sequence in the amino-terminal segment of Gp1bα.[74,75] Binding of thrombin via the exosite-2 domain permits thrombin full access to macromolecular substrates, such as PAR-1 via binding to exosite-1. Binding of thrombin to Gp1bα is also suggested to enhance the catalytic activity of thrombin. Jandrot-Perrus and coworkers[76] observed that there was a slight enhancement of the rate of inactivation of thrombin by DFP in the presence of glycocalicin. These investigators also observed that glycocalicin inhibits the binding of thrombin to fibrin and thrombomodulin. Several years later, De Marco and coworkers[75] identified a region on GpIbα that interacts with thrombin. Synthetic peptides derived from this region were evaluated for their ability to affect the catalytic activity of thrombin. Peptides comprising residues 265–285 and 271–285 had the greatest effect on calcium mobilization and aggregation. None of the peptides inhibited the hydrolysis of a tripeptide nitroanilide substrate, S-2238. Some stimulation of activity was observed with peptide 216–240 and 235–262. Peptides 216–240 had little effect on the binding of thrombin to platelets, while peptides 235–262 stimulated the binding of thrombin to platelets. These two peptides showed modest inhibition of fibrinogen clotting activity. In other work, Li and coworkers[77] observed that glycocalicin inhibited the release of FpA from fibrinogen. This effect was sensitive to sodium chloride concentration. Mutation in the exosite-2 domain (R89E, K248E) reduced the affinity of thrombin for glycocalicin, while exosite-1 mutations had little effect on the binding of thrombin to glycocalicin. Heparin, but not hirugen, inhibited the binding of thrombin to immobilized GpIb. De Candia and coworkers[78] reported that the binding of thrombin to GpIbα on intact platelets accelerated the rate of hydrolysis of PAR-1. These experiments did not actually measure a stimulation of PAR-1 hydrolysis by thrombin bound to GpIbα, but rather measured the inhibition of PAR-1 hydrolysis by ligands that bound to GpIbα (aptamer H-22; a monoclonal antibody to GpIb). The thrombin-binding sequence in Gp1bα contains sulfated tyrosine residues important for binding thrombin.[79,80] While binding to GpIbα seems to be the most reasonable explanation for the activation of platelets by thrombin, there is considerable work to suggest that the action of thrombin on platelets is more complex.[81]

Jandrot-Perrus and coworkers[82] compared the action of α-thrombin and γ-thrombin on normal platelets and platelets from patients with Bernard-Soulier syndrome; platelets from patients with Bernard-Soulier syndrome lack GpIbα glycoprotein. While the results showed a common mechanism for α-thrombin and γ-thrombin, activation of normal platelets was much faster with α-thrombin than with γ-thrombin. Previous work from this laboratory had shown that γ-thrombin could not be crosslinking to a glycoprotein, presumably GpIbα. Activation of Bernard-Soulier platelets with α-thrombin was slower than normal platelets but similar to that observed with γ-thrombin based on the increase in thromboxane B_2. The results with α-thrombin are consistent with the previous observations of Jamieson and coworkers[55] and with those of Cooper and coworkers[57] with platelets treated with the metalloproteinase from *S. marcescens*. These observations support the proposition that binding to GpIbα is not necessary for cleavage of PAR-1 but is likely physiologically essential. Dormann and coworkers[10] have shown that GpIbα is essential for thrombin-induced expression of platelet factor 3 (platelet procoagulant activity). De Marco and coworkers[83] have also shown the importance of GpIbα in the activation of platelets by thrombin. These investigators used a monoclonal antibody to GpIb that did not block binding of vWF but did block the binding of thrombin. The binding isotherm for thrombin to platelets treated with the antibody was similar to that observed with platelets from a patient with Bernard-Soulier syndrome.

While this is all satisfactory in providing an understanding of the role of GpIbα and the mechanism of thrombin binding to GpIbα, other observations need to be considered. Jandrot-Perrus and coworkers[84] presented evidence suggesting that thrombin could bind to GpIbα via exosite-1. Prior work from this group[76] had shown that glycocalicin could inhibit the binding of thrombin to fibrin or thrombomodulin. Thrombomodulin binds to both exosite-1 and exosite-2, as does fibrin, although high-affinity binding of thrombin occurs at exosite-2 via the γ'-chain. In a more recent work,[84] Jandrot-Perrus and coworkers also showed that in addition to fibrin(ogen) and thrombomodulin, the *C*-terminal fragment of hirudin inhibited the interaction of thrombin with GpIbα.

It would have been nice to write at this point that x-ray crystallography studies would clarify the binding of thrombin to GPIbα. I should that my first contact with x-ray crystallography was with Joe Kraut and Steve Freer at the University of

Washington. This was in the day of many images collated manually and isomorphous replacement of mercury. Joe and Steve left UW after my first year there to go to La Jolla, where they continued to do great work in this, at the time a very difficult area. Most of the first-year grad students were terrified to even think about asking Professor Kraut for an interview, let also consider working with him. It is not that Joe was not a nice guy—he was—but crystallography was intimidating to most of us. At that time (1961), most of us first-year graduate students had not had an undergraduate biochemistry course (Dave Tinker from the University of Toronto was an exception, and he came to the UW to work with Don Hanahan). Most of us had vague ideas about proteins and enzymes. I still reflect on the patience of the faculty, which included Hans Neurath, Ed Krebs, and Ed Fischer, on working with us. So, if my discussion of the following material is less than satisfactory to the more sophisticated, I apologize. Having written all of this, the crystallographic studies on the complex of GpIbα with thrombin have not provided the necessary clarification. Perhaps new imaging technologies such as cryoelectron microscopy[85] will be more useful. There were two crystal structures for the complex of GpIBα with thrombin published in *Science* in 2003[86,87] together with a commentary.[88] Celikel and coworkers[86] reported a structure for the complex of thrombin with GpIbα. This structure shows thrombin bound to GpIbα via exosite-1 with a second molecule of thrombin bound to GpIbα via exosite-1. It is suggested that the second binding site is cryptic and exposed following binding of the first molecule of thrombin. Dumas and coworkers[87] also produced a structure with two molecules of thrombin bound to one molecule of GpIbα, but there was no suggested cooperativity in the binding of the two thrombin molecules. As with Celikel and coworkers,[86] one of the two thrombin molecules is bound via exosite-2 and the second bound by exosite-1. While these two publications differed in the observed structures and interpretations, neither report provided the clarity necessary to resolve the differences reported in the solution chemistry of the interaction of thrombin with GpIbα. However, this comment should not be construed as a criticism of the quality of the work, but rather the difficulty of the problem. Evan Sadler[88] discussed the differences between the two structures but provided no resolution. Vanhoovelbeke and coworkers[89] note that both structures have one thrombin molecule bound to the same region of GpIbα, while there are differences in the binding of the second molecule of thrombin. As reported by other investigators, only about 5% of the GpIbα molecules on the platelet surface bind thrombin, and it is argued that GpIbα is only important at low thrombin concentrations, which are present at the early stage of coagulation.[90] Vanhoovelbeke and coworkers[89] also suggest the potential of noncovalent crosslinking of GpIbα by thrombin, providing a mechanism for signal transduction independent of PAR-1. Adams and Huntington[91] note the importance of sulfated tyrosine residues in the interaction with thrombin, focusing on the importance of the exosite-2 domain in the interaction of thrombin with GpIbα. Zarpellon and coworkers[92] also suggested that exosite-1 was important in the interaction of thrombin with GpIb. Some years later, Kobe and coworkers[93] presented a further discussion of the two structures. This was a useful discussion of technical details that might account for some of the differences between the two structures. These investigators also suggest that the differences in structure might be explained by the plasticity of thrombin.[94] Kobe and coworkers also suggested that the 1:1 complex of GpIbα and thrombin is the only strong interaction. The concentration of protein in the crystallization studies (25 mg/mL; 10 mg/mL) is high compared to physiological conditions or most solution protein chemistry. *In vivo*, thrombin is formed at nanomolar concentrations (5–10 nM) during the initiation of coagulation, but larger concentrations are formed during the propagation phase.[90,95,96] It would seem likely that platelet aggregation and release/exposure of PF3 would occur in the initiation phase and continue during propagation. It would also seem likely that thrombin acts on platelets at a very low concentration. Ruggeri and coworkers[97] also presented a discussion that the two structures could represent two different modes of binding with physiological consequences. In an additional discussion that follows, I do suggest that the role of platelets may expand after the propagation phase of blood coagulation. Lechtenberg and coworkers[74] used a combination of solution chemistry, x-ray crystallography, and NMR to study the interaction of thrombin with GpIbα and compared the interaction with other thrombin binding partners. The binding of a PAR-1 peptide (Ac-Ser$_{16}$-Arg$_{36}$-NH$_2$) to a thrombin exosite-2 mutant (R89E) was somewhat reduced (K_c = 11.86 μM) compared to a wild-type thrombin (K_d = 5.4 μM), while binding of an exosite-1 mutant (R68E) was greatly reduced (K_d > 70 μM). Binding of the exosite-1 mutant to a GpIbα peptide (Ac-Gly$_{268}$-Thr$_{294}$-NH$_2$) was somewhat reduced (K_d = 5.63 μM) compared to wild-type thrombin (K_d = 1.03 μM); binding of the exosite-1 mutant to the PAR-1 peptide was greatly reduced (K_d > 1000 μM). X-ray crystallography of the complex between thrombin and the GpIbα peptide showed exclusive interaction with exosite-2. NMR studies supported interaction of the GpIbα peptide with exosite-2 of thrombin, although interaction with exosite-1 could not be excluded. The bulk of the observations led Lechtenberg and coworkers[74] to suggest that the interaction of exosite-2 and GpIbα was the most important interaction of thrombin with blood platelets. In a more recent study,[98,99] Estevez and coworkers suggested that there was cooperativity between GpIb-IX and PAR-1 in calcium signaling at low thrombin concentrations (3 nM–10 nM). Estevez and coworkers[98] continued the effort to show that there was a receptor/signaling mechanism for thrombin as a ligand independent of PAR receptors. They suggest that there is cooperativity between the PAR/G-protein pathway and GPIb-IX based on the linkage between the GpIb cytoplasmic domain and putative 14–3–3[100,101]/ RAC1/ LIMK1. These studies used an engineered carbohydrate (CHO) cell line that had been transfected to express GPIb-IX or GPIb-IX mutants The level of surface GPIb-IX expression was determined by fluorescence-activated cell sorting analysis (FACS) analysis. Jerry Ware at the University of Arkansas had a comment[102] on the article by Estevez and coworkers. Ware raised concern about the absence of GPV (see later) in their construct, as well as the use of the CHO cell model. Regardless, the work of Estevez and coworkers is a valuable contribution to the identification of signaling pathway alternatives to PAR receptors. In subsequent work, Estevez and Du[99] reviewed the various mechanisms for platelet activation, including the possible synergism between the thrombin occupancy of GPIbα and the proteolysis of a PAR receptor.

GPV AND THROMBIN

An early study[68] suggested that the hydrolysis of GPV might be important in thrombin signaling. Characterization of the GpIb-IX-V complex has raised questions about whether GPV is an integral part of the complex, leading to a paper where the complex was referred to as GPIb-IX.[58] Early work by Berndt and Phillips[68] had shown that hydrolysis of GPV by thrombin could not be directly associated with platelet activation. Later work by Knupp and White[52] presented data showing that hydrolysis of GPV was not associated with the binding of thrombin to GPIbα. Earlier work by Mosher and coworkers[103] had shown that treatment of platelets with thrombin resulted in the loss of a platelet membrane glycoprotein with M_r = 57–68 kDa and the appearance of a glycoprotein with M_r=57–68 kDa in the supernatant fraction. Other agents that cause platelet activation (A23187k ADP, collagen) did not cause the release of the soluble protein seen with the treatment of thrombin. Comparable results were obtained earlier by Phillips and Agin.[104] Several years later, McGowan and coworkers[105] examined the role of thrombin-catalyzed hydrolysis of GPV and platelet activation. They reported that the hydrolysis of GPV was linear with time and thrombin concentration. Platelet activation was complete well before the complete hydrolysis of GPV. The authors said that complete activation occurred with about 1% hydrolysis of GPV. Although this number seems small, Vanhoorelbeke and coworkers[89] observed that only 5% of the GpIbα site binds thrombin, with the high affinity associated with platelet activation. This number is consistent with the number (500/platelet) suggested for thrombin high-affinity binding sites.[12,13,15] The authors observed that their results do not exclude hydrolysis of GPV as important for platelet activation, but that there was a not a clear relationship between GPV hydrolysis and platelet activation. Some years later, Dong and coworkers,[106] in the López laboratory in Houston, at that time showed the GPV was necessary for the high-affinity binding of thrombin to transfected L cells. PPACK-thrombin was also observed to bind to these transfected cells, suggesting that cleavage of GPV is requisite for thrombin binding to GpIbα and platelet activation. These investigators suggest that GPV is important for the conformation of GpIbα essential for high-affinity binding. Subsequently Ramakrishnan and coworkers[107] proposed a model for thrombin activation of platelets, where thrombin cleavage of GPV resulted in a change in the conformation of GpIbα, initiating a pathway for platelet activation separate from the cleavage of PAR-1 and/or PAR-4. In other words, the proteolytic action of thrombin is not required for platelet activation by thrombin. GPV null mouse platelets were aggregated by DIP-thrombin, while such platelets were not activated by wild-type thrombin. Wild-type platelets were not activated by DIP-thrombin unless pretreated with wild-type thrombin in the absence of calcium ions (calcium-free Tyrodes-HEPES). They also showed that S205A thrombin could activate GPV null platelets. DIP-thrombin can slowly reactivate in a reaction enhanced by the presence of a nucleophile (including basic conditions).[108] Likewise, S205A thrombin may retain a low level of activity, as shown by early studies on mutants of subtilisin.[109] This residual activity of S205A thrombin was noted by Estevez and coworkers[98] and could be abolished by reaction with PPACK. While the S205A thrombin mutant might have some activity, Wu and coworkers[110] reported that it did not aggregate human platelets at 3–5 times the concentrations of recombinant or plasma-derived thrombin. At the same time as the study by Ramakrishan and coworkers,[107] some 5–10 miles away at UCSF, Sambrano and coworkers[111] showed that mouse platelets that did not express PAR-4 (mouse platelets don't express PAR-1; PAR-4 is the activation pathway in mice) could not be activated (shape change, ATP secretion, aggregation) under conditions where wild-type mouse platelets can undergo full activation with thrombin. In other words, PAR-4 null mouse platelets could not be activated by thrombin under conditions where there was full activation of wild-type mouse platelets. Recent work has suggested that the action of thrombin on PAR-4 may be more important than the action on PAR-1. Wu and coworkers[112] showed that two exosite-I inhibitors, hirugen and aptamer HD1, enhanced the ability of a PAR-4 antagonist [YD-3, 1-benzyl-3-(ethoxycarbonylphenyl)-imidazole] but did influence PAR-1 inhibitors (PPACK, hirudin). There has been additional work supporting the importance of thrombin activation of PAR-4 in platelet function.[113,114] Other work from Lindahl's laboratory at Linköping[115] demonstrated the importance of exosite-2 in the thrombin action on PAR-4. These investigators showed that aptamer H22, which binds to exosite-2, blocked platelet activation by thrombin. They also showed that depletion of GpIbα either by treatment with a snake venom protease or blockage with a specific monoclonal antibody (SZ2) had a small, if any, effect, on the activation of platelets by either α-thrombin or γ-thrombin under conditions where the HD22 aptamer essentially blocks activation. PAR-4 receptors have received additional recent attention.[116]

The road to understanding the action of thrombin on platelets has been convoluted, and there are still questions to be answered. I have listed my list of significant observations on the action of thrombin on platelets in Table 7.2. I close this brief

---

TABLE 7.2
Significant Observations on the Interaction of Thrombin with Platelets

Observation	Reference
Thrombin causes a change in platelets in a calcium-dependent reaction independent of fibrinogen.	1
Proteolysis is required for platelet activation.	2
Physical binding to thrombin is required for platelet activation.	3
DIP-thrombin does not activate platelets but does bind to platelets with the same characteristics of native α-thrombin. The active site of thrombin is not required for binding to platelets, but the active site is required for platelet activation.	4
There might be two binding sites for thrombin on platelets: a high-affinity site (500/platelet, KD = 0.21 nM) and low-affinity site (50,000/platelet, KD = 30 nM).	4
Chemical crosslinking shows thrombin bound to a discrete platelet surface glycoprotein with Mr ≈160 kDa.	5
Chemical crosslinking at low and high thrombin concentrations shows a single product suggesting a single binding site for thrombin on platelets.	5
Platelet protease nexin as a secreted protein reacts with thrombin to form a 77-kDa SDS-stable product, which can subsequently react with thrombospondin via disulfide exchange.	6–9

(Continued)

TABLE 7.2
(Continued)

Observation	Reference
Glycoprotein Ib is missing from platelets obtained from patients with Bernard-Soulier syndrome, which also showed a delayed response to thrombin.	10,11
A protease from *Serratia marcescens* hydrolysis GpIb platelet surface membrane protein resulted in a delayed response to thrombin, while the response to vWF was totally absent.	12
Structural characterization of GpIb complex.	13,14
Glycocalicin as product of the proteolysis of GpIb.	15,16
Thrombin cleavage of platelet GPV.	17
PAR receptors (protease-activated receptors).	18,19
Glycoprotein Ibα modulates thrombin function.	20–22
Crystal structures of GpIbα-thrombin complex.	23,24
Evidence supporting essential interaction of thrombin exosite-2 in the interaction with glycoprotein Ibα on the platelet surface.	25

References
1. Shermer, R.W., Mason, R.G., Wagner, R.H., and Brinkhous, K.M., Studies on thrombin-induced platelet agglutination, *J. Expt. Med.* 114, 905–920, 1961.
2. Davey, M.G. and Luscher, E.F., Actions of thrombin and other proteolytic enzymes on blood platelets, *Nature* 216, 857–858, 1967.
3. Detwiler, T.C. and Feinman, R.D., Kinetics of the thrombin-induced release of calcium (II), *Biochemistry* 12, 282–289, 1973.
4. Tollefsen, D.M., Feagler, J.R., and Majerus, P.W., The binding of thrombin to the surface of human platelets, *J. Biol. Chem.* 249, 2645–2651, 1974.
5. Tollefsen, D.M. and Majerus, P.W., Evidence for a single class of thrombin-binding sites on human platelets, *Biochemistry* 15, 2144–2149, 1976.
6. Yeo, K.T., and Detwiler, T.C., Analysis of the fate of platelet-bound thrombni, *Arch. Biochem. Biophys.* 236, 399–410, 1985
7. Gronke, R.S., Bergman, B.I., and Baker, J.B., Thrombin interaction with platelets. Influence of a platelet protease nexin, *J. Biol. Chem.* 262, 3030–3036, 1987.
8. Gronke, R.S., Knauer, D.J., Veeraghearnan, S., and Baker, J.B., A form of protease nexin is expressed on the platelet surface during platelet activation, *Blood* 73, 472–478, 1989.
9. Detwiler, T.C., Chang, A.C., Speziale, R.R., Complexes of thrombin with proteins secreted by activated platelets, *Sem. Thromb. Hemost.* 18, 60–66, 1992.
10. Nurden, A.T. and Caen, J.P., Specific roles for platelet surface glycoproteins in platelet function, *Nature* 258, 720–722, 1975.
11. Jamieson, G.A. and Okamura, T., Reduced thrombin binding and aggregation in Bernard-Soulier platelets, *J. Clin. Invest.* 61, 861–864, 1978.
12. Cooper, H.A., Bennett, W.P., White, G.C., II, and Wagner, R.H., Hydrolysis of human platelet membrane glycoproteins with a *Serratia marcescens* metalloprotease: Effect on response to thrombin and von Willebrand factor, *Proc. Natl. Acad. Sci. USA* 79, 1433–1437, 1982.
13. Berndt, M.C., Gregory, C., Kobali, A., *et al.*, Purification and preliminary characterization of the glycoprotein Ib complex in the human platelet membrane, *Eur. J. Biochem.* 151, 637–649, 1985.
14. Wicki, A.N. and Clemetson, K.J., The glycoprotein Ig complex of human blood platelets, *Eur. J. Biochem.* 163, 43–50, 1987.
15. Jamieson, G.A., Okumura, T., and Hasitz, M., Structure and function of glycocalicin, *Thromb. Haemost.* 42, 1673–1678, 1979.

16. Loscalzo, J. and Handin, R.I., Platelet glycocalicin, *Methods Enzymol.* 215, 289–294, 1992.
17. Berndt, M.C. and Phillips, D.R., Interaction of thrombin with platelets: purification of the thrombin substrate, *Ann. N. Y. Acad. Sci.* 370, 87–95, 1981.
18. Vu, T.-K.H., Hung, D.T., Wheaton, V.I., and Couglin, S.R., Molecular cloning of a functional thrombin receptor reveals a novel proteolytic mechanism of receptor activation, *Cell* 64, 1057–1068, 1991.
19. Coughlin, S.R., How the protease talks to cells, *Proc. Natl. Acad. Sci. USA* 96, 11023–11027, 1999.
20. Jandrot-Perrus, M., Clemetson, K.J., Huisie, M.-G., and Guillin, M.-C., Thrombin interaction with platelet glycoprotein 1b: Effect of Glycocalicin on thrombin specificity, *Blood* 80, 2781–2786, 1992.
21. De Marco, L., Mazzucato, M., Masotti, A., and Ruggeri, Z.M., Localization and characterization of an α-thrombin binding site on glycoprotein GpIbα, *J. Biol. Chem.* 269, 6475–6484, 1994.
22. De Candia, E., Hall, S.W., Rutella, S., *et al.*, Binding of thrombin to glycoprotein Ib accelerates the hydrolysis of Par-1 on intact platelets, *J. Biol. Chem.* 276, 46924698, 2001.
23. Celikel, R., McClintock, R.A., Roberts, J.R., *et al.*, Modulation of α-thrombin function by distinct interactions with platelet glycoprotein Ibα, *Science* 301, 218–221, 2003.
24. Dumas, J.J., Kumar, R., Seehra, J., Somers, W.S., and Mosyak, L., Crystal structure of eh GpIbα complex essential for platelet aggregation, *Science* 301, 222–226, 2003.
25. Lechtenberg, B.C., Freund, S.M.V., and Huntington, J.A., GpIbα interacts exclusively with exosite II in thrombin, *J. Mol. Biol.* 426, 881–893, 2014.

discussion of the action of thrombin on platelets but direct the reader again to the excellent review by Erica De Candia.[73] There are also reviews by Nieman,[117] as well as by Wu and coworkers.[118] Nieman[117] focuses on the PAR receptors, while Wu and coworkers[118] discuss the effects of a number of proteases on platelet function.

REFERENCES

1. Wright, J.H. and Minot, J.R., The viscous metamorphosis of the blood platelet, *J. Expt. Med.* 26, 395–409, 1917.
2. Sherman, R.W., Mason, R.G., Wagner, R.H., and Brinkhous, K.M., Studies on thrombin-induced platelet agglutination, *J. Expt. Med.* 114, 905–920, 1961.
3. O'Brien, J.R., Platelet aggregation. Part 1. Some effects of adenosine phosphates, thrombin and cocaine on platelet adhesiveness, *J. Clin. Pathol.* 15, 446–452, 1962.
4. Davey, M.G., and Luscher, E.F., Actions of thrombin and other coagulant and proteolytic enzymes on blood platelets, *Nature* 216, 857–858, 1967.
5. Jennings, N.K., Fox, J.E.B., Edwards, H.H., and Phillips, D.R., Changes in the cytoskeletal structure of human platelets following thrombin activation, *J. Biol. Chem.* 256, 6927–6932, 1961.
6. Brass, L.F., Thrombin and platelet activation, *Chest* 143(Suppl) 18S–25S, 2003.
7. Joist, J.H., Dolezel, G., Lloyd, J.V., Kimlough-Rathbone, R.G., and Mustard, J.F., Platelet factor 3 availability and the platelet release reaction, *J. Lab. Clin. Med.* 84, 474–487, 1970.
8. Broekman, M.J., Handin, R.I., Derksen, A., and Cohen, P., Distribution of phospholipids, fatty acids, and platelet factor 3 activity among subcellular fractions of human platelets, *Blood* 47. 963–971, 1976.
9. Brox, J.H. and Østerud, B., Evidence for appearance of "lysis independent" procoagulant activity in collagen and thrombin-stimulated platelets, *Thromb. Res.* 30, 449–458, 1983.
10. Dormann, D., Clemetson, K.J., and Kehrel, B.E., The GpIb thrombin-binding site is essential for thrombin-induced procoagulant activity, *Blood* 96, 2469–2478, 2000.

11. Detwiler, T.C. and Feinman, R.D., Kinetics of the thrombin-induced release of calcium (II), *Biochemistry* 12, 282–289, 1973.
12. Tollefsen, D.M., Feagler, J.R., and Majerus, P.W., The binding of thrombin to the surface of human platelets, *J. Biol. Chem.* 249, 2645–2651, 1974.
13. Ganguly, P., Binding of thrombin to human platelets, *Nature* 247, 306–307, 1974.
14. Laganowsky, A., Reading, E., Hopper, J.T.S., and Robinson, C.V., Mass spectrometry of intact membrane protein complexes, *Nat. Protocols* 8, 639–651, 2013.
15. Ganguly, P. and Sonnichsen, W.J., Bindnig of thrombin to human platelets and possible significance, *Brit. J. Haematol.* 34, 291–301, 1976.
16. Martin, B.M., Feinman, R.D., and Detwiler, T.C., Platelet stimulation by thrombin and other proteases, *Biochemistry* 14, 1308–1314, 1975.
17. Neal, G.G. and Esnouf, M.P., Kinetic studies of the activation of factor X by factor IXa and VIIIC in the absence of thrombin, *Brit. J. Haematol.* 57, 123–131, 1984.
18. Martin, B.M., Wasiewski, W.W., Fenton, J.W., II, and Detwiler, T.C., Equilibrium binding of thrombin to platelets, *Biochemistry* 15, 4886–4893, 1976.
19. Tollefsen, D.M., and Majerus, P.W., Evidence for a single class of thrombin binding sites on human platelets. *Biochemistry* 15, 2144–2149, 1976.
20. Migneault, I., Dartiguehare, C., Bertrand, M.J., and Waldron, K.C., Glutaraldehyde: behavior in aqueous solution, reaction with proteins, and application to enzyme cross-linking, *Biotechniques* 17, 790–802, 2004.
21. Mohammed, S.F., Whitworth, C., Chuang, H.Y.K., Lundblad, R.L., and Mason, R.G., Multiple active forms of thrombin binding to platelets and effects on platelet function, *Proc. Natl. Acad. Sci. USA* 73, 1660–1663, 1976.
22. Workman, E.F., Jr., White, G.C., II, and Lundblad, R.L., High affinity binding of thrombin to platelets. Inhibition by tetranitromethane and heparin, *Bichem. Biophys. Res. Commun.* 75, 925–932, 1977.
23. Workman, E.F., Jr., White, G.C., II, and Lundblad, R.L., Structure-function relationships in the interaction of thrombin with blood platelets, *J. Biol. Chem.* 252, 7118–7123, 1977.
24. Ganguly, P. and Gould, V.L., Thrombin receptors on human platelets: thrombin binding and antithrombin properties of glycoprotein I, *Brit. J. Haematol.* 42, 137–145, 1979.
25. George, J.N., Lewis, P.C., and Sears, D.A., Studies on platelet plasma membranes. II Characterization of surface proteins rabbit platelets *in vitro* and during circulation *in vivo* using diazotized ^{125}I-diiodosulfanilic acid as a label, *J. Lab. Clin. Med.* 88, 247–260, 1976.
26. Nurden, A.T., A peripheral high molecular weight glycoprotein located at the surface of human platelets, *Experentia* 33, 331–332, 1977.
27. Mills, D.C.B., The basic biochemistry of the platelet, in *Haemostgsis and Thrombosis*, ed. A.L. Bloom and D.P. Thomas, Chapter 3, pp. 50–60, Churchill Livingstone, Edinburgh, Scotland, UK, 1981.
28. Watanabe, K., Chao, F.C., and Tullis, J.L., Antithrombin activity of intact human platelets, *Thromb. Haemost.* 34, 115–126, 1975.
29. Wu, H.-f., White, G.C., II, Workman, E.F., Jr., Jenzano, J.W., and Lundblad, R.L., Affinity chromatography of platelets on immobilized thrombin: retention of catalytic activity by platelet-bound thrombin, *Thromb. Res.* 67, 419–427, 1992.
30. Lundblad, R.L. and White, G.C., II, The interaction of thrombin with blood platelets, *Platelets* 16, 373–385, 2005.
31. Rotoli, B. and Majerus, P.W., Plasma antithrombins and thrombin binding to platelets, *Haematologica* 64, 537–551, 1979.
32. Ruggeri, Z.M., De Marco, L., Gatti, L., Bader, R., and Montgomery, R.R., Platelets have more that one binding site for von Willebrand factor, *J. Clin. Invest.* 72, 1–12, 1983.

33. Bennett, W.F. and Glenn, K.C., Hypersensivity of platelets to thrombin. Formation of a stable thrombin-receptor complex and the role of shape change, *Cell* 22, 621–627, 1980.

34. Yeo, K.-T. and Detwiler, T.C., Analysis of the fate of platelet-bound thrombin, *Arch. Biochem. Biophys.* 236, 399–410, 1985.

35. McGowan, E.B. and Detwiler, T.C., Modified platelet responsiveness to thrombin. Evidence for two types of coupling mechanisms, *J. Biol. Chem.* 261, 739–746, 1986.

36. Detwiler, T.P., Chang, A.C., Speziale, M.V., *et al.*, Complexes of thrombin with protein secreted by activated platelets, *Sem. Thromb. Hemost.* 18, 60–66, 1992.

37. Riewald, M., Morgenstern, K.A., and Schleef, R.R., Identification and characterization of the cytoplasmic antiproteinase (CAP) in human platelets, *J. Biol. Chem.* 271, 7160–7167, 1996.

38. Gronke, R.S., Knauer, D.J., Veeragheanan, S., and Baker, J.B., A form of protease nexin is expressed on the platelet surface during platelet activation, *Blood* 73, 472–478, 1989.

39. Gronke, R.S., Bergman, B.I., and Baker, J.B., Thrombin interaction with platelets. Influence of a platelet protease nexin, *J. Biol. Chem.* 262, 3030–3036, 1987.

40. Hagen, I., Gogstad, G.O., Brostad, F., and Solum, N.O., Demonstration of ^{125}I-labeled thrombin binding platelet protein by immunoelectrophoresis and autordiography, *Biochim. Biophys. Acta* 732, 600–606, 1983.

41. Knupp, C., The interaction of thrombin with platelet protease nexin, *Thromb. Res.* 56, 77–90, 1989.

42. Jung, S.M. and Moroi, M., Crosslinking of platelet glycoproteins by *N*-succinimidyl-(4-azidophenylthio)propioniate and 3,3'-dithiobis(succinimidyl)propiontate, *Biochim. Biophys. Acta* 76, 152–162, 1983.

43. Jandrot-Perrus, M., Didry, D., Guillin, M.-C., and Nurden, A.I., Cross-linking of α- and γ-thrombin to distinct binding site of human platelets, *Eur. U. Biochem.* 174, 359–367, 1988.

44. Larsen, N.E., Horne, W.E., and Simons, E.R., Platelet interaction with active and TLCK-inactivated α-thrombin, *Biochem. Biophys. Res. Commun.* 87, 403–408, 1979.

45. Larsen, N.E. and Simons, E.R., Preparation and application of photoreactive thrombin analogue: Binding to human platelets, *Biochemistry* 20, 4141–4147, 1981.

46. Simons, E.R., Davies, T.A., Greenberg, S.M., and Larsen, N.E., Thrombin receptors on human platelets, *Methods Enzymol.* 215, 155–176, 1997.

47. Danishefsky, K.J. and Detwiler, T.C., Photoaffinity labeling of thrombin binding protein, *Biochim. Biophys. Acta.* 801, 48–57, 1984.

48. Fenton, J.W., II., Fasco, J.J., Stackrom, A.B., *et al.*, Human thrombins Production, evaluation, and properties of α-thrombin, *J. Biol. Chem.* 252, 3587–3598, 1977.

49. Phillips, D.R. and Agin, P.P., Platelet plasma membrane glycoproteins, Evidence for the presence of nonequivalent disulfide bonds, *J. Biol. Chem.* 252m, 2121–2126, 1977.

50. Lopez, J.A., Chung, D.W., Fujikawa, K., *et al.*, Cloning of the α chain of human glycoprotein 1b. A transmembrane protein with homology to leucine-rich α_2-glycoprotein, *Proc. Natl. Acad. Sci. USA* 84, 5615–5619, 1987.

51. Horne, W.C. and Simons, E.R., Probes of transmembrane potential in plateletls: change in cyanine dye fluorescence in response to aggregation stimuli, *Blood* 51, 741–749, 1978.

52. Knupp, C.L. and White, G.C., II, Effect of active site-modified thrombin on the hydrolysis of platelet-associated glycoprotein V by native thrombin, *Blood* 65, 578–583, 1985.

53. Bernard, J. and Soulier, J.P., Sur une nouvelle variete de dystrophié thrombocytaire hémorragique congénitale, *Sem. Hop. Paris* 24, 367, 1948.

54. Bithel, T.C., Pareth, S.J., and Strong, R.R., Platelet-function studies in the Bernard-Soulier syndrome, *Ann. N. Y. Acad. Sci.* 201, 145–160, 1972.

55. Jamieson, G.A. and Okemura, T., Reduced thrombin binding and aggregation in Bernard-Soulier platelets, *J. Clin. Invest.* 61, 861–864, 1978.

56. Nurden, A.T. and Caen, J.P., Specific roles for platelet surface glycoproteins in platelet function, *Nature* 258, 720–722, 1975.
57. Cooper, H.A., Bennet, W.P., White, G.C., II., and Wagner, R.H., Hydrolysis of platelet membrane glycoproteins with *Serratia marcescens* metalloproteinase, *Proc. Natl. Acad. Sci. USA* 79, 1433–1438, 1982.
58. Quach, M.E., and Li, B., Structure-function of platelet glycoprotein Ib-IX, *J. Thromb. Haemost.* 18, 3131–3141, 2020.
59. Berndt, M.C., Gregory, C., Kobali, A., *et al.*, Purification and preliminary characterization of the glycoprotein 1b complex in the human platelet membrane, *Eur. J. Biochem.* 151, 637–649, 1985.
60. Wicki, A.N., and Clemetson, K.J., The glycoprotein 1b complex of human blood platelets, *Eur. J. Biochem.* 163, 43–50, 1987.
61. Modderman, P.W., Admiraal, L.g., Sonnenberg, A., and van dem Boone, A.E.C.Kr., GLycoproteins V and Ib-IX form a noncovalent complex in the platelet membrane, *J. Biol. Chem.* 267, 364–369, 1992.
62. Lopez, J.A., and Dong, J.-F., Structure and function of the glycoprotein Ib-IX-V complex, *Curr. Opin. Hematol.* 4, 323–329, 1997.
63. Mo, X., Liu, L., López, J.A., and Li, R., Transmembrane domains are critical for the interaction between platelet glycoprotein V and glycoprotein Ib-IX complex, *J. Thromb. Haemost.* 10, 1875–1886, 2012.
64. Berndt, M.C., Shen, Y., Dopheide, S.M., Gardiner, E.E., and Andrew, R.K., The vascular biology of the glycoprotein 1b-IX-V complex, *Thromb. Haemost.* 86, 178–188, 2001.
65. Jamieson, G.A., Okumura, T., and Hasitz, M., Structure and function of platelet glycocalcin, *Thromb. Haemost.* 42, 1673–1678, 1979.
66. Loscalzo, J. and Handin, R.L., Platelet glycocalicin, *Methods Enzymol.* 215, 289–294, 1992.
67. Detwiler, T., R., Hypothetial models for the thrombin-platelet interaction, *Ann. N. Y. Acad. Sci.* 370, 6767–6777, 1981.
68. Berndt, M.C. and Phillips, D.R., Interaction of thrombin with platelets: purification of the thrombin substrate, *Ann. N. Y. Acad. Sci.* 370, 87–95, 1981.
69. Vu, T.-K.H., Hung, D.T., Wheaton, V.I., and Coughlin, S.R., Molecular cloning of a functional thrombin receptor reveals a novel proteolytic mechanism of receptor activation, *Cell* 64, 1057–1068, 1991.
70. Coughlin, S.R., How the protease thrombin talks to cells, *Proc. Natl. Acad. Sci. USA* 96, 11023–11027, 1999.
71. Woulfe, D.S., Platelet G protein coupled receptors in hemostasis and thrombosis, *J. Thromb. Haemost.* 3, 2193–2200, 2005.
72. Offermanns, S., Activation of platelet function through G protein-coupled receptors, *Circ. Res.* 99, 1293–1304, 2006.
73. De Candia, E., Mechanisms of platelet activation by thrombin: a short history, *Thromb. Res.* 129, 250–256, 2012.
74. Lechtenberg, B.C., Freund, S.M.V., and Huntington, J.A., GpIBα interacts exclusively with exosite II in thrombin, *J. Mol. Biol.* 426, 881–893, 2014.
75. De Marco, L., Mazzucato, M., Masotti, A., and Ruggeri, Z.M., Localization and characterization of an α-thrombin-binding site on glycoprotein Ibα, *J. Biol. Chem.* 269, 6475–6484, 1994.
76. Jandrot-Perrus, M., Clemetson, K.J., Huisie, M.-G., and Guillin, M.-C., Thrombin interaction with platelet glycoprotein 1b: effect of Glycocalicin on thrombin specificity, *Blood* 80, 2781–2786, 1992.
77. Li, C.W., Vindigni, A., Sadler, J.E., and Wardell, M.R., Platelet glycoprotein Ibα binds to thrombi anion-binding exosite II inducing allosteric changes in the activity of thrombin, *J. Biol. Chem.* 276, 6161–6168, 2001.

78. De Candia, E., Hall, S.W., Rutella, S., *et al.*, Binding of thrombin to glycoprotein Ib accelerates the hydrolysis of Par-1 on intact platelets, *J. Biol. Chem.* 276, 4692–4698, 2001.

79. Dong, J.-C. L., Li, C.G., and López, J.A., Tyrosine sulfation of the glycoprotein Ib-IX complex. Identification of the sulfated residues and effect on ligand binding, *Biochemistry* 33, 13946–13953, 1994.

80. Marchese, P., Murato, M., Mazzucato, M., *et al.*, Identification of three tyrosine residues of glycoprotein Ibα with distinct roles in von Willebrand factor and thrombin binding, *J. Biol. Chem.* 270, 9571–9576, 1995.

81. Adam, F., Bouton, M.-C., Kukse, M.-G., and Jandrot-Perrus, M., Thrombin interaction with platelmembrane glycoprotein Ibα, *Trends. Mol. Med.* 9, 461–464, 2003.

82. Jandrot-Perrus, M., Rendu, F., Caen, J.P., Levy-Toledona, S., and Guillin, M.-C., the common pathway for alpha- and gamma-thrombin induced platelet activation independent of GPIb: a study of Bernard-Soulier platelets, *Brit. J. Haematol.* 75, 385–392, 1990.

83. De Marco, L., Mazzucato, M., Masotti, A., Fenton, J.W., II., and Ruggeri, Z.M., Function of glycoprotein Ibα in platelet activation induced by α-thrombin, *J. Biol. Chem.* 266, 23776–23783, 1991.

84. Jandrot-Perrus, M., Bouton, M.-C., Lanza, E., and Guillin, M.-C., Thrombin interaction with platelet membrane glycoprotein Ib, *Sem. Thromb. Haemost.* 22, 151–156, 1996.

85. Danev, R., Yanagisawa, H., and Kirkawa, M., Cryo-electron microscopy methodology: current aspects and future directions, *Trends Biochem. Sci.* 44, 837–848, 2019.

86. Celikel, R., McClintock, R.A., Roberts, J.R., *et al.*, Modulation of α-thrombin function by distinct interactions with platelet glycoprotein Ibα, *Science* 301, 218–221, 2003.

87. Dumas, J.J., Kumar, R., Seehra, J., Somers, W.S., and Mosyak, L., Crystal structure of the GpIbα complex essential for platelet aggregation, *Science* 301, 222–226, 2003.

88. Sadler, J.E., A ménage á trois in two configurations, *Science* 301, 177–179, 2003.

89. Vanhoovelbeke, K., Ulricht, H., Romijn, R.A., Huizinga, E.G., and Deckmyn, H., The GpIbα-thrombin interactions are far from crystal clear, *Trends Mol. Med.* 10, 33–39, 2004.

90. Mann, K.G., Brummel, K., and Butanas, S., What is all that thrombin for?, *J. Thromb. Haemost.* 1, 1504–1514, 2003.

91. Adams, T.F. and Huntington, J.A., Thrombin-cofactor interaction. structure insights not regulatory mechanism, *Arterioscler. Thromb. Vasc. Biol.* 26, 1738–1745, 2006.

92. Zarpellon, A., Celikel, R., Roberts, J.R., *et al.*, Binding of α-thrombin to surface-anchored platelet glycoprotein Ibα sulfotyrosines through a two-site mechanism invoving exosite I, *Proc. Natl. Acad. Sci. USA* 108, 8628–8633, 2011.

93. Kobe, B., Gunčor, G., Bucholz, R., Huber, T., and Maca, B., The many faces of platelet glycoprotein Ibα-thrombin interaction, *Curr. Prot. Pept. Sci.* 10, 551–558, 2009.

94. Huntington, J.A., Thrombin plasticity, *Biochim. Biophys. Acta* 1824, 246–252, 2012.

95. Brinkhous, K.M., Thrombin and the clinical application, *South. Surg.* 13, 297–407, 1947.

96. Butanos, S. and Mann, K.G., Blood coagulation, *Biochemistry* (Moscow) 67, 5–15, 2002.

97. Ruggeri, Z.M., Zarpellon, A., Roberts, J.R., *et al.*, Unravelling the mechanism and significance of thrombin binding to platelet glycoprotein Ib, *Thromb. Haemost.* 104, 894–902, 2010.

98. Estevez, B., Kim, K., Delaney, M.K., *et al.*, Signaling-mediated cooperativity between glycoprotein Ib-IX and protease-activated receptors in thrombin-induced platelet activation, *Blood* 127, 626–636, 2016.

99. Estevez, B. and Du, X., New concepts and mechanisms of platelet activation signaling, *Physiology* 32, 162–177, 2017.

100. Chen, Y., Ruggeri, Z.M., and Du, X., 14–3–3 proteins in platelet biology and glycoprotein Ib-IX signaling, *Blood* 131, 2436–2448, 2018.
101. Munier, C.C., Ottmann, C., and Perry, M.W.D., 14–3–3 modulation of the inflammatory response, *Pharmacol. Res.* 163, 105236, 2021.
102. Ware, J., A binding relationship with thrombin, *Blood* 127, 525–526, 2016.
103. Mosher, D.F., Vaheri, A., Choate, J.J., and Gahmberg, C.G., Action of thrombin ib on surface glycoproteins of human platelets, *Blood* 53, 437–445, 1979.
104. Phillips, D.R. and Agin, P.P., Plasma membrane glycoproteins. Identification of a proteolytic substrate for thrombin, *Biochem. Biophys. Res. Commun.* 75, 940–947, 1977.
105. McGowan, E.B., Ding, A.-h., and Detwiler, T.C., Correlation of thrombin-induced glycoprotein V hydrolysis and platelet activation, *J. Biol. Chem.* 258, 11243–11248, 1983.
106. Dong, J.-T., Sae-Tung, G., and López, J.A., Role of glycoprotein V in the formation of the high-affinity platelet binding site, *Blood* 89, 4355–4363, 1997.
107. Ramankrishnan, V., De Guzman, F., Bao, M., *et al.*, A thrombin receptor function for platelet glycoprotein Ib-IX unmasked by cleavage of glycoprotein V, *Proc. Natl. Acad. Sci. USA* 98, 1823–1828, 2001.
108. Reactivation of DIP-thrombin
109. Carter, P. and Wells, J.A., Dissecting the catalytic triad of a serine proteases, *Nature* 332, 564–568, 1988.
110. Wu, Q., Sheenan, J.P., Tsang, M., *et al.*, Single amino acid substitution dissociated fibrinogen-clotting and thrombomodulin-binding activities of human thrombin, *Proc. Natl. Acad. Sci. USA* 88, 6775–6779, 1991.
111. Sambrano, G.R., Weiss, E.J., Zheng, Y.-W., Huang, W., and Coughlin, S.R., Role of thrombin signaling in platelet in haemostasis and thrombosis, *Nature* 413, 74–78, 2001.
112. Wu, C-.C., Wong, W.-Y., Wei, C.-K., and Teng, C.-M., Combined blockage of thrombin anion-binding exosite I and PAR4 produces synergistic antiplatelet effect in human platelets, *J. Thromb. Haemost.* 105, 88–95, 2011.
113. LIndahl, T.L., Macwan, A.S., and Ramström, S., Protease-activated receptor 4 is more important than protease-activated receptor 1 for the thrombin-induced procoagulant effect on platelets, *J. Thromb. Haemost.* 141, 1639–1641, 2016.
114. French, S.L., Arther, J.F., Lee, H., *et al.*, Inhibition of protease-activated receptor 4 impairs platelet procoagulant activity during thrombin formation in human blood, *J. Thromb. Haemost.* 14, 1642–1654, 2016.
115. Boknäs, V., Faxäly, L., Sanchez-Centellas, D., *et al.*, Thrombin-induced platelet activation with PAR4: pivotal role for exosite II, *J. Thromb. Haemost.* 112, 558–565, 2014.
116. French, S.L., and Hamilton, J.R., Protease-activated receptror 4: from structure to function and back again, *Br. J. Pharmacol.* 173, 2952–2965, 2016.
117. Nieman, M.T., Protease-activated receptors in hemostasis, *Blood* 128, 169–177, 2016.
118. Wu, J., Heemskerk, J.W.M., and Baaten, C.F.M.J., Platelet membrane proteolsysi: implications for platelet function, *Front. Cardiovasc. Med.* 71, 608391, 2021.

8 Inhibitors of Thrombin

Thrombin can be considered the terminal effector of clot formation in the normal hemostatic response and in thrombus formation; thrombin also has a role in disseminated intravascular coagulation (DIC) and possibly in fibrotic disease.[1] It has been shown that dabigatran can influence the development of idiopathic pulmonary fibrosis.[2] Since thrombin is a potent effector in physiology, there are natural inhibitors that can modulate *in vivo* function in both the intravascular and extravascular space. There are also agents that can be used as therapeutics for controlling thrombin function. This chapter will be divided into two sections: natural inhibitors of thrombin, including hirudin, and synthetic therapeutic inhibitors of thrombin.

NATURAL INHIBITORS OF THROMBIN

The presence of an inhibitor of thrombin was recognized in early work on thrombin in the 1800s.[3] Several investigators described the inhibition of "fibrin ferment" by serum. Morowitz described the presence of a substance in serum that generated thrombin activity on treatment with a base.[3,4] It was suggested that this represented a second process for the formation of thrombin coagulation. Morowitz referred to the material in serum as metathrombin.[3] Several investigators subsequently studied the formation and properties of metathrombin. The results from these studies suggested that metathrombin was the complex of thrombin and antithrombin and that treatment of metathrombin with base did result in thrombin.[4,5] It is acknowledged that treatment with base of a thrombin-antithrombin complex formed with purified protein releases active thrombin.[6-8] In early work on thrombin, the thrombin-antithrombin complex was known as metathrombin, and dissociation of this complex was thought to represent a pathway for physiological thrombin formation,[9] but there was skepticism of this concept.[10] Gerber and Blanchard[11] reported that phenols accelerated the clotting of plasma by thrombin, suggesting the phenol increased the fibrinogen clotting activity of thrombin. It was later shown by Fantl[12] that the effect of phenols was to depress the antithrombin activity of plasma. Depletion of antithrombin in substrate plasma results in a decrease in the clotting time of an activated prothrombin complex concentrate.[13] Wilson reported that peptide material from human brain enhanced the antithrombin activity of defibrinated human plasma.[14] Ferguson and Glazko[15] observed that heparin alone did not inhibit thrombin but the addition of a small amount plasma resulted in inactivation of thrombin. Antithrombin-enzyme complexes are removed by the low-density lipoprotein receptor.[16] There have been many studies on the clearance of thrombin *in vivo*. One study examined this in a canine model.[17] The clearance occurred over a period of hours, and it was observed that the liver was importance in clearance. Vogel and coworkers[18] showed the clearance of thrombin in an autologous model (rabbit). The dominant clearance process of thrombin from the circulation occurs via formation of a complex with antithrombin,

DOI: 10.1201/b22204-8

TABLE 8.1
Protease Inhibitors in Plasmaa

Inhibitor	Concentration (µM)	k_{on}[a]	Relative Effectiveness	*In Vivo* Distribution
Antithrombin	4.46	3.36×10^5	15.64	79%
α_1-Antitrypsin	42.4	6.5×10^3	3.03	0
α_2-Macroglobulin	3.11	2.93×10^4	1	21%

Source: Downing, M.R., Bloom, J.W., and Mann, K.G., *Biochemistry* 17. Vogel, C.N., Kingdon, H.S., and Lundblad, R.L., *J. Lab. Clin. Med.* 93, 661–673, 1979

[a] Second-order rate constant based on the rate of loss of fibrinogen clotting activity

with a much lesser role for α_2-macroglobulin. Vogel and coworkers[18] did not find a complex of thrombin with α_1-antitrypsin They did show that TLCK-thrombin could form an *in vivo* complex with α_2-macroglobulin, but not with antithrombin. The distribution of the *in vivo* thrombin observed by Vogel and coworkers is consistent with *in vivo* human data[19] and with kinetic data.[20] Data for the various plasma protein inhibitors in plasma is shown in Table 8.1. The rate of reaction of thrombin with α_1-antitrypsin is considerably lower than that observed with the other plasma proteinase inhibitors, possibly providing an explanation for the results obtained by Vogel and coworkers.[18]

ANTITHROMBINS

ANTITHROMBIN I

Antithrombin I is described as the adsorption of thrombin by fibrin.[21] It was recommended that this description not be used to describe this property of fibrin, since fibrin was a derivative of an existing coagulation factor.[22] Personally, I find the term to be useful.

ANTITHROMBIN II (HEPARIN COFACTOR II)

The designation of antithrombin II is rarely used except as a typographical error or in combination with antithrombin III, such as antithrombin II/III. While I find it difficult to locate a specific citation, it is likely that antithrombin II is identical with heparin cofactor II.[23] There was considerable confusion in early studies about the relations of heparin cofactor II and antithrombin. Heparin cofactor was described as a heparin-dependent inhibitor, while antithrombin (antithrombin III) was described as a progressive antithrombin present in blood. Heparin cofactor II (HCII) is better known as an inhibitor of thrombin in the extravascular space after binding to dermatan sulfate on the extracellular matrix. As with other serpins, HCII is pleiotropic. One study shows a protective of HCII against angiotensin II–induced cardiac remodeling[24] to the extent of abolishing cardiac fibrosis in a murine model. It was suggested the effect of HCII was mediated through suppression of the NAD(P)H oxidase-transforming growth factor-β1 pathway. NAD(P) oxidase 4 has been shown to affect the transforming growth factor-β1 pathway in the differentiation of cardiac fibroblasts.[25]

ANTITHROMBIN (ANTITHROMBIN III)

Antithrombin is the major inhibitor of thrombin in plasma,[18–20] substantially more effective than either α_1-antitrypsin or α_2-macroglobulin. Antithrombin does exist as a population of stable conformers. Latent antithrombin is a stable conformer with markedly reduced activity as an inhibitor of thrombin but increased antiangiogenic activity.[26,27] In addition, cleaved forms of antithrombin that result from the dissociation of the enzyme (thrombin)-inhibitor complex,[28] as well as from proteolysis by other enzymes such as elastase,[29] have antiangiogenic activity. A latent form of antithrombin can also be obtained by heating the native protein in citrate.[30] A latent form of α_1-antitrypsin can be obtained by treating the native protein under similar conditions.[31] A variety of antithrombin mutations can cause thrombotic consequences.[27] There continue be advances in our understanding of the reaction between thrombin and antithrombin.[32,33]

Phage display technology has been demonstrated to be a useful approach for the synthesis and evaluation of a large number of mutant peptides and proteins.[34–39] Phage display technology has been used to develop more effective thrombin inhibitors. α_1-Trypsin inhibitor (α_1-proteinase inhibitor, serpin A1) has been an attractive target.[40,41] Native α_1-antitrypsin is not as effective as antithrombin in the inhibition of thrombin.[20] A mutant α_1-antitrypsin, α_1-antitrypsin Pittsburgh, was discovered to be a potent antithrombin,[42] where the "active site" methionine is replaced by arginine.[43] There has been some success in the use of phage display technology to engineer α_1-antitrypsin.[44] Other investigators have used a leech-derived tryptase inhibitor[45] or the first Kunitz domain of human lipoprotein–associated coagulation inhibitor[46] as the matrix.

THERAPEUTIC INHIBITORS OF THROMBIN

The term therapeutic inhibitor is used to describe a substance, organic or biological in nature, developed and used as a pharmaceutical/biopharmaceutical product. Antithrombin (antithrombin III) is both a natural inhibitor and a therapeutic inhibitor.[47] There are therapeutic preparations of α_1-antitrypsin (α_1-antiprotease inhibitor).[48] There are attempts to use α_2-macroglobulin as a therapeutic.[49]

HEPARIN

Heparin and other sulfated mucopolysaccharides are best known for the ability to accelerate the reaction of coagulation proteases such as thrombin with antithrombin and HCII. These anionic materials can directly affect thrombin function.[50–54] Several of these studies show an interaction of heparin and other sulfated oligosaccharides or related compounds with exosite-2.[38,55,56] There is a long history of direct interaction of anionic materials, including heparin with thrombin.[57] The ability of heparin, either unfractionated or low-molecular-weight heparin, is dependent on the presence of antithrombin. The absence of heparin results in a condition known as heparin-resistant coagulopathy, which can be treated with argatroban, as described later.

HIRUDIN

Hirudin is a mixture of polypeptides found in the salivary glands of the medicinal leech.[58] Although best known as a product of the European medicinal leech,[59] it is found in medicinal leeches in other areas.[58,60] The effort to develop a specific inhibitor of thrombin for therapeutic use is not new and can be traced back at least to the work of Markwardt and others on hirudin.[61–64] Recombinant DNA technology permitted the manufacture of hirudin and derivatives,[65–67] such as bivalirudin, for current clinical use.[68] Bivalirudin (Angiomax) started life off as hirugen[69] with Biogen and currently is a product of The Medicines Company. While I do not know of a direct connection, it is interesting to note that clinical use of leeches has increased markedly in the past 20 years.[70–74] Peptide mimetics of hirudin, hirunorms, have been developed as potential therapeutic agents,[75] and the crystal structure of the complex of one mimetic, hirunorm IV, with thrombin has been determined.[76] These mimetics have nanomolar K_i values in the inhibition of the hydrolysis of Tos-Gly-Pro-Arg-pNA. These mimetics are approximately one-third the size of hirudin and have a novel design in the inclusion of a space which permits hirudin-like binding at the active size and a fibrinogen binding site with a smaller mass. While there does not seem to be any further development of hirunorms as therapeutics, hirunorm IV has been used in NMR measurement to further define the fibrinogen binding site in thrombin.[77] Hirugen is a small derivative fragment of hirudin that inhibits fibrinogen clotting but has no effect on the hydrolysis of small molecule substrates.[78] The crystal structure of the complex of hirugen and thrombin has been determined to show binding to the fibrinogen binding site (exosite-1),[79] leaving the active site free to react with different inhibitors.[80] Radiolabeled (^{131}I) hirudin was suggested as a probe for active thrombin in tissue (aorta wall).[81] ^{131}I-labeded hirudin was also suggested as a probe for thrombi, but there are issues, as hirudin can displace a substantial amount of thrombin bound by fibrin.[82] As with other radiolabeled material, heterogeneity of product can occur with iodination.[83]

Other anticoagulant proteins directed at thrombin can be obtained from other species that rely upon keeping blood fluid during the digestive process.[84–88] There has been interest in the development of some of these proteins as therapeutics. However, with the exception of heparin and low-molecular-weight heparin (LMWH), the novel oral anticoagulants (NOACs) have come to dominate the therapeutic anticoagulant marketplace.

THE DEVELOPMENT OF SYNTHETIC ORGANIC NONPEPTIC INHIBITORS OF THROMBIN

There is a long history of developing inhibitors to thrombin using synthetic organic chemistry. This work was driven in part by a desire to understand the specificity of thrombin as a serine protease, but much more by the need to develop better anticoagulants. Thrombin has a distinct preference for the cleavage of peptide bonds where the carboxyl group is contributed by arginine. Cleavage of ester and amide bonds where lysine contributes the carboxyl group is seen in low-molecular weight synthetic substrates but rarely in proteins. Arginine occupies a site referred to as P_1 in many substrates, binding to the anionic site S_1 on thrombin. The presence of arginine in the P_1 site drove early interest in guanidine derivatives, including benzamidines. The nomenclature of P_1 and S_1 to define protease specificity was developed by Schechter and Berger in 1967[89] and was applied to the development of protease substrates and synthetic organic inhibitors for thrombin.[90–93]

THE EARLY DAYS (1960–1980)

It is essential to review the early work on the development of synthetic organic inhibitors of thrombin before a discussion of the NOACs. This permits an understanding of the factors that were useful in the design of a successful oral inhibitor for thrombin. Much of the work came from two groups, that of Professor Fritz Markwardt in what was then East Germany and the laboratories of Dieter Geratz and Rick Tidwell in the Department of Pathology at the University of North Carolina at Chapel Hill. Markwardt led a prolific group of researchers working on the development of enzyme inhibitors based on benzamidine.[94,95] His group also made major contributions to the development of inhibitors based on hirudin. Markwardt and Walsmann[96] initially worked with some simple derivatives of benzamidine (Figure 8.1). As far as I could find, this was the first work showing that benzamidine could inhibit thrombin (Table 8.2).

It was noted that 4-amidinophenylpyruvic acid was also a potent inhibitor of the hydrolysis of TosArgOMe and the clotting of fibrinogen. Markwardt and coworkers also published a larger study in the same year[97] describing the systematic development of inhibitors for several arginine-specific regulatory proteases based on benzylamine and benzamidine (Figure 8.2, Table 8.3). p-Chlorobenzylamine was

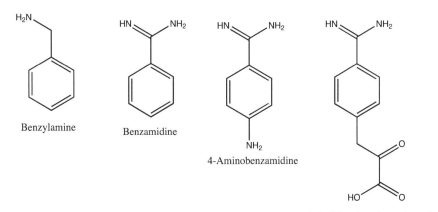

Benzylamine

Benzamidine

4-Aminobenzamidine

4-Amidinophenylpyruvic acid

FIGURE 8.1 Benzyamine, benzamidine, and related compounds.

Source: Adapted from Markwardt, F. and Walsmann, P., Uber die Hemmung des Gerinnungsferment durch benzamidinderivate, *Experientia* 24, 25–26, 1968.

TABLE 8.2
The Inhibition of Thrombin by Benzamidine and Related Compounds

Compound	K_i (mM)[a]	pKa/log P
Benzylamine	12	9.3
4-Methoxybenzylamine	4	
Phenylguanidine	9	10.88
Benzamidine	0.2	11.6
4-Aminobenzamidine	0.08	
4-amidinophenylpyruvic acid	0.03	

Source: Markwardt, F. and Walsmann, P., *Experientia* 24, 25–26, 1968
[a] Determined with Bz*dl*-ArgpNA in 0.1 M Tris, pH 8.4, at 25°C using the method of Dixon (Dixon, M., The determination of enzyme inhibitor constants, *Biochem. J.* 55, 70–71, 1953)

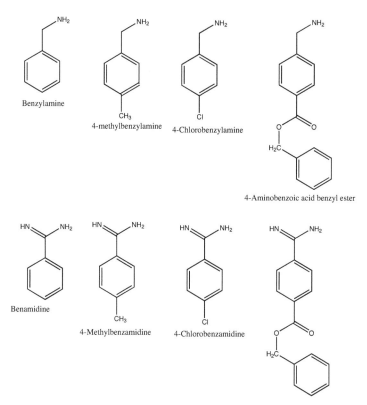

FIGURE 8.2   Benzylamine, benzamidine, and halogenated derivatives.

Source: Adapted from Markwardt, F., Landsmann, H., and Walsmann, P., Comparative studies on the inhibition of trypsin, plasmin, and thrombin by derivatives of benzylamine or benzamidine, *Eur. J. Biochem.* 6, 502–508, 1968,

TABLE 8.3
Inhibition of Trypsin, Plasmin, and Thrombin by Benzylamine, Benzamidine, and Derivatives

Compound	K_i (mM)[a]		
	Bovine Trypsin	Human Plasmin	Bovine Thrombin
Benzylamine	0.3	20	12.3
4-Methylbenzylamine	1.5	3.0	1.6
4-Chlorobenzylamine	0.7	2.0	0.23
4-Aminomethylbenzoic acid, benzyl ester	0.03	0.2	1.4
Benzamidine	0.035	0.35	0.22
4-Methylbenzamidine	0.03	0.30	0.40
4-Chlorobenzamidine	0.04	0.25	0.40
4-Amidinobenzoic acid, benzyl ester	0.02	20	4.2

Source: Markwardt, F., Landmann, H., and Walsmann, P., *Eur. J. Biochem.* 6, 502–508, 1968

[a] Determined by the effect on the hydrolysis of BzdlArgPNA at pH 8.0/25°. The K_k was determined as described by Dixon (Dixon, M., The determination of enzyme inhibitor constants, *Biochem. J.* 55, 70–71, 1953)

later developed by Art Thompson[98,99] as a ligand for the affinity chromatography of thrombin. This is a halogenated aromatic structure that may bind to the S_1 pocket in the binding site, but more likely to other regions in the extended active site. Baum and coworkers[100] have presented a thermodynamic analysis of the binding of this and similar structures of the S_1 binding site in thrombin.

The results obtained from the early work of Markwardt and colleagues showed that it is possible to build specificity into the structures of either benzylamine or benzamidine, but further work was required to build specificity and affinity, although the latter is not as critical. A decade after the work cited earlier, Markwardt reviewed the work from his laboratory.[101,102] Markwardt and coworkers continued their work focusing on new derivatives of benzamidine (amidinophenyl) (Figure 8.3 and Figure 8.4), which were effective in the micromolar range (Table 8.4).[103]

The Erfurt group continued their work on developing more increasing complex derivatives of benzylamine (4-amidinophenylalanine),[104] although Markwardt's interest also included useful work on hirudin and hirudin derivatives.[105,106] Some of the results obtained with inhibitors based on N_α-arylsufonyl-4-amidinophenylalanine (Figure 8.5 and Figure 8.6) are given in Table 8.5. Values of K_i below 2 µM were obtained with S-2238 tripeptide nitroanilide substrate in 0.154 M NaCl. These studies were performed before our current understanding of the structural role of sodium ions in thrombin. It is my sense that this does not affect the conclusions of the study. Stürzebecher and coworkers also compared the action of N_α-arylsulfonyl-4-amidinophenylalanine on other proteinases (Table 8.6 and Figure 8.7). Bz-*dl*-ArgpNA was also used for plasmin.

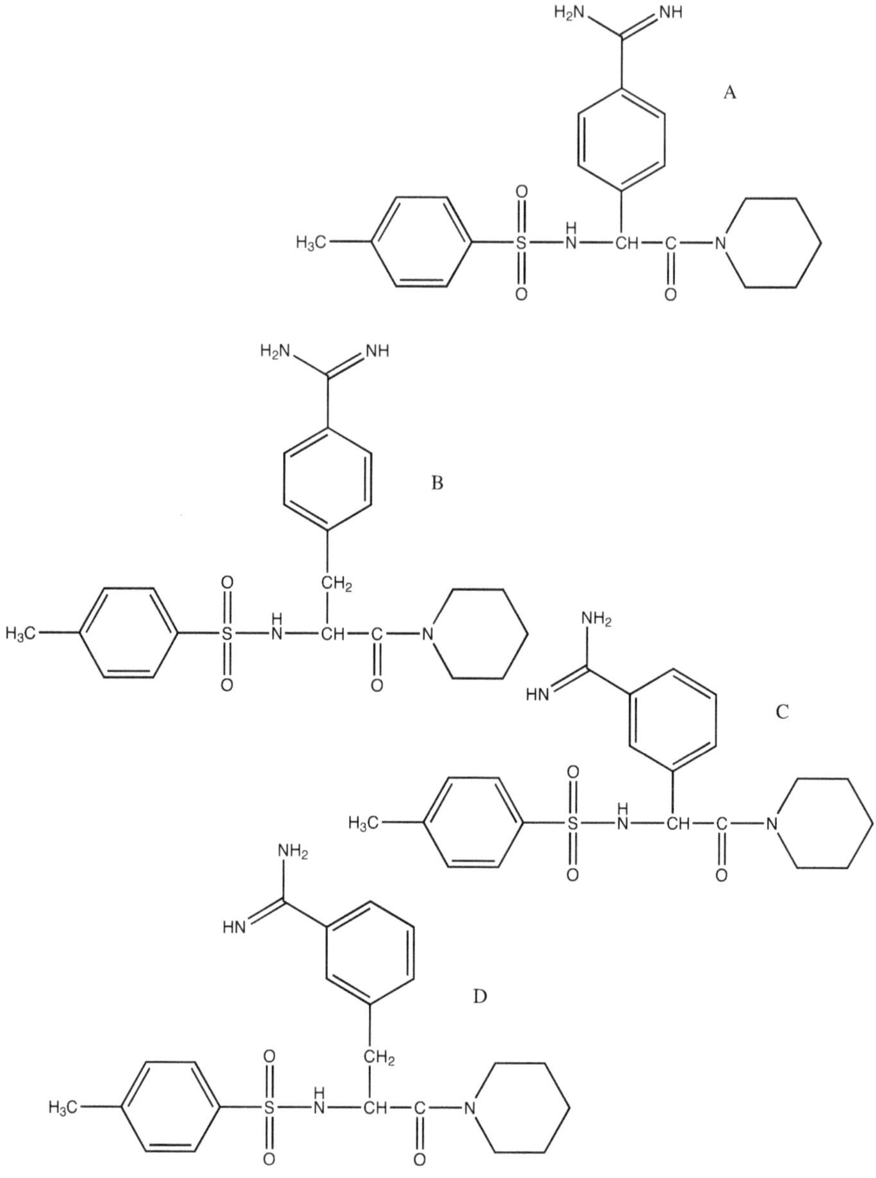

FIGURE 8.3 From Table 8.4 (A) Piperidino N^α-tosyl-2-amino-2′-(4-amidinophenyl)acetate; (B) Piperidino N^α-tosyl-4′-amidinophenylalanine; (C) Piperidino N^α-tosyl-2-amino-2′-3-amidinophenylalanine; (D) Piperidino N^α-tosyl-2-amino-3′-amidinophenylalanine.

Source: Adapted from Markwardt, F., Wagner, G., Stürzebecher, J., and Walsmann, P., N^α-Arylsulfonyl-ω-(4-amidinopheny) α-aminoalkylcarboxylic acid amides, *Thromb. Res.* 17, 425–431, 1980.

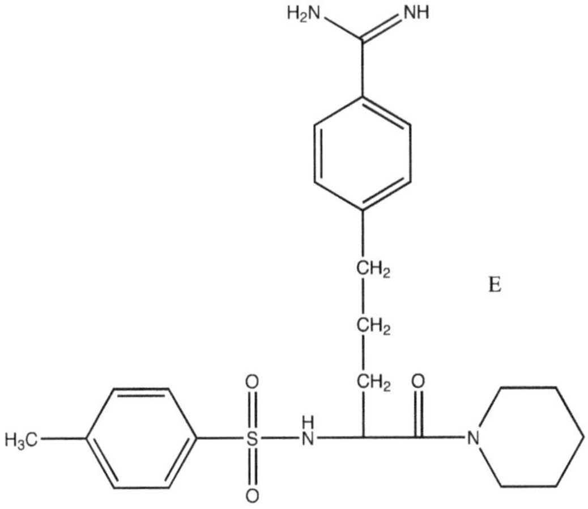

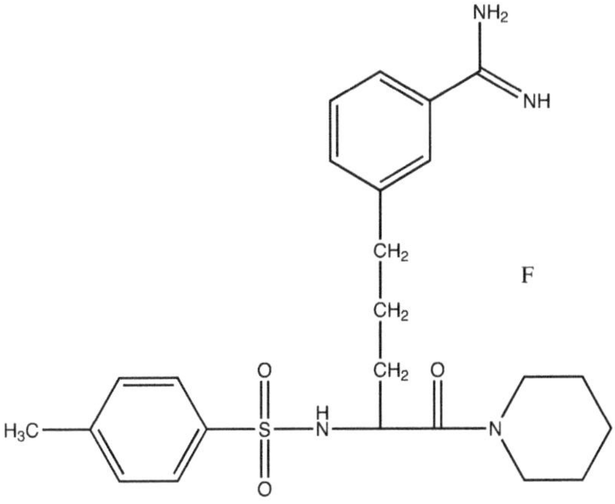

FIGURE 8.4 From Table 8.4 (E) 1-Piperidino-N^{α}-tosyl-2-amino-5′-(4-amidinophenyl)vale-ric acid and (F) 1-Piperidino-N^{α}-tosyl-2-amino-5′-(3-amidinophenyl)valeric acid.

Source: Adapted from Markwardt, F., Wagner, G., Stürzebecher, J., and Walsmann, P., N^{α}-Arylsulfonyl-ω-(4-amidinopheny) α-aminoalkylcarboxylic acid amides, *Thromb. Res.* 17, 425–431, 1980.

TABLE 8.4
Inhibition of Thrombin by Derivatives of N^{α}-Toluenesulfonylamidinophenyl Piperides

Compound	K_i (μM)[a]
(A) Piperidino N^{α}-tosyl-2-amino-2′-(4-amidinophenyl)acetate	61
(B) Piperidino N^{α}-tosyl-4′-amidinophenylalanine	2.3
(C) Piperidino N^{α}-tosyl-2-amino-2′-3-amidinophenylalanine	1.7
3(D) Piperidino N^{α}-tosyl-2-amino-3′-amidinophenylalanine	1.2
(E) Piperidino N^{α}-tosyl-2-amino-5′-(4-amidinophenyl)valerate	1.7
(F) PIperidino N^{α} -2-amino-5′-(3′-Amidinophenyl)valerate	0.6

(Continued)

TABLE 8.4
(Continued)

Source: Markwardt, F., Wagner, G., Stürzebecher, J., and Walsmann, P., *Thromb. Res.* 17, 425–431, 1980
[a] Measurement of activity was performed in 0.1 M Tris, pH 8.0., at 25°C with Bz-*dl*-ArgpNA. The K_k was determined as described by Dixon ((Dixon, M., The determination of enzyme inhibitor constants, *Biochem. J.* 55, 70–71, 1953)

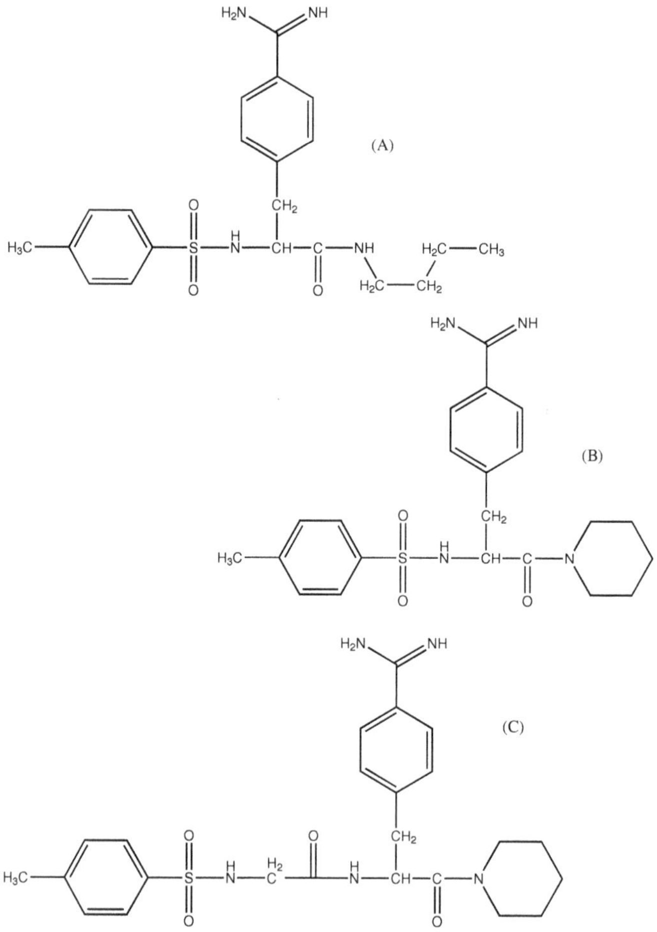

FIGURE 8.5 Cyclic amide derivatives of N_α-arylamidinophenyalanine as inhibitors of thrombin. (A) N^α-tosyl-(4-amidinophenylalanyl butylamide. (B) Piperidino N^α-tosyl-4-amidinophenylalanine. (C) Piperidino N^α-tosyl-glycyl-4-amidinophenylalanine.

Source: Adapted from Stürzebecher, J., Markwardt, F., Voight, B., Wagner, G., and Walsmann, P. Cyclic amides of N$_\alpha$-arylsulfonylaminoacylated 4-amidinophenylalanine-tight binding inhibitors of thrombin, *Thromb. Res.* 29, 635–642, 1983.

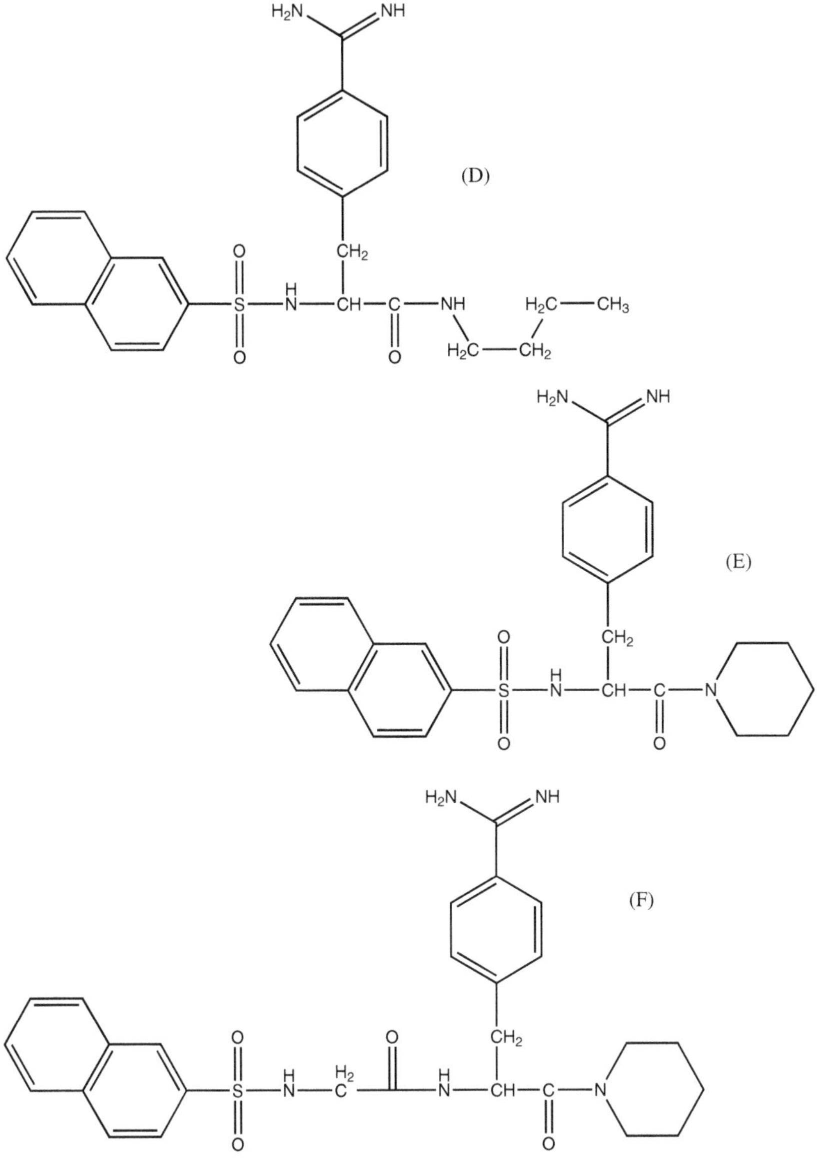

FIGURE 8.6 Cyclic amide derivatives of N_α-arylamidinophenyalanine as inhibitors of thrombin. (D) N^α-2-naphthylsulfonyl-(4-amidinophenyl)alanyl butylamide. (E) Piperidino-2-naphthylsulfonyl-4-amidinopphenylalanine. (F) Piperidino-2-naphthylsulfonylglycyl-4-amidinopphenylalanine.

Source: Adapted from Stürzebecher, J., Markwardt, F., Voight, D., Wagner, G., and Walsmann, P., Cyclic amides of N^α-arylsulfonylaminoiacylated 4-amidinophenylalanine tight-binding inhibitors of thrombin, *Thromb. Res.* 29, 635–642, 1983.

TABLE 8.5
Cyclic Amide Derivatives of N_α-Arylamidinophenyalanine as Inhibitors of Thrombin

Compound	K_i (µM)[a]
(A) N^α-tosyl-(4-amidinophenylalanyl butylamide	23
(B) Piperidino N^α-tosyl-4-amidinophenylalanine	2.3
(C) Piperidino N^α-tosyl-glycyl-4-amidinophenylalanine	0.048
(D) N^α-2-naphthylsulfonyl-(4-amidinophenyl)alanyl butylamide	3.6
(E) Piperidino-2-naphthylsulfonyl-4-amidinopphenylalanine	0.42
(F) Piperidino-2-naphthylsulfonylglycyl-4-amidinopphenylalanine	0.006

Source: Stürzebecher, J., Markwardt, F., Voight, B., Wagner, G., and Walsmann, P., *Thromb. Res.* 29, 635–642, 1983

[a] Bz-*dl*ArgPNa was the substrate in 0.1 M Tris, pH 8.0, when $K_i \geq 2$ µM; $K_i < 2$µM, S-2238 (H-D-Phe-Pip-Arg-pNA) was the substrate in 50 mM Tris, pH 8.0, containing 0.154 M NaCl and 1.25%(V/V). Reactions were performed at 25°C. The values for K_i were obtained by graphical methods after the method of Dixon (Dixon, M., The determination of enzyme inhibitor constants, *Biochem. J.* 55, 70–71, 1953)

TABLE 8.6
Comparison of N_α-Arylsulfonyl-4-Amidinophenylalanine Cyclic Amides in the Inhibition of Thrombin, Trypsin, Plasmin, and Factor Xa

R	K_i (µM)[a]			
	Thrombin	Trypsin	Plasmin[b]	Factor X
Tosyl	2.3	64	>1000	73
α-Napthylsulfonyl	2.8	110	>1000	240
β-Napthylsulfonyl	0.42	2.3	800	15
Tosylglycyl	0.048	1.7	57	20
α-Napthylsulfonylglycyl	0.014	1.6	34	18.
β-Napthylsulfonylglycyl (NAPAP)	0.006	0.69	30	7.9

Source: Stürzebecher, J., Markwardt, F., Voight, B., Wagner, G., and Walsmann, P., *Thromb. Res.* 29, 635–642, 1983. Castro, M.L., Kingston, M.L., and Anderson, F., *Anal. Biochem.* 226, 225–231, 1995

[a] The assay for purified bovine thrombin were performed with Bz-*dl*- ArgpNA as the substrate in 0.1 M Tris, pH 8.0, when $K_i \geq 2$ µM; $K_i < 2$µM, S-2238 (H-D-Phe-Pip-Arg-pNA) was the substrate in 50 mM Tris, pH 8.0, containing 0.154 M NaCl and 1.25%(V/V). Reactions were performed at 25°C. Assays for trypsin and plasmin were performed with BzArgNA in 0.1 M Tris, pH 8.0, at 25°C. The assay for factor Xa was performed with S-2222(Bz-Ile-Glu-Gly-Arg-pNA in 50 mM Tris, pH 8.0, containing 0.154 M NaCl and 1.25%(V/V). Reactions were performed at 25°C. The values for K_i were obtained by graphical methods after the method of Dixon (Dixon, M., The determination of enzyme inhibitor constants, *Biochem. J.* 55, 70–71, 1953)

[b] Plasmin has a greater specificity for the hydrolysis of bonds containing lysine residues but will catalyze the hydrolysis of BzArgpNA. Therefore, the use of BzArgpNA for the measurement of plasmin should not affect the observations of the relative effectiveness of the several inhibitors

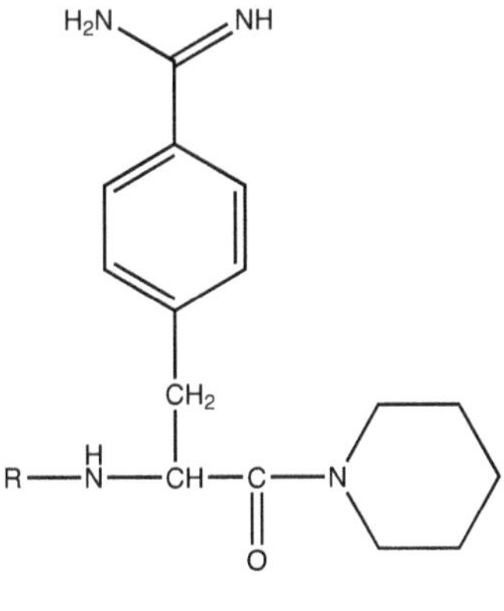

FIGURE 8.7   Base structure for Table 8.6.

*N*p-(β-naphthyl-sulfonyl-glycyl)-4-amidinophenylalanyl-piperidine (*N*$^\alpha$-(2-naphthyl-sulfonyl-glycyl)-*p*-amidinophenylalanyl-piperidine, NAPAP) has been used for continuing studies on the structure of the extended active site of thrombin. The crystal structure of the trypsin-NAPAP complex was used to model the thrombin-NAPAP complex.[107] These investigators note that the model predicts an excellent fit of the amidinophenyl group into the S_1 specificity pocket and that the naphthyl functions into the S_2 aromatic (aryl) binding site. There has been considerable work to improve the quality of NAPAP as an inhibitor of thrombin. Subsequent work from this laboratory[108] presented the crystal structures for the complex of NAPAP and bovine thrombin, confirming and extending the modeling studies. NAPAP has been shown to only bind as the D-isomer, with the amidino group binding to Asp199 and the naphthyl group interacting with Trp227 in the D-pocket. Di Cristofero and Landolfi[109] have studied the thermodynamics of NAPAP to thrombin, describing it as process of a chemical compensation. NAPAP binds to thrombin, eliciting a conformation change similar to that observed with the interaction of PPACK with thrombin. These investigators suggest that the binding of NAPAP involves entropy-enthalpy compensation,[110] which was also observed by Treuheit and coworkers.[111] One attempt prepared a conjugate of a heparin-derived pentasaccharide with NAPAP.[112] This derivative inhibited thrombin and accelerated the inactivation of factor Xa by antithrombin. Rewinkel and coworkers,[113] at Organon in The Netherlands, replaced the benzamidine moiety in the P_1 segment on NAPAP with a less basic moiety (Figure 8.8), which improved selectivity and also improved membrane transport (CaCO-2 system[114,115]). Hilbert and coworkers[116] noted that NAPAP and the OM-inhibitor described later which led to the development of argatroban (MD-805),[117] were based on the structure of TosArgOMe, one of the first, if not the first, synthetic substrates described for thrombin (Figure 8.9). Hilpert and coworkers modified the central building block, as well as the putative S_1 binding component to develop a new inhibitor Ro 46–6240 (Napsagatron) (Figure 8.9). This was a potent inhibitor of thrombin[118] but was not approved for clinical use. Stürzebecher and coworkers[119] had another paper that evaluated the 3-amidinophenylalanine derivatives as inhibitors of bovine thrombin and bovine factor Xa. There appears to be more specificity for thrombin than for factor

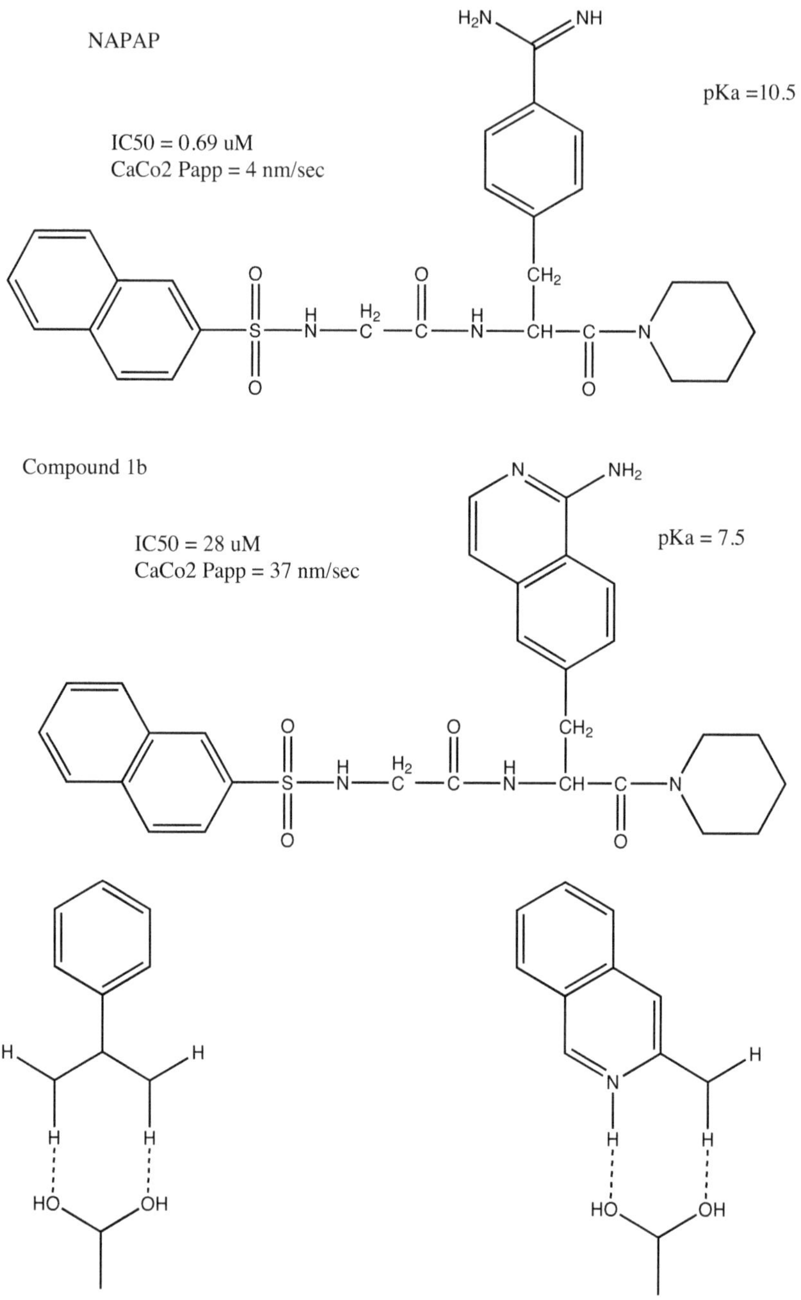

NAPAP

IC50 = 0.69 uM
CaCo2 Papp = 4 nm/sec

pKa =10.5

Compound 1b

IC50 = 28 uM
CaCo2 Papp = 37 nm/sec

pKa = 7.5

FIGURE 8.8 Structure of a novel thrombin inhibitor. Also shown is a model for the interaction between the P_1 residue and S_1 site on thrombin.

Source: Adapted from Rewinkel, J.B.M., Lucas, H., van Galen, P.J.M., *et al*., 1-Aminoisoquinoline as benzamidine isosteric in the design and synthesis of orally active thrombin inhibitors, *Bioorg. Med. Chem. Lett.* 9, 685–690, 1999.

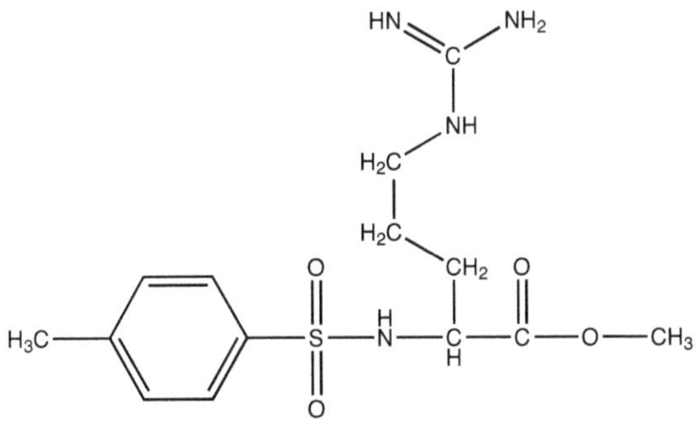

Tos ArgOMe

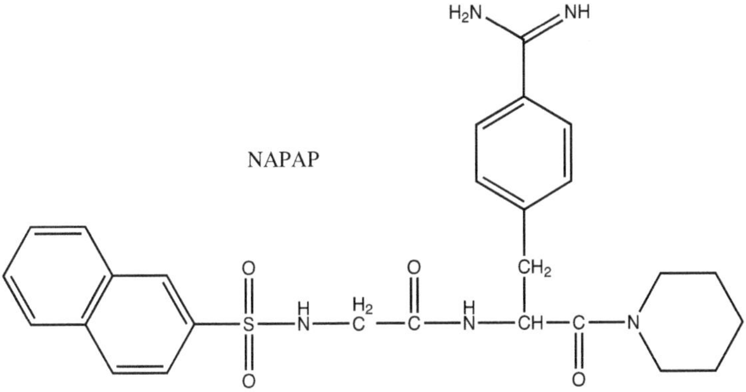

NAPAP

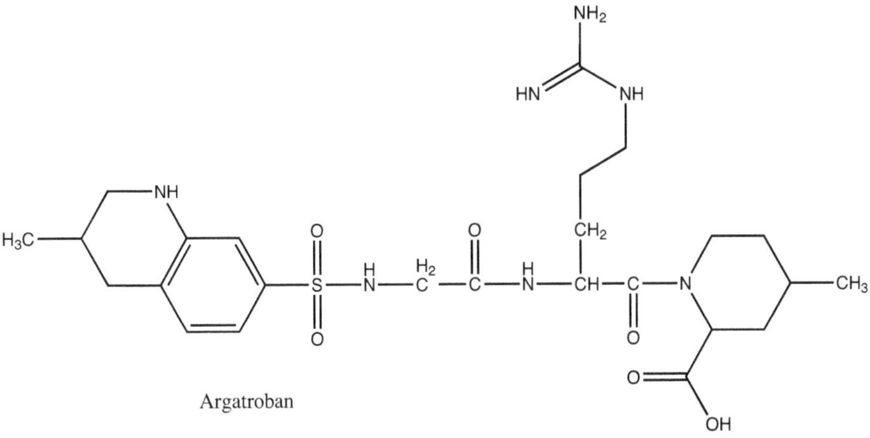

Argatroban

FIGURE 8.9 Structural relationships between TosArgOMe, NAPAP, argatroban, and napsagratran.

Source: Adapted from Hilpert, K., Ackermann, J., Banner, D.W., *et al.*, Design and synthesis of potent and highly selective thrombin inhibitors, *J. Med. Chem.* 37, 3889–3901, 1994.

Xa. I could find no further work from this group in this area other than the 2002 review by Professor Markwardt.[95]

The other major group developing synthetic organic inhibitors during the period of 1970–1990 was led initially by Joachim Dieter Geratz and later in collaboration with Richard Tidwell, both in the Department of Pathology at the University of North Carolina at Chapel Hill. I was fortunate to be in the same department at that time to hear about their work in progress. The work of the UNC group, as with the Erfurt group, was based on a derivative of benzamidine. As far as I can tell, although there was earlier work on trypsin,[120] Markwardt and Walsmann[96] were the first to demonstrate inhibition of thrombin by benzamidine. Geratz and coworkers[121] focused on α,ω-diphenoxy diamidines in their early work. In general, these compounds (Figure 8.10) proved to be better inhibitors of pancreatic kallikrein (tissue

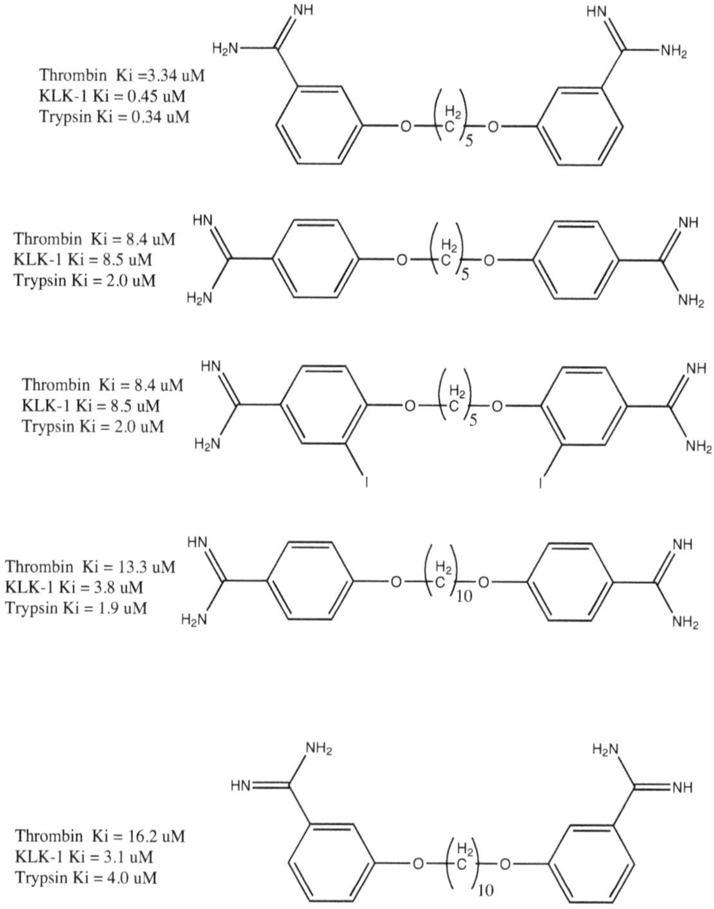

Thrombin Ki = 3.34 uM
KLK-1 Ki = 0.45 uM
Trypsin Ki = 0.34 uM

Thrombin Ki = 8.4 uM
KLK-1 Ki = 8.5 uM
Trypsin Ki = 2.0 uM

Thrombin Ki = 8.4 uM
KLK-1 Ki = 8.5 uM
Trypsin Ki = 2.0 uM

Thrombin Ki = 13.3 uM
KLK-1 Ki = 3.8 uM
Trypsin Ki = 1.9 uM

Thrombin Ki = 16.2 uM
KLK-1 Ki = 3.1 uM
Trypsin Ki = 4.0 uM

FIGURE 8.10   Some diamidino-α,ω-diphenylalkanes.

Source: Adapted from Geratz, J.D., Whitmore, A.C., Cheng, M.C.-F., and Piantidosi, C., Diamidino-α,ω-diphenylalkanes—Structure-activity relationships for the inhibition of thrombin, pancreatic kallikrein, and trypsin, J. Med. Chem. 16, 976–975, 1973.

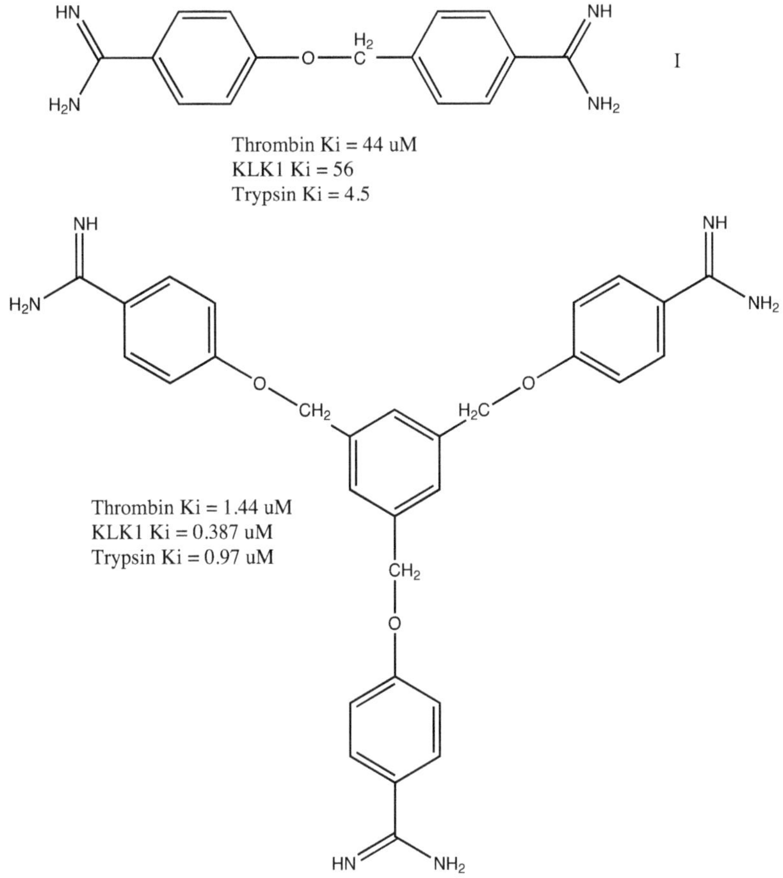

Thrombin Ki = 44 uM
KLK1 Ki = 56
Trypsin Ki = 4.5

Thrombin Ki = 1.44 uM
KLK1 Ki = 0.387 uM
Trypsin Ki = 0.97 uM

FIGURE 8.11 Aromatic di- and tri-amidines connected by a central 2-oxyalkanes as inhibitors of thrombin. Shown at the top is an aromatic diamidine with a central oxyalkane. At the bottom is shown a tris amidine coupled by monooxyalkane chains.

Source: Adapted from Geratz, J.D., Chang, M.C.-F., and Tidwell, R.R., New aromatic amidines with central α-oxyalkanes or α,ω-dioxyalkane chains. Structure-activity relationships for the inhibition of trypsin, pancreatic kallikrein, and thrombin and for inhibition of the overall coagulation process, *J. Med. Chem.* 18, 477–481, 1975 and Tidwell, R.R., Fox, L.L., and Geratz, J.D., Aromatic tris amidines. A new class of highly active inhibitors of trypsin-like proteases, *Biochim. Biophys. Acta* 445, 729–735, 1996.

kallikrein, now known as KLK1[122]) and trypsin.[121] There was continued work by this group on α-oxyalkane diamidines (Figure 8.11). There were also more effective inhibitors of KLK1 and trypsin than thrombin.[123] Subsequent work from the Chapel Hill group presents data for more complex tris-amidines (8–11).[124] Tidwell and coworkers[125] made more progress with compounds based on the fusion of the benzamidine ring with indole rings (Figure 8.12). Geratz and Tidwell reviewed the work of their groups in 1978.[126] Tidwell and Geratz made the comment that "the development of such agents would be greatly simplified if not only the primary structure, but the

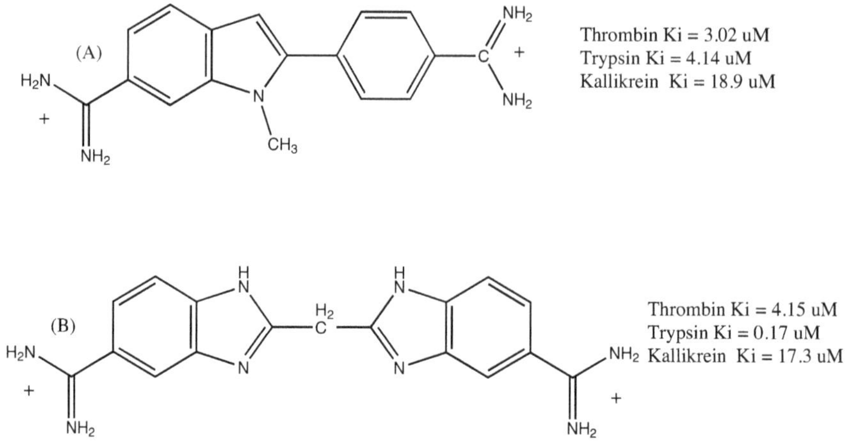

FIGURE 8.12   Some arylamidines with one or more indole rings. (A) 6-Amidino–2-(4-amidinophenol-1-methyl indole. (B) Bis-(5-amidinobenzimidoyl)methane.

Source: Adapted from Tidwell, R.R., Geratz, J.D., Dann, O., *et al.*, Diarylamidine derivatives with one or both of the aryl moieties consisting of indole or indole-like ring. Inhibition of arginine-specific estero-proteases, *J. Med. Chem.* 21, 613–623, 1978.

exact topography of thrombin was established." I am reminded of a discussion that I had with Stanford Moore over lunch at the Rockefeller University when I was privileged to be a research associate in the Moore-Stein laboratory. I cannot recall his exact words, but it was something to the effect that when they had solved the primary structure of pancreatic ribonuclease, they thought they would know the mechanism. Later, when they had the crystal structure from Fred Richards's laboratory at Yale, they thought they would know the mechanism. He said that they were disappointed that the mechanism of ribonuclease continued to evade them. Some years later, a collection of investigators in the Department of Pharmaceutical Chemistry at the University of California at San Francisco used emerging computation approaches to identify an inhibitor of human immunodeficiency virus 1 protease.[127] The computer makes a negative image of the active site cavity and then screens a library of chemicals for the best steric fit. It was (as told to me by one of the authors) a bit of a surprise that the program identified haloperidol as a candidate. This discovery did lead to work developing inhibitors based on haloperidol.[128] As we well see later, the process of drug discovery has markedly changed over the past 60 years.[129] Geratz and Tidwell[126] do suggest that their data obtained with *bis* and *tris* amidines provide support for the importance of a secondary binding side in thrombin. Their results also showed the importance of an indole amidine structure over a benzamidine-based structure. Geratz and coworkers[130] reported results with very simple compounds (Figure 8.13), showing the importance of an indole structure. There is a 30- to 40-fold increase in affinity for thrombin with 5-amidinoindole compared to benzamidine. Geratz and Tidwell[126] also presented some data suggesting that the inhibition of factor Xa was more important than the inhibition of thrombin in the

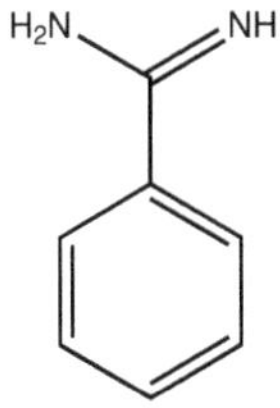

Benzamidine

bovine thrombin Ki = 332 uM
human thrombin Ki = 235 uM

5-amidinoindane

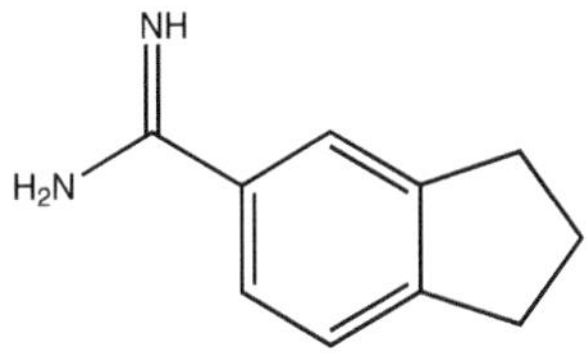

bovine thrombin Ki = 199 uM
human thrombin Ki = 1032 uM

5-amidinoindole

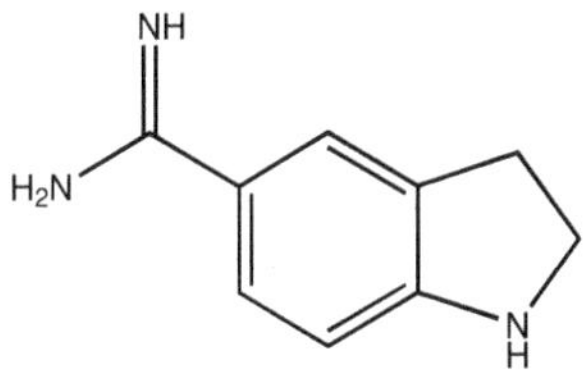

bovine thrombin Ki = 7.68 uM
human thrombin Ki = 7.9 uM

FIGURE 8.13 Comparison of the inhibition of thrombin by three simple aromatic diamidines.

Source: Adapted from Geratz, J.D. Stevens, F.M., Polakoski, K.L., Parrish, R.F., and Tidwell, R.R., Amidino-substituted aromatic heterocycles as probes of the specificity pocket of trypsin-like proteases, *Arch. Biochem. Biophys.* 197, 551–559, 1979.

prolongation of the activated partial thromboplastin time. However, such data is somewhat limited in that there are other arginine-specific proteases involved in the intrinsic coagulation pathways. There was continuing work from these laboratories in the development of cyclic aromatic amidine inhibitors.[131,132]

While the Markwardt group and the laboratories of Geratz and Tidwell were productive in the early development of inhibitors of thrombin, other groups later made advances in this area building on this early work. In addition to the work cited later, there was other work occurring in nonacademic laboratories that did not see the light of day. While companies vary in their approach to publications of research data, investigators usually do not have the time or incentive to publish. Then, there is gauntlet of corporate law. The intellectual property (IP) groups want you to save every restaurant napkin that you scribbled on, while the litigation group wants you to eat every scrap of paper at the end of the day.

Proflavin (proflavine) (Figure 8.14) is polycyclic aromatic acridine-like fluorescent dye that is used for nucleic acids[133] and proved useful for the study of the enzyme-active site of chymotrypsin.[134,135] Proflavin was shown to be a competitive inhibitor of chymotrypsin,[134] and the binding results in a hyperchromic red shift similar to that seen with the dye in dioxane.[135,136] Feinstein and Feeney at the University of California at Davis[137] showed that proflavin could be displaced from the enzyme-active site by avian orosomucoid inhibitor. These assorted studies showed that proflavin bound to the active site of chymotrypsin (and trypsin) and that the affinity of binding could be determined by the spectra changes associated with the binding of proflavin. Several investigators[138–140] then reported on the inhibition of thrombin by proflavin for the use of dye binding to measure active-site function. The Ki for the inhibition of thrombin esterase activity is approximately 10 µM and competes with benzamidine for the enzyme-active site. Conti and coworkers[141] reported on the crystal structure of a proflavin--thrombin complex, which shows binding to the S$_1$ specificity pocket. Proflavin is not as basic (pKa 9.5) as benzamidine (pKa 11.7) but would be positively charged at neutral pH. The spectral change observed on the binding of proflavin to thrombin suggesting a nonpolar environment can be explained, in part, by the hydrophobic nature of the S$_1$ specificity site.[141] Nienaber and coworkers[142] developed 6-fluorotryptamine (Figure 8.14) based on previous studies in their laboratory on the binding of proflavin to thrombin.[140] 6-Fluorotryptamine is a weak inhibitor of thrombin (Ki = 6.9 mM) with a binding site in the S$_1$ pocket. This work also supported the previous crystallography analysis of Conti and coworkers.[141] Much, if not all, of the early work on the development of synthetic thrombin inhibitors was done before the use of either solution-based or virtual high-throughput screening for drug discovery.[129] The Ohio State work and that from Markwardt, Stürzbecher, Hauptmann, Geratz, and Tidwell emphasized the importance of various component parts of their inhibitor, a concept to become known as fragment-based drug discovery.[143]

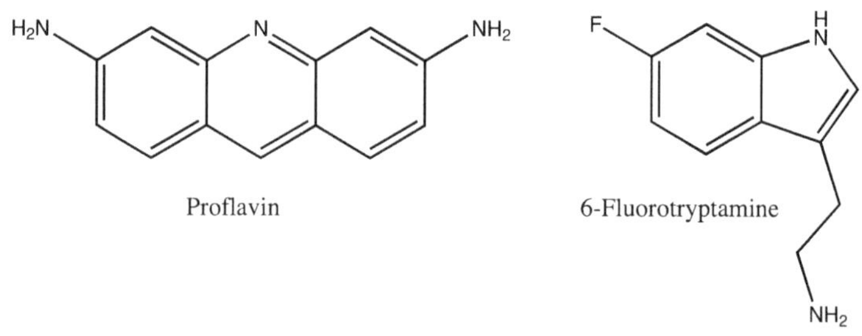

FIGURE 8.14 Structure of proflavine and 5-fluorotryptamine.

Source: Adapted from Nienaber, V.L., Boxrud, P.D., and Berliner, L.J., *J. Prot. Chem.* 19, 327–333, 2000.

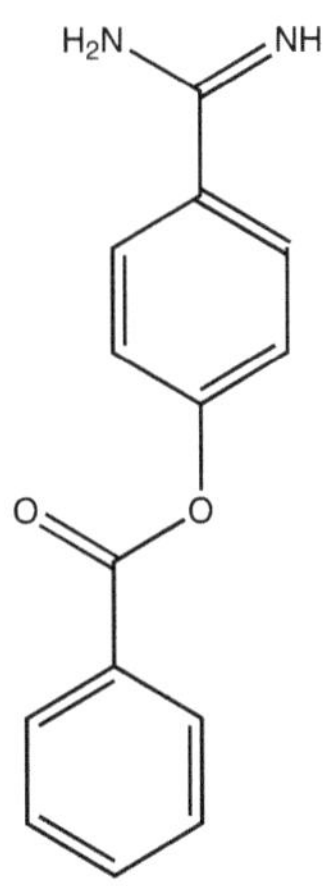

p-aminophenyl benzoate

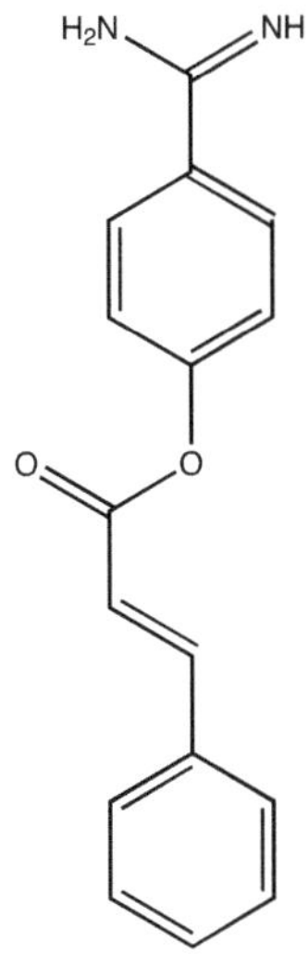

p-aminophenyl cinnamate 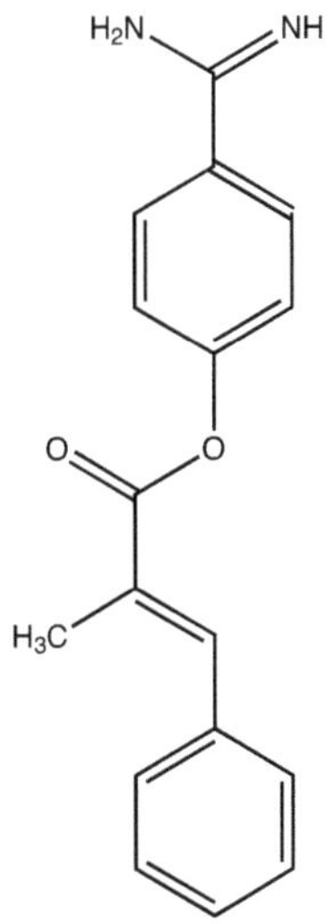   p-aminophenyl methylcinnamate

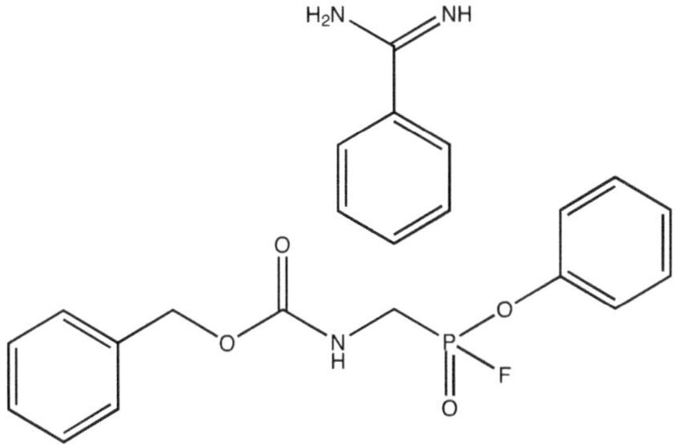

Phenyl N-benzyloxycarbonylamido-(4-amidinophenyl)methane phosphoflouridate

FIGURE 8.15 Structure of some novel amidino compounds which modify the active-site serine in thrombin. Shown at the top are some esters of 4-amidinophenol, which inactivate thrombin. At the bottom is a novel phosphofluoride inhibitor of thrombin.

Source: Adapted from Turner, A.D., Monroe, D.M., Roberts, H.R., Porter, N.A., and Pizzo, S.V., *p*-Amid-ino esters as irreversible inhibitors of factor IXa, and Xa, and thrombin, *Biochemistry* 25, 4929–4935, 1986 and Ni, L.-M. and Powers, J.C., Synthesis and kinetic studies on amidino-containing phosphofluo-ridate: Novel potent inhibitors of trypsin-like enzymes, *Bioorg. Med. Chem.* 6, 1767–1763, 1998.

Other researchers have used amidine function as an active site-directed inhibitor for thrombin. Turner and coworkers,[144] in a collaboration between the two blues, reported on the reaction of a series of aromatic esters of *p*-amidinophenol (Figure 8.15) with thrombin, factor Xa, and factor IXa. These are compounds that presumably react with the active serine to form an acyl-enzyme derivative that would undergo

TABLE 8.7
**Rate Constants for the Inactivation of Thrombin, Factor Xa, and Factor IXa
by Several Aromatic Esters of *p*-Amidinophenol**

	k_2 (M^{-1}min^{-1}) × 10^{-4}		
Enzyme	PAB[a]	PAC[b]	PAMC[c]
Human α-thrombin	108[d]	7.2	16
Bovine α-thrombin	144	4.5	32
Human factor Xa	78	110	9.5
Bovine factor Xa	22	105	3.5
Human factor IXa	4.8	61	16

Source: Turner, A.D., Monroe, D.M., Roberts, H.R., Porter, N.A., and Pizzo, S.V., *Biochemistry* 29, 4929–4935, 1986

[a] *p*-Amidinophenyl benzoate
[b] *p*-Amidinophenyl cinnamate
[c] *p*-Amidinophenyl methylcinnamate
[d] Data obtained in reactions at 23°C in 0.05 M Tris, pH 7.2. Deacylation does occur, but not rapid enough to affect the measurement of the rate of acylation

slow diacylation. Rate constants for the inactivation of thrombin and some other regulatory serine proteases by these compounds are shown in Table 8.7.

Ni and Powers[145] subsequently reported on the reaction of amidino phosphorofluoridates (Figure 8.15) with thrombin and other proteases. Their compound, phenyl-*N*-carbobenzoxyamido-(4-amidinophenyl)methane phosphofluoridate, showed a second-order rate constant of 1 × 10^5 M^{-1}s^{-1} with human thrombin (80 mM HEPES, pH 7.5, containing 485 mM NaCl and 8% DMF at 25°C). More complex phosphono esters were used by Pan and coworkers[146] as activity-based probes.[147] The use of phosphono esters for the study of proteases has been presented by Grzywa and Sieńczyk.[148]

A group of Japanese investigators developed inhibitors which they described as OM inhibitors and were also described as MQPA inhibitors. Kikomoto and coworkers[149] showed that the stereochemistry of a substituted piperidine ring was important in their arginine-based inhibitors (Figure 8.16).[150] It is suggested that the development of dansyl-arginine-*N*-(3-ethyl-1,5 pentandiol)amide (dansyl-arginine-4-ethylpiperidine, DAPA)[150–152] was based on prior work by Okamota, Hijikata, and coworkers (Figure 8.17). The interaction of MQPA ((2R,4R)-4-methyl-[N²[(3-methyl-1,2,3,4-tetrahydro-8-quinolinyl)sulfonyl]-L-arginine]-2-piperidinecarboxylic acid with thrombin was not affected by heparin or prior modification of lysine residues with pyridoxal phosphate. This work resulted in the development of argatroban as an inhibitor of thrombin for clinical use. This development process is described in a comprehensive work by Hijikata-Okunomiya and Okamoto.[153]

Obst and coworkers[154,155] continued the work on improving the selectivity of thrombin inhibitors, focusing on the S$_1$ binding site and the S$_2$ aromatic(aryl) binding site (D-pocket involving Trp227 and Tyr49). One compound designated roc-13d had an improved selectivity. (Ki trypsin/Ki thrombin; the higher the number, the greater

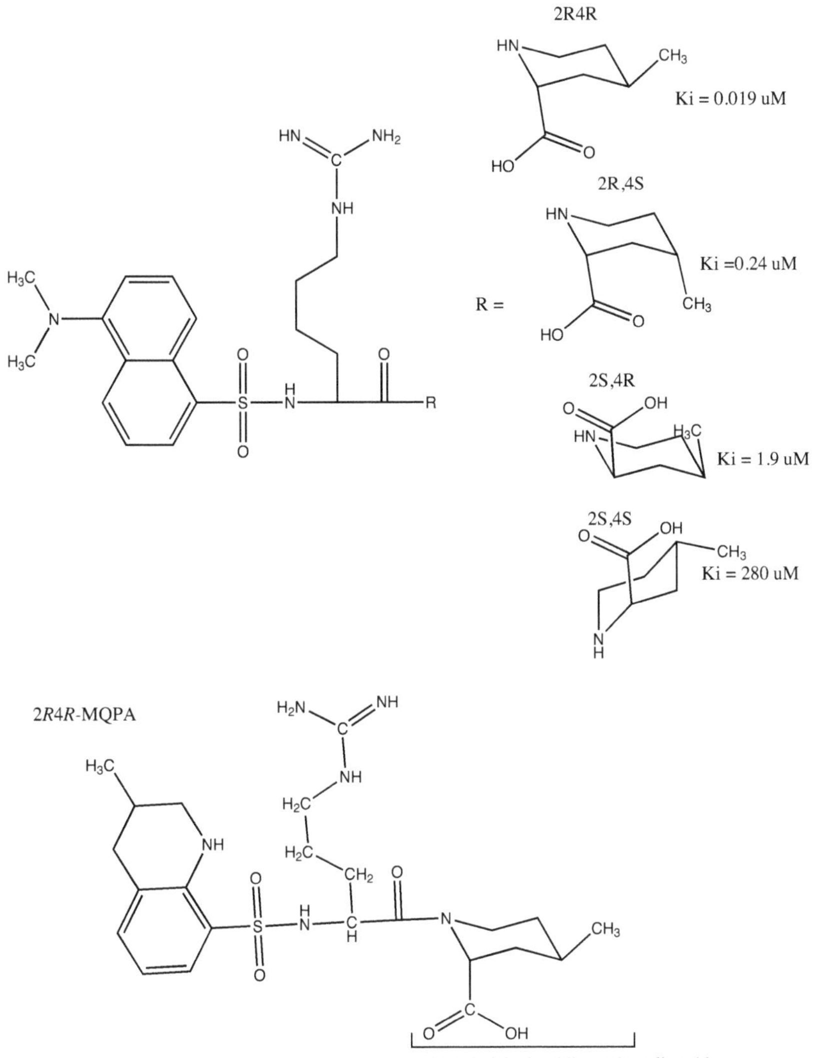

FIGURE 8.16 The structure of some OM-inhibitors and NAPAP. Shown at the top are stereoisomers of two OM-inhibitors. Shown at the bottom is 2R4R-MQPA(2R4R-4-methyl-1-[N²-[(3-methyl-1,2,3,4-tetrahydro-8-quinolinyl)sulfonyl]-L-arginine)]-2-piperidine carboxylic acid (NAPAP).

Source: Adapted from Okamoto, S. and Hijikaka-Okunomiya, A., Synthetic selective inhibitors of thrombin, *Methods Enzymol.* 222, 328–340, 1993.

the selectivity as assessed by this ratio. As will be seen later, much better numbers were obtained in later work by Merck.) Nevertheless, this work was very significant in its own right. Osbt and coworkers worked with the crystal structure of thrombin-inhibitor complexes to develop compounds (Figure 8.18) that maximized interaction

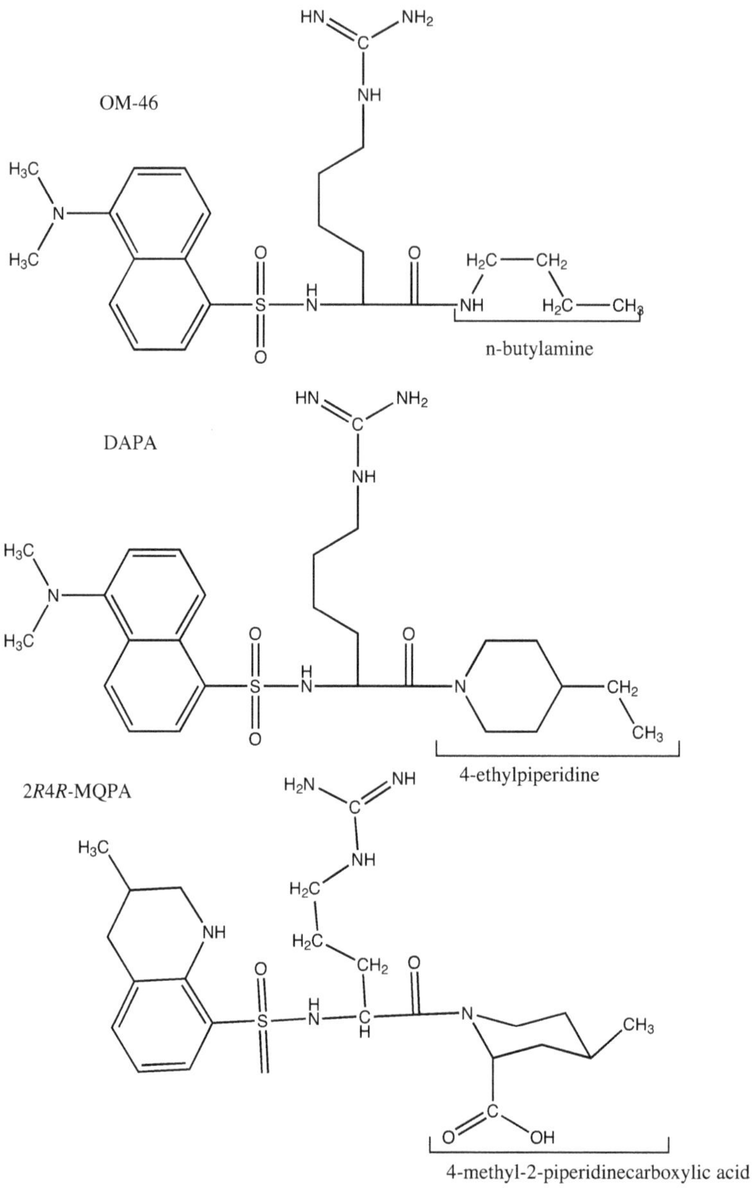

FIGURE 8.17   The development of DAPA. Shown is OM-46, DAPA, and 2R4R-MQPA(2R4R-4-methyl-1-[N²-[(3-methyl-1,2,3,4-tetrahydro-8-quinolinyl)sulfonyl]-L-arginine)]-2-piperidine carboxylic acid (NAPAP).

Source: Adapted from Hijikata-Okunomiya, A. and Okamoto, S., A strategy for a rational approach to designing synthetic selective inhibitors, *Sem. Thromb. Hemost.* 18, 135–149, 1992; Nesheim, M.E., Prendergast, F.G., and Mann, K.G., Interaction of a fluorescent active-site-directed inhibitor of thrombin: Dansylarginine N-(3-ethyl-1,5-pentandiyl)amide, *Biochemistry* 18, 996–1003, 1979; and Okamoto, S. and Hijikaka-Okunomiya, A., Synthetic selective inhibitors of thrombin, *Methods Enzymol.* 222, 328–340, 1993.

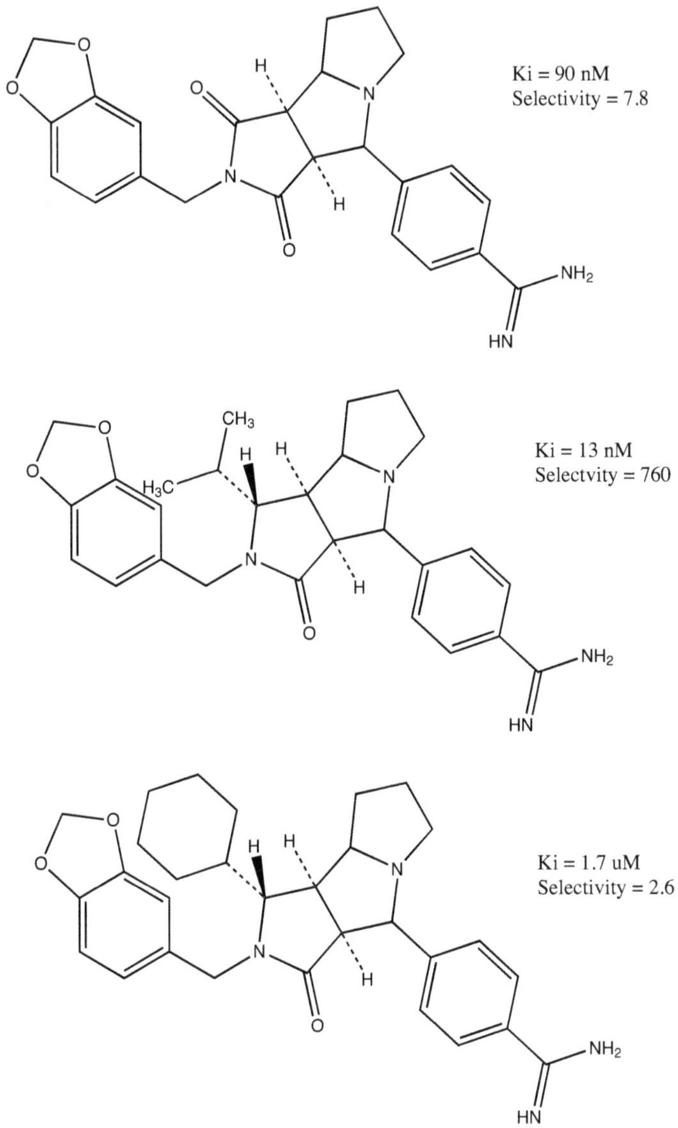

FIGURE 8.18 Some stages in the development of an orally available selective thrombin inhibitor. Shown at the top is a lead compound that was developed from the crystal structure of thrombin The middle structure was developed from the lead compound and shows both a better affinity and selectivity (Ki trypsin/Ki thrombin), while the lower compound has a poor fit in the P-pocket and presents a better Ki, but poor selectivity.

Source: Adapted from Adapted from Obst, U., Grenwich, U., Diedrich, F., Weber, L., and Banner, D.W., Design of novel nonpeptidic thrombin inhibitors and structure of a thrombin-inhibitor complex, *Angew. Chem. Int. Edn.* 34, 1739–1742, 1995; and Obst, U., Banner, D.W., Weber, L., and Diedrich, Molecular recognition at the thrombin active site: Structure-based design and synthesis of potent and selective thrombin inhibitors and the X-ray crystal structure of two thrombin-inhibitor complexes, *Chem. Biol.* 4, 287–295, 1997).

with the S_1 site (primary binding site), S_2 site (D-pocket binding the P_2 residue), and S_3 site (P-pocket).[155]

While many groups have worked to develop selective inhibitors for thrombin, I have selected work from Merck (West Point, Pennsylvania) for discussion. These investigators and others have noted the limited bioavailability of drugs based on strong basic components binding to the S_1 specificity site. These highly charged drugs, while potent inhibitors, were not effectively transported from the gastrointestinal tract into the vascular space.[156,157] As was the case with other investigators, the Merck investigators developed peptidomimetics based on the tripeptide, D-Phe-Pro-Arg,[158,159] which had been shown to be an excellent inhibitor of thrombin either as an aldehyde[157] or chloromethylketone.[158] Sanderson and coworkers[160] reported the synthesis and evaluation of a candidate for an orally available inhibitor of thrombin (Compound A in Figure 8.19). Replacement of the methyl group on the pyridinone core with a propyl group by Issacs and coworkers[161] (Compound B in Figure 8.22) permitted a better fit with the D-pocket, while a cyclopropyl group was too bulky. The cyclopropyl derivative had a Ki of 16 nm compared to 0.85 for the propyl derivative. In another work published at the same time, Sanderson and coworkers[162] extended the half-life of L-374,087 by substituting a 3-alkylpyrazinone for the 3-sulfoamino-pyrazinone (methylene group substituted for sulfo group (Compound C in Figure 8.19). Several years later Raffel and coworkers[163] showed the compound (L-375,378) developed by Sanderson and coworkers[162,] was metabolized to derivatives (Figure 8.20) that retained anticoagulant activity. Rittle and coworkers[164] discovered that placement of an o-aminoalkylbenzamide in the P_1 position improved potency and selectivity (Figure 8.21). These investigators suggested that the improved potential of the 2-methylamino-5-clorophenyl derivative was due to hydrogen bonding to Glu202. In vivo pharmacokinetic studies at Merck revealed that the functional half-life was limited by metabolism of the inhibitor.[165] These observations led to the development of an inhibitor more resistant to oxidation and alkylation (Figure 8.22). Young and coworkers[166] followed this work with the P_1 segment. This work focused on the development of a weakly basic or neutral P_1 segment with increasing the binding efficacy of the P_2P_3 segments (Schecter and Berger nomenclature[89]). Young and coworkers[166] developed compounds with o-triazole and tetrazole rings in the P_1 position (Figure 8.23). Nantermet and coworkers[167] built on the previous work at Merck with a pyridine N-oxide derivative. Nfuh and Larionov[168] reviewed the development of heterocyclic N-oxide derivatives. While the Merck work was excellent chemistry, it did not lead to a product. A group at AstraZeneca[169] developed an inhibitor, designated AR-H067637 (Figure 8.24), which is obtained from a precursor form, AZ00837. This drug was taken into clinical trials,[170] but to the best of my knowledge, has not proceeded to the marketing stage. Having worked in the biopharm industry, I understand that many factors go into the development and eventual marketing of a product. Failure to go to market can have nothing to do with the quality of the product. Other approaches to the development of less basic inhibitors have been advanced,[171,172] where work has focused on the modification of the P_1 group in the inhibitor.

The identification of natural products as inhibitors of thrombin extends beyond heparins and saliva-based anticoagulants such as heparin. de Andrade Moura and

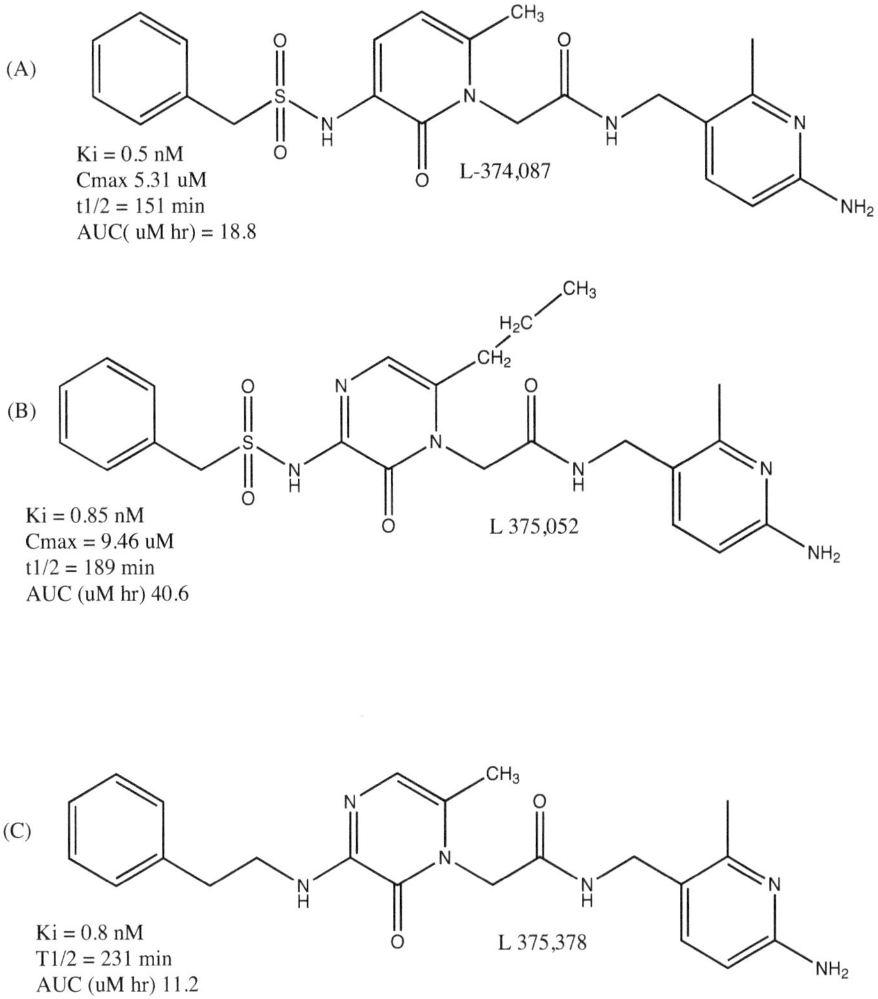

(A)

Ki = 0.5 nM
Cmax 5.31 uM
t1/2 = 151 min
AUC(uM hr) = 18.8

L-374,087

(B)

Ki = 0.85 nM
Cmax = 9.46 uM
t1/2 = 189 min
AUC (uM hr) 40.6

L 375,052

(C)

Ki = 0.8 nM
T1/2 = 231 min
AUC (uM hr) 11.2

L 375,378

FIGURE 8.19 Development of nonpeptidic orally available inhibitors of thrombin. Shown at the top (A) is a candidate compound (L-374,087) for a thrombin inhibitor (L-374,087, an efficacious, orally bioavailable, pyridinone acetamide thrombin inhibitor. Shown in the middle (B) is a second generation compound (L-375,052). Shown at the bottom (C) is another second-generation product (L-375,378) which addressed one of the weaknesses of (L-374,087)

Source: Sanderson, P.E.J., Cutrona, K.J., Dorsey, B.D., Dyer, C.L., and Vacca J., L-374,087, an efficacious, orally bioavailable pyridinone acetamide thrombin inhibitor, *Bioorg. Med. Chem. Lett.* 8, 817–822, 1998; Isaacs, R.C.A., Cutrona, K.J., Newton, C.L., *et al.*, C6 modification of the pyridinone core of thrombin inhibitor L-374,087 as a means of improving its oral absorption, *Bioorg. Med. Chem. Lett.* 8, 1719–1724, 1998; and Sanderson, P.E.J., Lyle, T.A., Cutrona, K.J., Efficacious, orally bioavailable thrombin inhibitors based on 3-aminopyridinone or 3-aminopyrazinone acetamide peptidomimetic templates, *J. Med. Chem.* 41, 4466–4474, 1998.

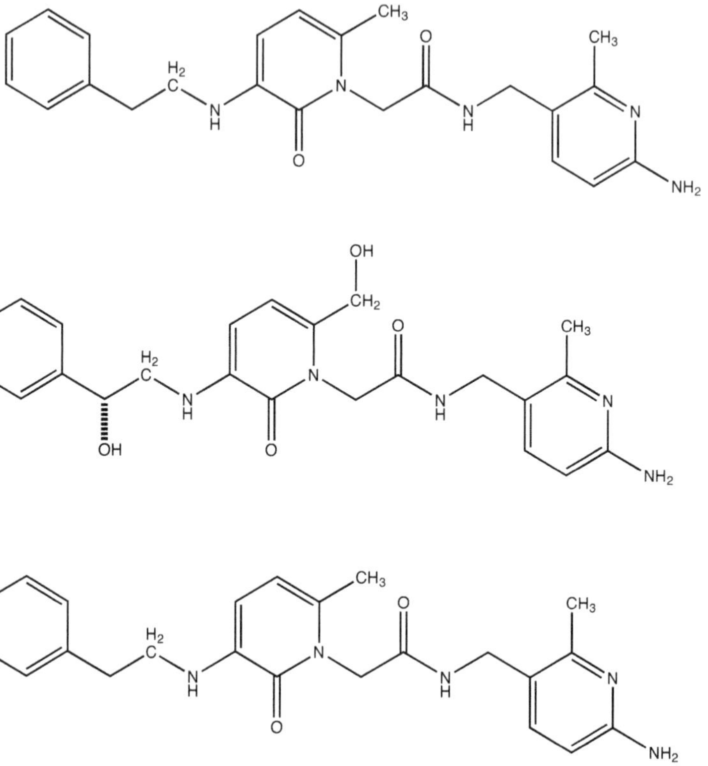

FIGURE 8.20 Metabolic development of orally available thrombin inhibitors. Shown are the products obtained by the *in vivo* degradation of L-375,378.

Source: Adapted from Riffel, K.A., Song, H., Go, X., Yan, K., and Lo, M.-W., Simultaneous determination of a novel thrombin inhibitor and its two metabolites in human plasma by liquid chromatography/tandem mass spectrometry, *J. Pharm. Biomed. Anal.* 23, 607–616, 2000.

coworkers[173] reported on the inhibition of thrombin by diterpenes obtained from a marine algae (Figure 8.25). These compounds are weak inhibitors, as measured by the effect on fibrinogen clotting activity or the hydrolysis of S-2238. Subsequent work by Pereira and coworkers[174] presented modeling studies for the mechanism of interaction of these compounds with thrombin. Dwivedi and Pomin[175] have reviewed the various inhibitors of blood coaguation which can be obtained from marine organisms.

As can been seen from this discussion, there has been some interesting chemistry on the way to the development of an approved synthetic organic chemical inhibitor of thrombin. There are several recent reviews of value that consider the later aspect of the development process.[176–179] prior to approval for clinical use. There are two direct thrombin inhibitors that are currently approved for clinical use in the United States. One inhibitor was removed from clinical use as a result of toxicity. I am also not including hirudin or hirudin derivatives such as bivalirudin (Angiomax). There is one direct thrombin inhibitor, dabigatran etexilate (Pradaxa) (Figure 8.26), which is approved for oral use.[180] The other synthetic inhibitor, argatroban (Argatroban Novoplus) (Figure 8.27), is approved for *in vivo* use (IV injection).[181,182] Argatroban

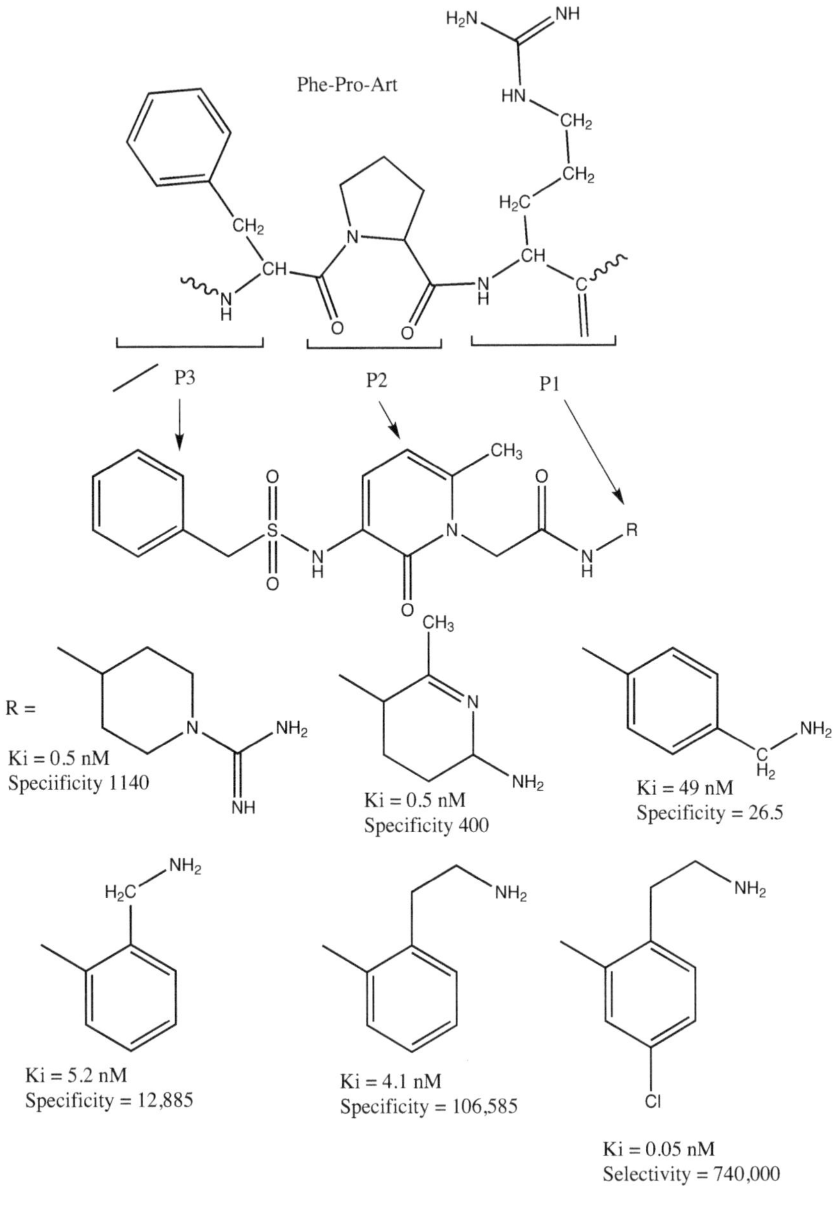

FIGURE 8.21 Modification of the P_1 position in a candidate orally available thrombin inhibitor. Shown at the top is the Phe-Pro-Arg sequence, which has been used as the basis for many substrates and inhibitors for thrombin. Below that is a template for an orally available thrombin inhibitor. At the bottom are several fragments evaluated for performance at the P_1 position.

Source: Adapted from Rittle, K.E., Barrow, J.C., Cutrona, K.J., *et al.*, Unexpected enhancement of thrombin inhibitor potency with *o*-aminoalkylbenzylamides in the P_1 position, *Biooorg. Med. Chem. Lett.* 13, 3477–3482, 2003.

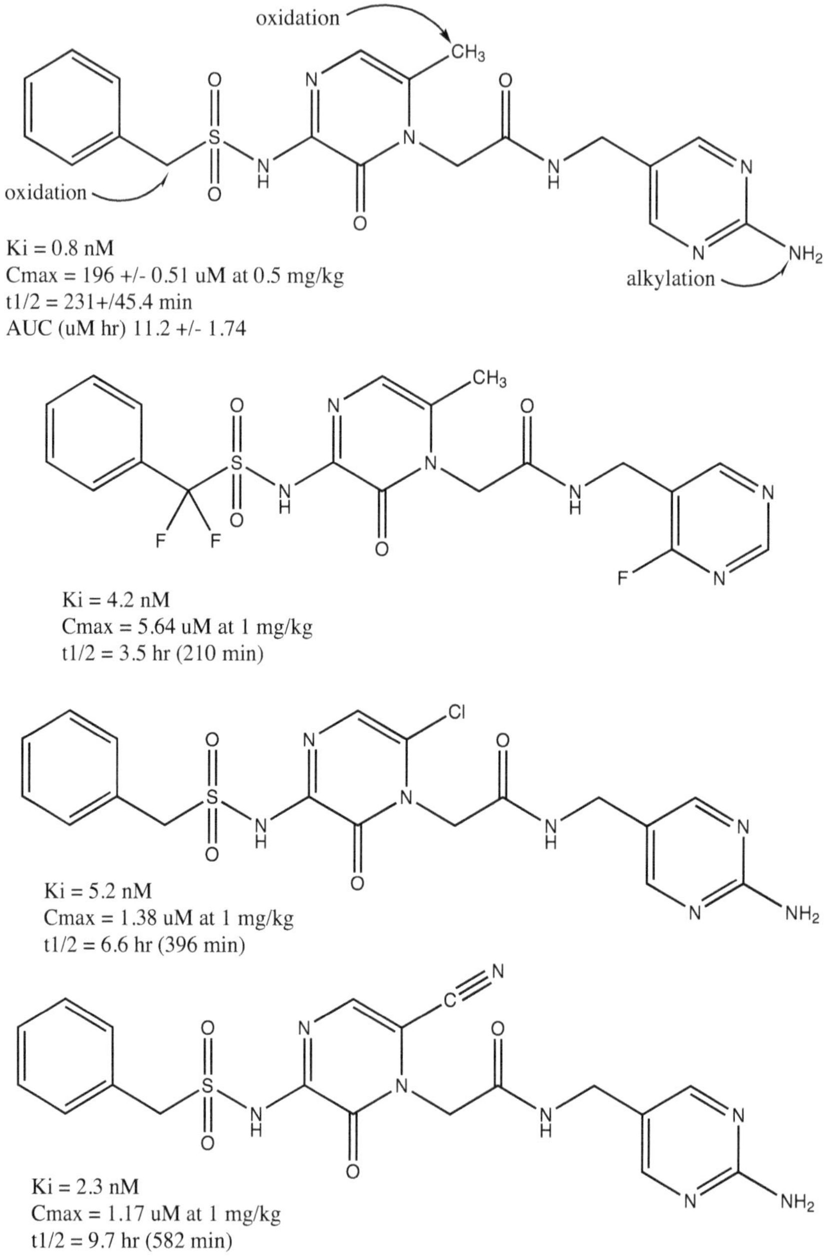
oxidation
CH3
oxidation
alkylation
NH2
Ki = 0.8 nM
Cmax = 196 +/- 0.51 uM at 0.5 mg/kg
t1/2 = 231+/45.4 min
AUC (uM hr) 11.2 +/- 1.74
CH3
F
F
F
Ki = 4.2 nM
Cmax = 5.64 uM at 1 mg/kg
t1/2 = 3.5 hr (210 min)
Cl
NH2
Ki = 5.2 nM
Cmax = 1.38 uM at 1 mg/kg
t1/2 = 6.6 hr (396 min)
C
N
NH2
Ki = 2.3 nM
Cmax = 1.17 uM at 1 mg/kg
t1/2 = 9.7 hr (582 min)

FIGURE 8.22 Metabolic product development. Shown are the primary sites of *in vivo* modification of an orally available thrombin inhibitor and structural modifications to obviate such *in vivo* modifications.

Source: Adapted from Burgey, C.S., Robinson, K.A., Lyle, T.A., *et al.*, Metabolism-directed optimization of 3-aminopyrazinone acetamide thrombin inhibitors. Development of an orally bioavailable series containing P_1 and P_3 pyridines, *J. Med. Chem.* 46, 461–473, 2003.

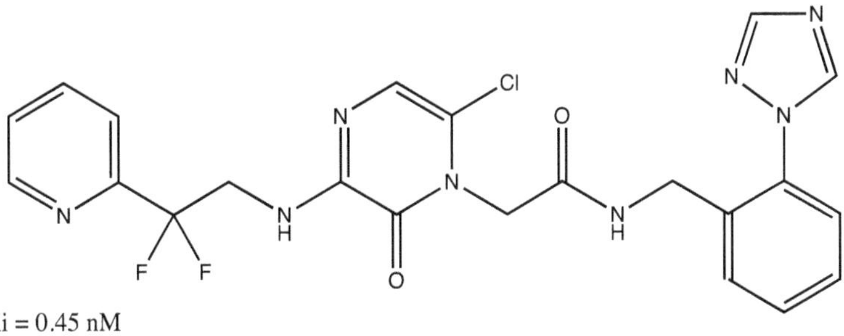

Ki = 0.45 nM
Tnn1/2 = 3.5 hrs
Cmax = 1.1 uM with 0.5 kg dose

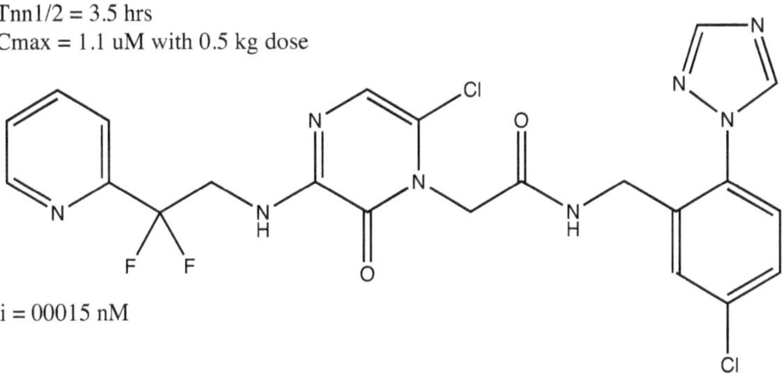

Ki = 00015 nM

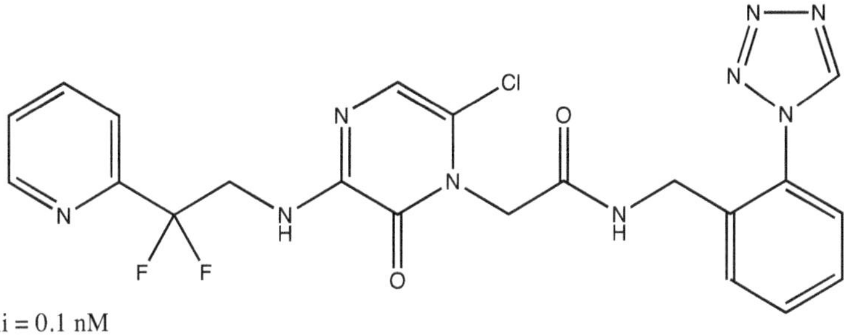

Ki = 0.1 nM

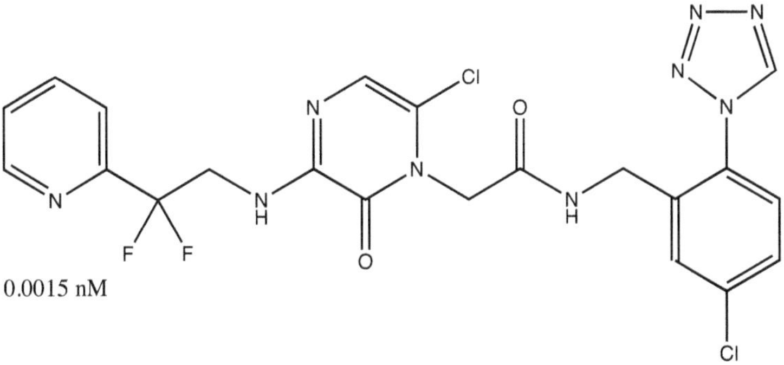

Ki = 0.0015 nM

FIGURE 8.23 Development of an orally available thrombin inhibitor with a triazole or tetrazole in the P$_1$ position. Shown at the top are two inhibitors with a 1,2,4 triazole in the P$_1$ position and below two inhibitors with 1,2,3,4 tetrazole in the P$_1$ positions.

Source: Adapted from Young, M.B., Barrow, J.C., Glass, K.L., *et al.*, Discovery and evaluation of potent P$_1$ aryl heterocycle-based thrombin inhibitors, *J. Med. Chem.* 47, 2995–3008, 2004.

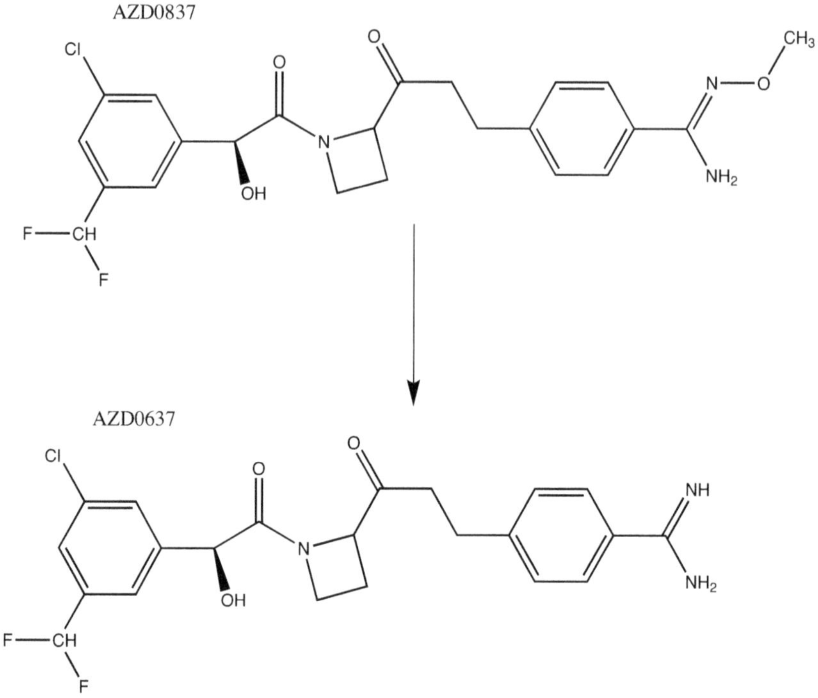

AZD0837

AZD0637

FIGURE 8.24   An orally available thrombin inhibitor that is a prodrug. Shown at the top is AZD0837, which is transformed *in vivo* to the active drug, AZD0637. Activation is accomplished by *in vivo* hydrolysis of the *N*-methoxl function.

Source: Adapted from Deinum, J., Mattsson, C., Inghardt, T., and Elg, M., Biochemical and pharmacological effects of the direct thrombin inhibitor AR-H067637, *Thromb. Haemost.* 101, 1051–1059, 2009.

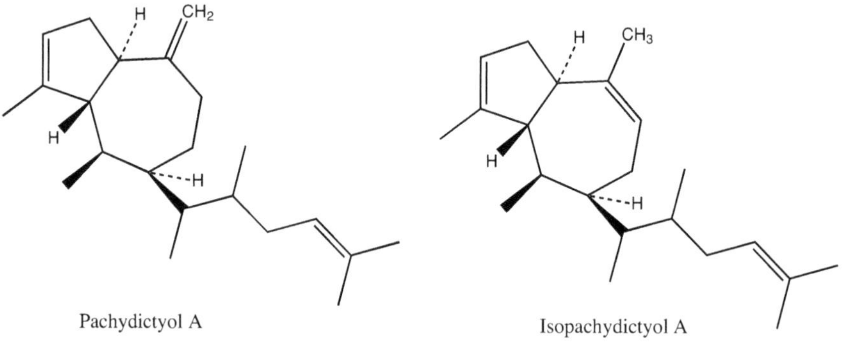

Pachydictyol A   Isopachydictyol A

FIGURE 8.25   Natural product inhibitors of thrombin from a marine algae. Shown are the structures of two thrombin inhibitors obtained from the marine algae *Dictyota mentruales*.

Source: Adapted from de Andrade Moura, L., Marque de Almeida, A.C., Souza, T.F., *et al.*, Antiplatelet and anticoagulant effects of diterpenes isolated from the marine alga, *Dictyota menstrualis*, *Marine Drugs* 12, 2471–2484, 2014.

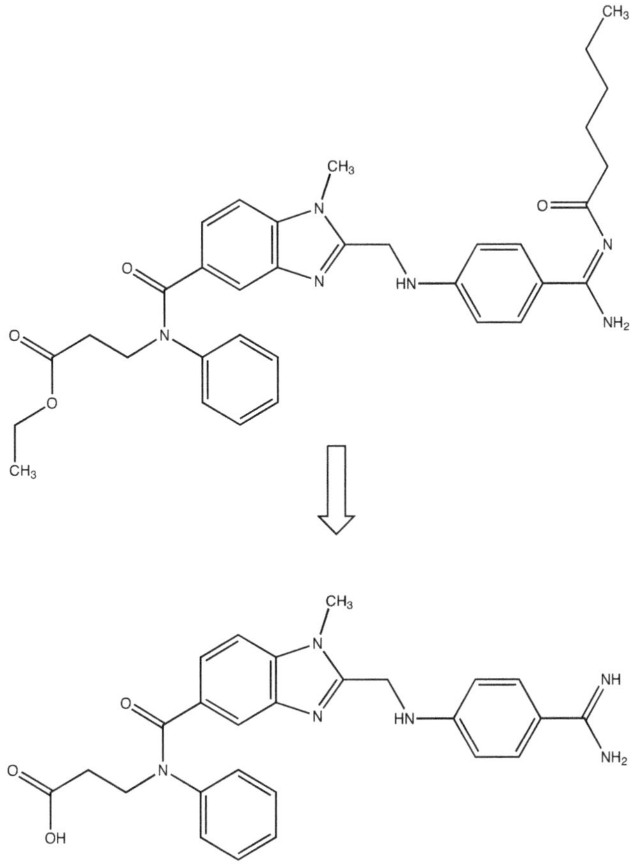

FIGURE 8.26 The structure of dabigatran and its prodrug form. Shown is the prodrug, dabigatran etexilate, and its conversion by *in vivo* hydrolysis to the active form. The *in vivo* hydrolysis removes the alkylcarboxylic acid, exposing the amidine function.

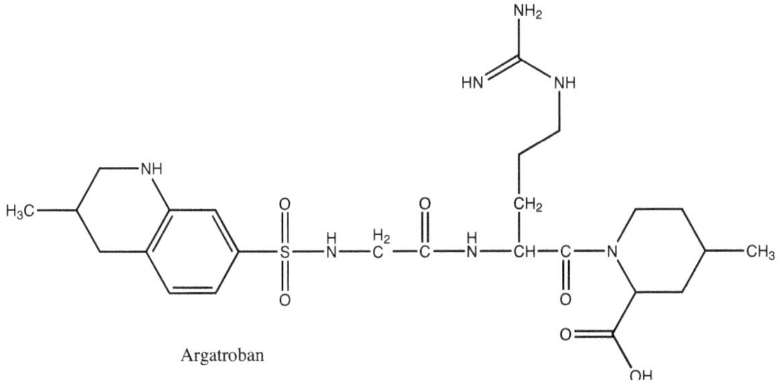

FIGURE 8.27 The structure of argatroban. Argatroban has an exposed guanidine function and is an active drug.

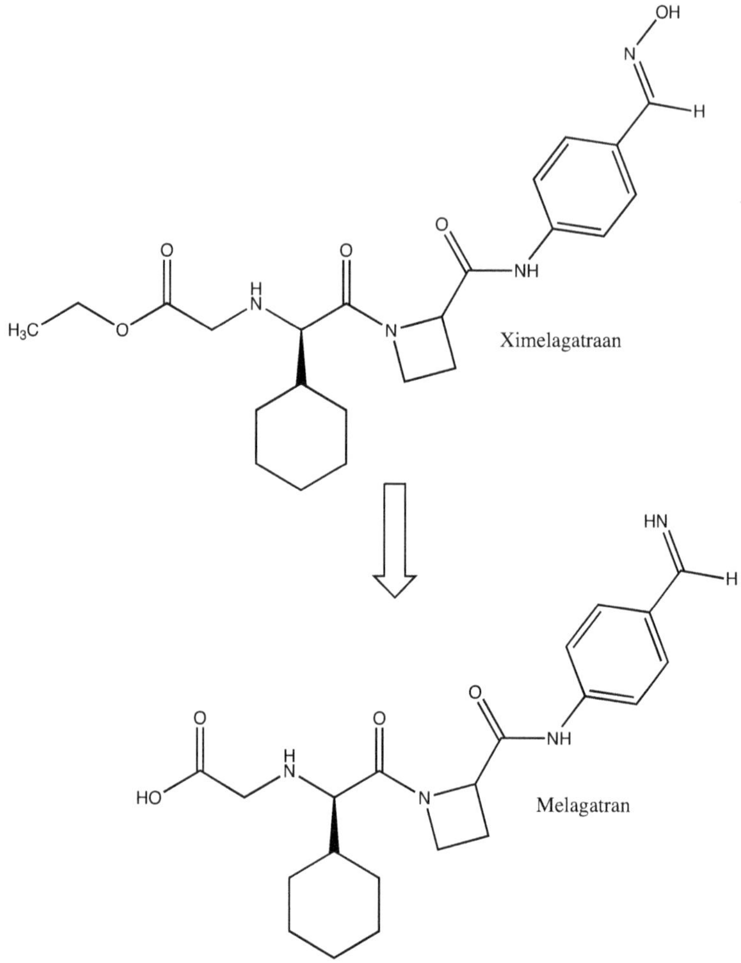

FIGURE 8.28 The structure of ximelagatran and melagatran. Ximelagatran is the inactive precursor drug, which is converted to melagatran, the active drug form, by *in vivo* hydrolysis.

is a highly charged compound containing arginine, which will not cross cell membranes and requires intravenous administration and which was approved for use in heparin-induced thrombocytopenia,[182,183] but is also used in situations with acquired antithrombin deficiency,[183] including COVID-19.[184,185] Ximelagatran/melagatran (Exanta) (Figure 8.28)[186–188] was developed as an oral direct thrombin inhibitor but was withdrawn as a result of hepatic toxicity.[188,189]

REFERENCES

1. de Ridder, G.G., Lundblad, R.L., and Pizzo, S.V., Actions of thrombin in the interstitium, *J. Thromb. Haemost.* 14, 40–47, 2016.

2. Lumpkin, T.B., Southern, B.D., Scheraga, H.A., *et al.*, Direct thrombin inhibition with dabigatran alters the matix-induced pro-fibrotic phenotype of fibroblasts, *Am. J. Resp. Crit. Care Med.* 197, A5794, 2018.

3. Morowitz, P., Die Chemie der Blutgerinnung, *Ergibn. Physiol.* 4, 307–422, 1905.

4. Mellanby, J., The coagulation of blood, *J. Physiol.* 38, 28–112, 1908.

5. Weymouth, F.W., The relation of metathrombin to thrombin, *Am. J. Physiol.* 32, 266–285, 1913.

6. Rich, A.R., The nature and properties of metathrombin, *Am. J. Physiol.* 43, 549–570, 1917.

7. Jesty, J., Dissociation of complexes and their derivatives formed during inhibition of bovine thrombin and activated factor X by antithrombin III, *J. Biol. Chem.* 254, 1044–1049, 1979.

8. Jesty, J., The kinetics of formation and dissociation of the bovine thrombin.antithrombin III complex, *J. Biol. Chem.* 254, 10044–10050, 1979.

9. Fish, W.W. and Björk, I., Release of a two-chain form of antithrombin from the antithrombin-thrombin complex, *Eur. J. Biochem.* 101, 31–38, 1979.

10. Rich, A.R., The nature and properties of metathrombin, *Am. J. Physiol.* 43, 549–570, 1917.

11. Gerber, C.F. and Blanchard, E.W. The effect of certain substances on clotting times in vitro, *Am. J. Physiol.* 144, 447–456, 1945.

12. Fantl, P., The use of substances depressing antithrombin activity in the assay of prothrombin, *Biochem. J.* 57, 416–421, 1954.

13. White, G.C., 2nd., Roberts, H.R., Kingdon, H.S., and Lundblad, R.L., Prothrombin complex concentrates: potentially thrombgenic materials and clues to the mechanism of thrombosis in vivo, *Blood* 49, 159–170, 1977.

14. Wilson, W.C., Action on blood clotting of peptide material in human brain: with special reference to enhancement of antithrombin activity, *Scot. Med. J.* 2, 351–358, 1952.

15. Ferguson, J.H. and Glazko, A.J., Heparin, *J. Lab. Clin. Med.* 26, 1559–1564, 1941.

16. Strickland, D.K. and Kounnas, M.E., Mechanisms of cellular uptake of thrombin-antithrombin II complexes. Role of the low-density lipoprotein receptor-related protein as a serpin-enzyme complex receptor, *Trends Cardiovasc. Med.* 7, 9–16, 1997.

17. Gans, H., Stern, R., and Tan, B.H., On the *in vivo* clearance of thrombin, *Thromb. Diath. Haemhr.* 22, 1–12, 1969.

18. Vogel, C.N., Kingdon, H.S., and Lundblad, R.L., Correlation of in vivo and in vitro inhibition of thrombin by plasma inhibitors, *J. Lab. Clin. Med.* 93, 661–673, 1979.

19. Abildgaard, Fagerol, M.K., and Egeberg, O., Comparison of progressive antithrombin activity and concentrations of these inhibitors in human plasma, *Scand. J. Clin. Lab. Invest.* 26, 349–354, 1970.

20. Downing, M.R., Bloom, J.W., and Mann, K.G., Comparison of the inhibition of thrombin by three plasma proteinase inhibitors, *Biochemistry* 17, 2649–2653, 1978.

21. Mosesson, M.W., Antithrombin I. Inhibition of thrombin generation in plasma by fibrin formation, *Thromb. Haemost.* 89, 9–12, 2003.

22. Verstraete, M., Antithrombins, *Thromb. Diath. Haemorrh.* Suppl 20, 386–389, 1966.

23. Griffith, M.G., Carraway, T., White, G.C., and Dombrose, F.A., Heparin cofactor activities in a family with hereditary antithrombin deficiency: evidence for a second heparin cofactor in human plasma, *Blood* 61, 111–118, 1983.

24. Sumitomo-Ueda, Y., Aihara, K.-i., Tes, T., *et al* Heparin cofactor II protects against angiotensin II-induced cardiac remodeling via attenuation oxidative stress in mice, *Hypertension* 56, 430–436, 2010.

25. Cucoranu, I., Clempus, R., Dikalova, A., *et al.*, NAD(P)H oxidase 4 mediates transforming growth factor -β1-induced differentiation of cardiac fibroblasts into myofibroblasts, *Circ. Res.* 97, 900–907, 2005.

26. Larsson, H., Sjoblom, T., Dixelius, J., *et al.*, Antiangiogenic effects of latent antithrombin through perturbed cell-matrix interactions and apoptosis of endothelial cells, *Cancer Res.* 60, 6723–6729, 2000.

27. de la Morena-Barrio, M., Sandoval, E., Llamas, P., *et al.*, High levels of latent antithrombin in plasma from patients with antithrombin deficiency, *Thromb. Haemost.* 117, 880–888, 2017.

28. Griffith, M.J. and Lundblad, R.L., Dissociation of antithrombin III-thrombin complex. Formation of active and inactive antithrombin III, *Biochemistry* 20, 105–110, 1981.

29. Wardell, M.R., Chang, W.S., Bruce, D., *et al.*, Preparative induction and characterization of L-antithrombin: a structural homologue of latent plasminogen activator inhibitor-1, *Biochemistry* 42, 13133–13142, 1997.

30. Lomas, D.A., Elliott, P.R., Chang, W.S., Wardell, M.R., and Carrell, R.W., Preparation and characterization of latent α_1-antitrypsin, *J. Biol. Chem.* 270, 5282–5288, 1995.

31. Aguila, S., Igaguirre, G., Martinez-Martínez, I., *et al.*, Disease-causing mutations in the serpin antithrombin reveal a key domain critical for inhibiting protease activity, *J. Biol. Chem.* 292, 165513–16520, 2017.

32. Verhamme, I.M., A novel antithrombin domain dictates the journey's end of a proteinase, *J. Biol. Chem.* 292, 16521–16522, 2017.

33. Mourier, P.A.J., Guichard, O.Y., Herman, F., Sizun, P., and Viskov, C., New insights in thrombin inhibition structure-activity relationships by characterization octadecasaccharides from low molecular weight heparin, *Molecules* 228, 428, 2017.

34. *Phage Display in Biotechnology and Drug Discovery*, ed. S.S. Sidhu and C.R. Geyer, CRC Press, Boca Raton, FL, USA, 2015.

35. Li, K., Zettlitz, K.A., Lipianskaya, J., *et al.*, A fully human scFv phage display library for rapid antibody fragment reformatting, *Protein Eng. Des. Sel.* 28, 307–315, 2015.

36. de Souza, L.R., Scott, B.M., Bhatka, V., *et al.*, Serpin phage display: the use of a T7 system to probe reactive center loop libraries with different serine proteases, *Methods Mol. Biol.* 1826, 41–64, 2018.

37. Smith, G.P. and Petrenka, V.A., Phage display, *Chem. Rev.* 97, 391–410, 1997.

38. Azzazy, H.M.E. and Highsmith, W.E., Jr., Phage display technology: clinical applications and recent innovations, *Clin. Biochem.* 35, 425–435, 2002.

39. Pande, J., Magdelana, M., Szewczyk, M., and Gloven, A.K., Phage display: concept, applications, and future, *Biotechnol. Adv.* 28, 849–858, 2010.

40. Zani, M.-L. and Moreau, T., Phage display as a powerful tool to engineer protease inhibitors, *Biochimie* 92, 1689–1714, 2010.

41. Scott, B.M. and Sheffield, W.P., Engineering the serpin α_1-antitrypsin: a diversity of goals and techniques, *Protein Sci.* 29, 856–871, 2020.

42. Lewis, J.L., Iammanino, R.M., Spero, J.A., and Hasiba, V., Antithrombin Pittsburgh, an α_1-antitrypsin variant causing hemorrhagic disease, *Blood* 51, 129–137, 1978.

43. Owen, M.C., Brennan, S.O., Lewis, J.H., and Carrell, R.W., Mutation of antitrypsin to antithrombin. α_1-antitrypsin Pittsburgh (358Met →Arg), a fatal bleeding disorder, *New Engl. J. Med.* 309, 694–698, 1983.

44. Scott, B.M., Metochko, W.L., Gierczak, R.F., *et al.*, Phage display of the serpin alpha-1-proteinase inhibitor randomized at consecutive residues in the reactive centre lolop and biopannned with or without thrombin, *PLoS One* 9(11), e84491, 2014.

45. Tanaka, A.S., Silva, M.M., Torquato, R.J.S., *et al.*, Functional phage display of leech-derived tryptase inhibitor (LDTI): construction of a library and selection of thrombin inhibitors, *FEBS Lett.* 458, 11–16, 1999.

46. Markland, W., Ley, A.C., and Lodner, R.C., Interactive optimization of high-affinity protease inhibitors using phase display. 2. Plasma kallikrein and thrombin, *Biochemstry* 35, 8058–8067, 1996.

47. Wiedermann, C.J., Anticoagulant therapy for septic coagulopathy and disseminated intravascular coagulation: where do KyberSept and SCARLET leave us?, *Acute Med. Surg.* 7(1), e477, 2020.
48. Wewers, M.D., Casolaro, M.A., Seller, S.E., *et al.*, Replacement therapy for alpha₁-antrypsin deficiency associated with emphysema, *New Eng. J. Med.* 316, 1055–1062, 1987.
49. Jordan, S., Iovine, J., Kuhn, J., and Gelabert, H., Neurogenic outlet syndrome and other forms of cervical brachial syndrome treated with a plasma concentrate enriched for alpha-2-macroglobulin, *Pain Physician* 34, 229–233, 2002.
50. Sheehan, J.P. and Sadler, J.E., Molecular mapping of the heparin-binding exosite of thrombin, *Proc. Natl. Acad. Sci. USA* 91, 5518–5522, 1994.
51. Henry, B.L., Monien, B.H., Bock, P.E., and Desai, U.R., A novel allosteric pathway of thrombin inhibition: exosite II mediated potent inhibition of thrombin by chemo-enzymatic, sulfated dehydropolymers of 4-hydroxycinnamic acids, *J. Biol. Chem.* 282, 31891–31899, 2007.
52. Sidhu, P.S., Abdel-Ariz, M.H., Sarkar, A., *et al.*, Designing allosteric regulators of thrombin. Exosite 2 features multiple subsites that can be targeted by sulfated small molecules for inducing inhibition, *J. Med. Chem.* 56, 5059–5070, 2013.
53. Carter, W.J., Cama, E., and Huntington, J.A., Crystal structure of thrombin bound to heparin, *J. Biol. Chem.* 280, 2745–2749, 2005.
54. Lyttleton, J.W., The antithrombin activity of heparin, *Biochem. J.* 58, 15–23, 1954.
55. Desai, B.J., Boothello, R.S., Mehta, A.Y., *et al.*, Interaction of thrombin with sucrose octasulfate, *Biochemistry* 50, 6973–6982, 2011.
56. Thompson, R.E., Liu, X., Ripoli-Rozada, J., *et al.*, Tyrosine sulfation modulates activity of tick-derived thrombin inhibitors, *Nat. Chem.* 9, 909–917, 2–17.
57. Fernández, P.V., Quintana, I., Cerezo, A.S., Anticoagulant activity of a unique sulfated pyranosic (1→3)-β-L-arabinan through direct interaction with thrombin, *J. Biol. Chem.* 288, 223–233, 2013.
58. Miller, C., Mescke, K., Liebig, S., *et al.*, More than just one: multiplicity of hirudins and hirudin-like factors in the medicinal leech, *Hirudo medicinalis, Mol. Genet. Genomics* 291, 227–240, 2016.
59. Müller, C., Lukas, P., Lemke, S., and Hildebrandt, J.P., Hirudin and decorsins of the North American medicinal leech *Macrobdella decora*: gene structure reveals homology to hirduins and hirudin-like factors of Eurasian medicinal leeches, *J. Paristol.* 105, 423–431, 2019.
60. Lukas, P., Wolf, R., Rauch, B.H., Hildebrandt, J.P., and Müller, C., Hirudins of the Asian medicinal leech, *Hirudinaria manillensis*; the same, but different, *Paristol. Res.* 118, 2223–2233, 2019.
61. Markwardt, F., The comeback of hirudin—an old-established anticoagulant agent, *Folia Haematol. Int. Mag. Klin. Morphol. Blutforsch.* 115, 10–23, 1988.
62. Markwardt, F., Development of hirudin as an antithrombotic agent, *Semin. Thromb. Hemost.* 15, 269–282, 1989.
63. Eisenberg, P.R., Role of new anticoagulants as adjunctive therapy during thrombolysis, *Am. J. Cardiol.* 67, 19A–24A, 1991.
64. Beretz, A. and Cazenave, J.P., Old and new natural products as the source of modern antithrombotic drugs, *Planta Med.* 57, S68–S72, 1991.
65. Harvey, R.P., Degryse, E., Stefani, L., *et al.*, Cloning and expression of a cDNA coding for the anticoagulant hirudin from the blood sucking leech, *Hirudo medicinalis, Proc. Natl. Acad. Sci. USA* 83, 1084–1088, 1986.
66. Degryse, E., Acker, M., Defreyn, G., *et al.*, Point mutations modifying the thrombin inhibition kinetics and antithrombotic activity *in vivo* of recombinant hirudin, *Protein Eng.* 2, 459–465, 1989.

67. Märki, W.E. and Wallis, R.B., The anticoagulant and antithrombotic properties of hiru-dins, *Thromb. Haemost.* 64, 344–348, 1990.
68. Gladwell, T.D., Bivalirudin: a direct thrombin inhibitor, *Clin. Ther.* 24, 38–58, 2002.
69. Bourdon, P., Jablonski, J.A., Chao, B.H., and Maraganore, J.M., Structure-function relationships of hirulog peptide interactions with thrombin, *FEBS Lett.* 294, 163–166, 1991.
70. Nagler, M. and Wuillemin, W.A., Leeching as a substitute for phebotomy, *Br. J. Haematol.* 153, 420, 2011.
71. Proshinsky, B.S., Saha, S., Grossman, M.D., *et al.*, Clinical use of the medicinal leech: a practical review, *J. Postgrad. Med.* 57, 65–71, 2011.
72. Gröbe, A., Michalsen, A., Hanken, H., *et al.*, Leech therapy in reconstructive maxiloofa-cial surgery, *J. Oral Maxillofac. Surg.*, 70(1), 221–227, 2012.
73. Köse, R., Mordeniz, C., and Sanli, C., Use of expanded reverse sural artery flap in lower extremity reconstruction, *J. Foot Ankle Surg.*, 50(6), 695–698, 2011.
74. Oh, T.S., Hallock, G., and Hong, J.P., Freestyle propeller flaps to reconstruct defects of the posterior trunk: A simple approach to a difficult problem, *Ann. Plast. Surg.*, 68(1), 79–82, 2012.
75. Lombardi, A., Nastri, F., Della Morte, R., *et al.*, Rationale design of true hirudin mimet-ics: synthesis and characterization of multisite-directed α-thrombin inhibitors, *J. Med. Chem.* 39, 2008–2017, 1996.
76. Lombardi, A., De Simone, G., Nasti, F., *et al.*, The crystal structure of α-thrombin-hirunorm IV complex reveals a novel specificity site recognition mode, *Protein Sci*, 8, 91–95, 1999.
77. Kovach, I.M., Kakalis, L., Jordan, F., and Zhang, D., Proton bridging in the interactions of thrombin with hirudin and its mimics, *Biochemistry* 52, 2472–2481, 2013.
78. Zavyalova, E. and Kopylov, A., Multiple inhibitory kinetics reveal an allosteric interplay among thrombin functional sites, *Thromb. Res.* 135, 212–216, 2015.
79. Skrzypczak-Jankun, E., Carperos, V.E., Ravichandran, K.G., *et al.*, Structure of the hiru-gen and hirulog 1 complexes of α-thrombin, *J. Mol. Biol.* 221, 1379–1393, 1991.
80. Malley, M.F., Tabernero, L., Chang, C.Y., *et al.*, Crystallographic determination of the structure of human α-thrombin complexed with BMS-186282 and BMS-189090, *Protein Sci.* 5, 221–228, 1996.
81. Hatton, M.W.C. and Ross-Ouellet, B., Radiolabeled r-hirudin as a measure of thrombin activity at, or within, the rabbit aorta wall in vitro and in vivo, *Thromb. Haemost.* 71, 499–506, 1994.
82. Rubens, F.D., Ross-Ouellet, B., Dennie, C., *et al.*, Displacement of fibrin-bound throm-bin by r-hirudin precludes the use of ^{131}I-r-hirudin for detecting pulmonary emboli In the rabbit, *Thromb. Haemost.* 72, 212–238, 1994.
83. Komatsu, Y. and Hayashi, H., Reevaluating the effects of tyrosine iodination of recom-binant hirudin on its thrombin inhibition kinetics, *Thromb. Res.* 87, 343–352, 1997.
84. Francischetti, I.M., Valenzuela, J.G., and Ribiero, J.M., Amphelin: kinetics and mecha-nism of thrombin inhibition, *Biochemistry* 38, 16678–16685, 1999.
85. Assumpcão, T.C., Ma, D., Mizurini, D.M., *et al.*, *In vitro* mode of action and anti-thrombotic activity of boophilin, a multifunctional Kunitz protease inhibitor from the midgut ofa tick vector of Babesioosis, *Rhipicephalus microplus*, *PLoA Negl. Trop. Dis.* 10(1):0004298, 2016.
86. Pirone, L., Ripoll-Rozada, J., Leone, M., *et al.*, Functional analyses yield detailed insight into the mechanism of thrombin inhibition by the antihemostatic salivary protein cE5 from *Anopheles gambiae*, *J. Biol. Chem.* 292, 12632–12642, 2017.
87. Watson, E.E., Liu, X., Thrompson, R.E., *et al.*, Mosquito-derived Anophelin sulfopro-teins are potent antithrombotics, *ACS Cent. Sci.* 4, 468–476, 2018.

88. Ibrahim, M.A. and Masoud, H.M.M., Thrombin inhibitor from the salivary gland of the camel tick, *Hyalomma dromedarii*, *Exp. Appl. Acarol.* 74, 85–97, 2018.

89. Schechter, I. and Berger, A., On the active site in proteases. I. Papain, *Biochem. Biophys. Res. Commun.* 27, 157–162, 1967.

90. Leung, D., Abbenante, G., and Fairlie, D.P., Protease inhibitors: current status and future prospects, *J. Med. Chem.* 43, 305–341, 2000.

91. Liang, G. and Bowen, J.P., Development of trypsin-like serine protease inhibitors as therapeutic agents: opportunities, challenges and their unique structure-based rationales, *Curr. Top. Med. Chem.* 16, 1506–1509, 2016.

92. Okamotok, S. and Hijikata-Okunumiya, A., Synthetic selective inhibitors of thrombin, *Methods Enzymol.* 222, 328–340, 1993.

93. Srivastava, S., Goswami, L.N., and Dikshit, D.K., Progress in the design of low molecular weight thrombin inhibitors, *Med. Res. Rev.* 25, 86–92, 2005.

94. Markwardt, F., Pharmacological approaches to thrombin regulation, *Ann. N. Y. Acad. Sci.* 485, 204–214, 1986.

95. Markwardt, F., Historical perspectives of the development of thrombin inhibitors, *Patholophysiol. Haemost. Thromb.* 32(Suppl. 3), 15–22, 2002.

96. Markwardt, F. and Walsmann, P., Uber die Hemmung des Gerrinungsferment thrombin durch benamidinderivate, *Experientia* 24, 25–26, 1988.

97. Markwardt, F., Landmann, H., and Walsmann, P., Comparative studies on the inhibition of trypsin, plasmin and thrombin by derivatives of benzylamine and benzamidine, *Eur. J. Biochem.* 6, 502–506, 1968.

98. Thompson, A.R. and Davie, E.W., Affinity chromatography of thrombin, *Biochim. Biophys. Acta* 250, 210–215, 1971.

99. Thompson, A.R., High affinity binding of human and bovine thrombin to *p*-chlorobenzylamido-ε-aminocaproyl agarose, *Biochim. Biophys. Acta* 422, 200–209, 1976.

100. Baum, B., Mohammed, M., Zayed, M., *et al.*, More than a simple lipophilic contact: a detail thermodynamic analysis of nonbasic residues in the S1 pocket of thrombin, *J. Mol. Biol.* 390, 56–69, 2009.

101. Markwardt, F., Pharmacological control of blood coagulation by synthetic, low molecular weight inhibitors of clotting enzymes, *Trends Pharmacol. Sci.* 1, 153–157, 1980.

102. Markwardt, F., Pharmacological control of blood coagulation by synthetic, low molecular weight inhibitors of clotting enzymes, *Ann. N. Y. Acad. Sci.* 370, 757–764, 1980.

103. Markwardt, F., Wagner, G., Stürzebecher, J., and Walsmann, P., N_α-Arylsulfonyl-ω-(4-amidinophenyl-α-aminoalkylcarboxylic acid amides-novel selective inhibitors of thrombin, *Thromb. Res.* 17, 425–431, 1980.

104. Stürzebecher, J., Markwardt, F., Voight, B., Wagner, G., and Walsmann, P., Cyclic amides of N_α-arylsulfonylaminoacylated 4-amidinophenylalanine—tight binding inhibitors of thrombin, *Thromb. Res.* 29, 635–642, 1983.

105. Markwardt, F., Pharmacological approaches to thrombin regulation, *Ann. N. Y. Acad. Sci.* 485, 204–214, 1988.

106. Markwardt, F., Historical perspectives of the development of thrombin inhibitors, *Pathophysiol. Haemost. Thromb.* 32 (Suppl 3), 15–22, 2002.

107. Bode, W., Turk, D., and Stürzebecher, J., Geometry of binding of the benzamidine and arginine-based inhibitors N^α-(2-naphthylsuphonyl-glycyl-DL-*p*-amidinophenylalanyl-piperidine (NAPAP) and (2*R*,2*R*)-4-methyl-1-[(N^α-(3-methyl-1,2,3,4-tetrahydro-8-quinolinesulphonyl)-L-arginine]-2-piperidine carboxyl acid (MQPA) to human α-thrombin, *Eur. J. Biochem* 193, 175–182, 1990.

108. Brandstetter, H., Turk, D., Heoffken, H.W., *et al.*, Refined 2.3 Å x-ray crystal structure of bovine thrombin crystals formed with benzamidine and ariginine-based inhibitors NAPAP, 4-TAPAP and MQPA, *J. Mol. Biol.* 226, 1086–1089, 1992.

109. De Cristofero, R. and Landolfi, R., Thermodynamics of substrates and reversible inhibitors binding to the active site cleft of thrombin, *J. Mol. Biol.* 239, 569–577, 1994.
110. Chodera, J.D., and Mobley, D.L., Entropy-enthalpy compensation: role and ramifications in biomolecular recognition and design, *Ann. Rev. Biophys.* 42, 121–142, 2013.
111. Treuheit, N.A., Beach, M.A., and Komives, E.A., Thermodynamic compensation upon binding to exosite 1 and the active site of thrombin, *Biochemistry* 50, 4590–4596, 2011.
112. Buijsman, R.C., Boston, J.E.M., van Dinther, T.-G., *et al.*, Design and synthesis of a novel synthetic NAPAP-pentasaccharide conjugate displaying a due antithrombotic activity, *Bioorg. Med. Chem. Lett.* 9, 263–268, 1991.
113. Rewinkel, J.B.M., Lucas, H., van Galen, P.J.M., *et al.*, 1-Aminoiso quinoline as benzamidine isostere in the design and synthesis of orally active thrombin inhibitors, *Bioorg. Med. Chem. Lett.* 9, 685–690, 1999.
114. Artursson, P., Palm, K., and Luthman, K., CaCo-2 monolayers as experimental and theoretical predictors of drug transport, *Adv. Drug. Deliv. Rev.* 46, 27–43, 2001.
115. Sun, H., Chan, E.D., Liu, S., Du, Y., and Pang, K.S., The CaCo-2 cell monolayer: usefulness and limitations, *Expert. Opin. Drug Metab. Toxicol.* 4, 395–411, 2008.
116. Hilpert, K.F., Ackermann, J., Banner, D.W., *et al.*, Design and synthesis of potent and highly selective thrombin inhibitors, *J. Med. Chem.* 37, 3889–3901, 1994.
117. Aliter, K.F., and Al-Horani, R.A., Thrombin inhibition by argatroban: potential therapeutic benefits in COVID-19, *Cardiovasc. Drug Ther.* 35(2), 195–203, 2021.
118. Hauptmann, J., Pharmacokinetics of an emerging class of anticoagulant/antithrombotic drugs, *Eur. J. Clin. Pharmacol.* 57, 751–758, 2002.
119. Stürzebecher, J., Stürzebecher, U., Viewag, H., *et al.*, Synthetic inhibitors of bovine thrombin and bovine factor Xa. Comparison of their anticoagulant efficiency, *Thromb. Res.* 54, 245–252, 1989.
120. Geratz, J.D., *p*-aminobenzamidine as inhibitors of trypsinogen activation, *Experentia* 22, 90–91, 1966.
121. Geratz, J.D., Whitmore, A.C., Cheng, M.C.F., and Piantadosi, C., Diamidino-α,ω-diphenoxyalkanes. Structure-activity relationships for the inhibition of thrombin, pancreatic kallikrein and trypsin, *J. Med. Chem.* 16, 970–975, 1973.
122. Diamandes, E.P., Yousef, G.M., Clements, J., *et al.*, New nomenclature for the human tissue kallikrein gene family, *Clin. Chem.* 46, 1855–1858, 2000.
123. Geratz, J.D., Cheng, M.C.-F., and Tidwell, R.R., New aromatic diamidines with central α-oxyalkanes or α,ω-dioxyalkane chains. Structure-activity relationships for the inhibition of trypsin, pancreatic kallikrein, and thrombin and for the inhibition of the overall coagulation process, *J. Med. Chem.* 18, 477–481, 1975.
124. Tidwell, R.R., Fox, L.L., and Geratz, J.D., Aromatic tris amidines. A new class of highly active inhibitors of trypsin-like proteases, *Biochim. Biophys. Acta* 445, 729–738, 1976.
125. Tidwell, R.R., Geratz, J.D., Dann, O., *et al.*, Diarylamidine derivatives with one or both aryl moieties consisting of an indole or indole-like ring. Inhibition of arginine-specific esteroproteases, *J. Med. Chem.* 21, 613–623, 1978.
126. Geratz, J.D. and Tidwell, R.R., Current concepts on action of synthetic thrombin inhibitors, *Haemostasis* 7, 170–176, 1978.
127. DesJarlais, R.L., Seibel, G.L., Kuntz, J.D., *et al.*, Structure-based design of nonpeptide inhibitors specific for the human immunodeficiency virus 1 protease, *Proc. Natl. Acad. Sci. USA* 87, 6644–6648, 1990.
128. De Voss, J.J., Sui, Z., DeCamp, D.L., *et al.*, Haloperidol-based irreversible inhibitors of the HIV-1 and HIV-2 proteases, *J. Med. Chem.* 37, 665–673, 1994.
129. Lundblad, R.L., Drug design, in *Encyclopedia of Cell Biology*, Vol. 1, ed. R.A. Bradshaw and P. Stahl, pp. 135–140, Academic Press, Waltham, MA, 2016.

130. Geratz, J.D., Stevens, F.M., Polakaski, K.L., Parish, R.F., and Tidwell, R.R., Amidine-substituted aromatic heterocycles as probes of the specificity pocket of trypsin-like proteases, *Arch. Biochem. Biophys.* 197, 551–559, 1979.
131. Dubovi, E.J., Geratz, J.D., and Tidwell, R.R., Enhancement of respiratory syncytial virus-induced cytopathology by trypsin, *Infect. Immun.* 40, 351–358, 1983.
132. Tidwell, R.R., Geratz, J.D., and Dubavi, E.J., Aromatic amidiines: comparison of their ability to block respiratory syncytial virus-induced cell fusion and to inhibit plasmin, urokinase, thrombin, and trypsin, *J. Med. Chem.* 26, 294–298, 1983.
133. Nedu, M.E., Testis, M., Cristea, C., and Georgescu, A.N., Comparative study regarding the properties of methylene blue and proflavine and their optimal concentrations in *in vitro* and *in vivo* applications, *Diagostics* (Basal) 10(4), 223, 2020.
134. Wallace, R.A., Kurtz, A.N., and Niemann, C., Interaction of aromatic compounds with α-chymotrypsin, *Biochemistry* 2, 824–836, 1963.
135. Glazer, A.N., Spectral studies on the interaction of α-chymotrypsin and trypsin with proflavine, *Proc. Natl. Acad. Sci. USA* 54, 171–176, 1965.
136. Bernard, S.A., Lee, B.T., Tashjian, Z.H., Proflavin binding and enzyme conformation during catalysis, *J. Mol. Biol.* 18, 405–420, 1966.
137. Feinstein, G. and Feeney, R.E., Binding of proflavin to α-chymotrypsin and trypsin and its displacement by avian orosomucoid, *Biochemistry* 6, 749–753, 1967.
138. Koehler, K.A. and Magnusson, S., The binding of proflavin to thrombin, *Arch. Biochem. Biophys.* 160, 17584, 1974.
139. Li, E.H.H., Orton, C., and Feinman, R.D., Interaction of thrombin and heparin, *Biochemistry* 13, 5012–5017, 1974.
140. Berliner, L.J. and Shen, Y.Y.L., Physical evidence o fan apolar binding site near the catalytic αcenter of human α-thrombin, *Biochemistry* 16, 4622–4626, 1977.
141. Conti, E., Rivetti, C., Wanacott, A., and Brick, P., X-ray and and spectrophotometric studies of the binding of proflavin to the S1 specificity pocket of human α-thrombin, *FEBS Lett.* 425, 229–233, 1998.
142. Nienaber, V.L., Boxrud, P.D., and Berliner, L.J., Thrombin inhibitor design: x-ray and solution studies, *J. Prot. Chem.* 19, 327–333, 2000.
143. Erlanson, D.A., McDowell, R.S., and O'Brien, T., Fragment-based drug discovery, *J. Med. Chem.* 17, 3463–3482, 2004.
144. Turner, A.D., Monroe, D.M., Roberts, H.R., Porter, N.A., and Pizzo, S.V., *p*-Amidino esters as irreversible inhibitors of factors IXa, and Xa and thrombin, *Biochemistry* 29, 4929–4935, 1986.
145. Ni, L.-M. and Powers, J.C., Synthesis and kinetic studies of an amidine-containing phosphonofluoridate: a new potent inhibitor of trypsin-like enzymes, *Bioorg. Med. Chem.* 6, 1767–1773, 1998.
146. Pan, Z., Jeffery, D.A., Chehade, K., *et al.*, Development of activity-based probes for trypsin-family serine proteases *Bioorg. Med. Chem.* 16, 2882–2885, 2006.
147. Fonović, M. and Bogyo, M., Activity-based probes as a tool for the functional proteomics analysis of proteases, *Expert Rev. Proteomics* 5, 721–730, 2008.
148. Grzywa, R. and Sieńczyk, M., Phosphonic esters and their application of proteases control, *Curr. Pharm. Design* 19, 1154–1178, 2013.
149. Kikomoto, R., Tamao, Y., Tezuka, I., *et al.*, Selective inhibition of thrombin by (2R,4R)-4-methyl-[N^2[(3-methyl-1,2,3,4-tetrahydro-8-quinolinyl)sulfonyl]-L-arginine]-2-piperidinecarboxylic acid, *Biochemistry* 23, 85–90, 1984.
150. Nesheim, M.E., Prendergast, F.G., and Mann, K.G., Interaction of a fluorescent active site-directed of thrombin-Dansylarginine-*N*-(3-ethyl-1,5-pentanedial) amide, *Biochemistry* 18, 996–1003, 1979.

151. Okamoto, S., Hijikata, A., Ikezawa, K., *et al.*, A new series of synthetic thrombin inhibibitors (OM-inhibitors) having extremely potent and selective action, *Thromb. Res.* 8(Suppl II), 77–82, 1976.

152. Hijikawa, A., Okamoto, S., Mori, E., *et al.*, In vitro and in vivo studies of a new series of synthetic thrombin inhibitors (OM-inhibitors), *Thromb. Res.* 8(Suppl II), 83–89, 1976.

153. Hijikata-Okunomiya, A. and Okamoto, S., A strategy for a rational approach to designing synthetic selective inhibitors, *Sem. Thromb. Hemots.* 18, 135–149, 1992.

154. Obst, U., Gramlich, V., Diederich, E., Weber, L., and Banner, D.W., Design of novel nonpeptidic thrombin inhibitors and structure of the thrombin-inhibitor complex, *Angew. Chem. Int. Edn.* 34, 1739–1742, 1995.

155. Obst, U., Banner, D.W., Weber, L., and Diederich, F., Molecular recognition at the thrombin active site: structure-based design and synthesis of potent and selective thrombin inhibitors and the x-ray crystal structure of the thrombin-inhibitor complex, *Chem. Biol.* 4, 287–295, 1997.

156. Hauptmann, J. and Stürzebecher, J., Synthetic inhibitors of thrombin and factor Xa: from bench to bedside, *Thromb. Res.* 93, 203–241, 1999.

157. Reeh, C., Wundt, J., and Clement, B., N,N'-Dihydroxyamidines: a new prodrug principle to improve he oral availability of amidines, *J. Med. Chem.* 50, 6730–6734, 2007.

158. Bajusz, S., Barabás, E., Tolnay, P., Széll, E., and Bagdy, D., Inhibition of thrombin and trypsin by tripeptide aldehydes, *Int. J. Pept. Prot. Res.* 12, 217–221, 1978.

159. Ketter, C. and Shaw, E., D-Phe-pro-argCHCl$_2$: a selective affinity label for thrombin, *Thromb. Res.* 14, 969–973, 1979.

160. Sanderson, P.E.J., Cutrona, K.J., Dorsey, B.D., *et al.*, L-374,087, an efficacious, orally available, pyridinone acetamide thrombin inhibitor, *Bioorg. Med. Chem. Lett.* 8, 817–822, 1998).

161. Issacs, R.C.A., Cutrano, K.I., Newton, C.I., *et al.*, C6 modification of the pyridinone core of thrombin inhibitor L-374,087 as a means of enhancing oral absorption, *Bioorg. Med. Chem. Letter* 8, 1710–1724, 1998.

162. Sanderson, P.E.J., Lyle, T.A., Cutrona, K.I., *et al.*, Efficacious orally available thrombin inhbitiors based on 3-aminopyridinone or 3-aminopyrazinone acetamid peptidomimetic templates, *J. Med. Chem.* 41, 4466–4474, 1998.

163. Riffel, K.A., Song, H., Go, X., and Lo, M.-N., Simultaneous determination of a novel thrombin inhibitor and its metabolites in human plasm a by liquid chromatography/ tandem mass spectrometry, *J. Pharm. Biomed. Anal.* 23, 607–616, 2000.

164. Rittle, K.E., Barrow, J.E., Cutrona, K.I., *et al.*, Unexpected enhancement of thrombin inhibitor potency with *o*-aminoalkylbenzylamides in the P$_1$ position, *Bioorg. Med. Chem. Lett.* 13, 3477–3482, 2003.

165. Burgey, C.S., Robinson, K.A., Lyle, T.A., *et al.*, Metabolism-directed optimization of 3-aminopyrazinone acetamide thrombin inhibitors. Development of an oral bioavailable series containing P1 and P3 pyridines, *J. Med. Chem.* 46, 461–473, 2003.

166. Young, M.B., Barrow, J.C., Glass, K.L., *et al.*, Discovery and evaluation of potent P$_1$ and heterocycle-based thrombin inhibitors, *J. Med. Chem.* 47, 2995–3008, 2004.

167. Nantermet, P.G., Burgey, C.S., Robinson, K.A., *et al.*, P$_2$ pyridine *N*-oxide thrombin inhibitors: a novel peptidomimetic scaffold, *Bioorg. Med. Chem. Lett.* 15, 2771–2775, 2005.

168. Mfuh, A.M. and Larionov, O.V., Heterocyclic *N*-oxides An emerging class of therapeutics agents, *J. Med. Chem.* 47, 2991–3008, 2006.

169. Deinum, J., Mattsson, C., Inghardt, T., and Eig, M., Biochemical and pharmacological effects of the direct thrombin inhibitor, AR-H067637, *Thromb. Haemost.* 101, 1051–1059, 2009.

170. Lip, G.Y.H., Rasmussen, L.H., Olsson, S.B., *et al.*, Exposure-reponse for biomarkers of anticoagulant effects by the oral direct thrombin inhibitor AZD0837 in patients with atrial fibrillation, *Brit. J. Clin. Pharmacol.* 80, 1363–1373, 2015.

171. De Simone, G., Menchise, V., Omaggio, S., Design of weakly basic thrombin incorporating novel P1 binding function: molecular and X-ray crystallographic studies, *Biochemistry* 42, 9013–9021, 2003.

172. Wu, E.L., Han, K., and Zhang, J.Z.H., Selectivity of neutral/weakly basic P1 group inhibitors of thrombin and trypsin by a molecular dynamics study, *Chem. Eur. J.* 14, 8704–8714, 2008.

173. de Andrade Moura, L., de Almeida, B.C., Dominiguez, T.E.S., *et al.*, Antiplatelet and anticoagulant effects of diterpenes isolated from the marine alga, *Dictyota menstruralis*, *Marine Drugs* 12, 2471–2484, 2014.

174. Pereira, R.C.C., Lourenge, A.L., Terra, L., *et al.*, Marine diterpenes: molecular modeling of thrombin inhibitors with potential biotechnological applications as an antithrombic, *Marine Drugs* 15, 79, 2017.

175. Dwivedi, R. and Pomin, V.H., Marine antithrombotics, *Marine Drugs* 18, 514, 2020.

176. Kikelj, D., Peptidomimetic thrombin inhibitors, *Pathophys. Haemost. Thromb.* 33, 487–491, 2003/2004.

177. Gustafsson, D., Oral direct thrombin inhibitors in clinical development, *J. Int. Med.* 254, 322–334, 2003.

178. Steinmetzer, T. and Stürzebecher, J., Progress in the development of synthetic thrombin inhibitors as new orally active anticoagulants, *Curr. Med. Chem.* 11, 2297–2321, 2004.

179. Di Niaio, M., Middeldorp, S., and Büller, H.R., Direct thrombin inhibitors, *New Engl. J. Med.* 353, 1028–1040, 2005.

180. Hankey, G.J. and Eikelboom, J.W., Dabigatran etexilate A new oral thrombin inhibitor, *Circulation* 123, 1436–1450, 2011.

181. Serebruany, M.V., Malinin, A.J., and Serebruany, V.L., Argatroban, a direct thrombin inhibitor for heparin-induced thrombocytopaenia: present and future perspectives, *Expert Opin. Pharmocther.* 7, 81–89, 2006.

182. McKeage, K. and Plosker, G.L., Argatoban, *Drugs* 61, 515–522, 2001.

183. Dingman, J.S., Smith, Z.R., Coba, V.E., Peter, M.A., and To, L., Argatroban dosing requirements in extracorporeal life support and other critically ill populations, *Thromb. Res.* 159, 69–76, 2020.

184. Aliter, K.F. and Al-Horani, R., Thrombin inhibition by argatroban: potential therapeutic benefits in COVID-19, *Cardiovasc. Drug Ther.*, 35(2), 195–203, doi: 10.1007/s10557-020-07066-x.

185. Rosenberg, A., Xu, AT., Passariello, M., *et al.*, Anticoagulation with argatroban in patients with acute antithrombin deficiency in severe COVID-19, *Br. J. Haematol.* 190, e286–e288, 2020.

186. Gustafsson, D., Nyström, J.E., Carlsson, S., *et al.*, The direct thrombin inhibitor melagatran and its prodrug H376/95: intestinal absorption properties, biochemical and pharmacodynamic effects, *Thromb. Res.* 101, 171–181, 2001.

187. Evans, H.C., Perry, C.M., and Faulds, D., Ximelagatran/melagatran A review of its use in the prevention of venous thromboembolism in orthopaedic surgery, *Drugs.* 64, 649–678, 2004.

188. Gurewich, V., Ximelagatran—promises and concerns, *JAMA* 293, 736–739, 2005.

189. Southworth, H., Predicting potential liver toxicity form phase 2 data: a case study with ximlagatran, *Stat. Med.* 33, 2914–2923, 2014.

Index

A

acetic anhydride, 170
2-acetoxy-5-nitrobenyl bromide, 102, 162
N-acetylimidazole, 168–169, 174–175
active site mutants, 132–133, 150
affinity labeling, 62–63, 69, 162
alanine scanning, 145–146
alpha-1(α_1)-antitrypsin, 84, 95, 135–136, 224–225
alpha-1(α_1)-antitrypsin Pittsburgh, 86, 134–135, 141
Amberlite CG-50, 11, 71, 174
amidino phosphorofluoridates, 244
anhydrothrombin, 62, 160
anion-binding site, 71–77, 87, 190
antithrombin, 72–78, 223–226
antithrombin I, 191–192, 224
antithrombin II, 192, 224
antithrombin III, 192, 224–225
antithrombin IV, 192
antithrombin V, 192
antithrombin VI, 192
antithrombin Denver, 135
apolar binding site, 71, 101–103
aptamer, 85–91, 173, 190, 211
argatroban, 81, 226, 235–237, 244, 250, 255
ATP, 71, 90–91, 103, 140, 204
autolysis Loop, 137–145

B

benzamidine, 66, 71, 134, 169, 227–242
Bernard-Soulier syndrome, 209, 211
beta (β)-thrombin, 43, 63, 98–99, 161, 164–166
biothrombin, 9, 11, 33
bivalirudin, 226, 250
bothrojaracin, 77, 79, 85
bovine pancreatic trypsin inhibitor, 134
N-bromosuccinimide, 175–176
N-butyrylimidazole, 169, 175

C

catalytic triad, 58, 161
chondroitin sulfate, 91, 104, 171
citrate activation of prothrombin, 9, 42–43
citrate thrombin, 9, 11, 24, 187
CLiPS, 84

D

dabigatran, 24, 29, 38, 223, 250, 255
dansylarginine *N*-(3-ethyl-1, 5-pentadiyl)amide, 34, 143, 150–151, 244, 246
DAPA, *see* dansylarginine *N*-(3-ethyl-1, 5-pentadiyl)amide
denaturase, 7, 186
DEPC, *see* diethyl pyrocarbonate
Der p 1, 26, 36
DFP, 61, 157, 160, 164, 188
diethyl pyrocarbonate, 63, 161
digestive protease, 57, 61
diisopropylphosphorofluoridate, 11, 139

E

ecarin, 5, 29, 131–132
endosite, 68
esterase thrombin, 170
exosites, 11, 33, 58, 68–72
extended binding site, 61, 64, 150
extracellular matrix, 37, 69, 88, 91, 103–104, 194

F

factor Xa, 24–25, 35–36, 43–45
factor XII, 5–6, 95
factor XIII, 7, 63, 94, 160, 187, 193

G

gamma (γ)-thrombin, 43, 71–73, 76–78, 137, 205, 211
glycocalcin, 122, 144, 206, 209–211, 216

H

haloperidol, 240
HD-1, 85–87, 190
HD-22, 86, 91, 190
hirulog-1, 73
hironorms, 226
HNB, *see* 2-hydroxy-5-nitrobenzyl bromide
hydrogen peroxide, 161, 175
hydrophobic, 68, 80, 97, 101–102, 242
2-hydroxy-5-nitrobenzyl bromide, 102, 151, 175–176

For Product Safety Concerns and Information please contact our EU
representative GPSR@taylorandfrancis.com
Taylor & Francis Verlag GmbH, Kaufingerstraße 24, 80331 München, Germany

www.ingramcontent.com/pod-product-compliance
Ingram Content Group UK Ltd.
Pitfield, Milton Keynes, MK11 3LW, UK
UKHW021011180425
457613UK00020B/892